Recent Developments and Knowledge in Intelligent and Safe Marine Navigation

Recent Developments and Knowledge in Intelligent and Safe Marine Navigation

Editors

Mingyang Zhang
Xinyu Zhang
Shanshan Fu
Lei Dai
Qing Yu

MDPI

Basel • Beijing • Wuhan • Barcelona • Belgrade • Novi Sad • Cluj • Manchester

Editors

Mingyang Zhang
Aalto University
Espoo, Finland

Xinyu Zhang
Dalian Maritime University
Dalian, China

Shanshan Fu
Shanghai Maritime University
Shanghai, China

Lei Dai
Shanghai Jiao Tong University
Shanghai, China

Qing Yu
Jimei University
Xiamen, China

Editorial Office
MDPI
St. Alban-Anlage 66
4052 Basel, Switzerland

This is a reprint of articles from the Special Issue published online in the open access journal *Journal of Marine Science and Engineering* (ISSN 2077-1312) (available at: https://www.mdpi.com/journal/jmse/special_issues/OW599B24NN).

For citation purposes, cite each article independently as indicated on the article page online and as indicated below:

Lastname, A.A.; Lastname, B.B. Article Title. *Journal Name* **Year**, *Volume Number*, Page Range.

ISBN 978-3-03928-623-2 (Hbk)
ISBN 978-3-03928-624-9 (PDF)
doi.org/10.3390/books978-3-03928-624-9

Cover image courtesy of Mingyang Zhang

Contents

About the Editors

Mingyang Zhang

Mingyang Zhang is a postdoctoral researcher from the Marine and Arctic Technology research group at Aalto University, Finland. His primary focus lies in ship safety and sustainable maritime transportation systems, utilizing emerging technologies. The objective of his work is to contribute to the development of intelligent decision-making systems that hold the potential to become critical enablers for future intelligent ship operations or sustainable maritime transportation systems. Additionally, he is keenly interested in implementing collaborative ship control and ship motions through the application of big data analytics and machine learning methods in ship operations modeling. He has about 50 publications in the corresponding fields. He rightfully earned a place on the 2023 list of the Top 2% Scientists Worldwide by Stanford University. Currently, he is actively involved in the RETROFIT55 project, collaborating with stakeholders from industry, academia, and policymakers in the maritime industry.

Xinyu Zhang

Xinyu Zhang is the Head of the Maritime Intelligent Traffic Research Team at the College Navigation of Dalian Maritime University in Dalian, China. He holds an M.Sc. in Applied Mathematics (2003) and is a Doctor of Geo-Information Exploration Technology (2006). He has been selected for various talent programs, including the Talent Program of One Hundred, Thousand, and Ten Thousand in Liaoning Province, the "Golden Mountain Talent" Leading Talents Program in Zhenjiang City, and the Outstanding Talent Support Program in Liaoning Province's universities. His research interests include sustainable green port development; safe and sustainable vessel traffic; autonomous collision avoidance for MASS; and maritime big data mining and applications. He has about 150 publications in the corresponding fields, obtained more than 20 national invention patents, and received over 10 national and provincial awards for outstanding achievements in scientific research. He holds positions as a researcher in the China Shipping 50-Person Forum, is a special research fellow in the China Maritime Think Tank, and an editorial board member of the "*Chinese Journal of Ship Research*." He has led the development of engineering applications such as the Vessel Accurate Riding Tidal System, the Ship Collision Avoidance Decision Support System, and the Port Water Area Layout Planning Support System. This cumulative achievement transformation has earned more than CNY 200 million.

Shanshan Fu

Shanshan Fu is an associate professor in the College of Transport and Communications, Shanghai Maritime University, China. She acquired her BS in June 2010, majoring in logistics engineering at Wuhan University of Technology, China. She concluded her post-graduate studies (M.Sc. in June 2013 and Ph.D. in January 2017) at the Intelligent Transport System Center, Wuhan University of Technology, China. In August 2017, she joined Shanghai Maritime University. She is the leader of two National Natural Science Foundation of China (NSFC) projects and three Shanghai Science and Technology Innovation Action Plan projects. She has published more than 60 scientific papers in international journals and conferences, such as *Reliability Engineering and System Safety*, *Safety Science*, and the International Conference on Ocean, Offshore, and Arctic Engineering. Her research interests include the quantitative risk assessment of maritime accidents and the safety analysis of ship operations in ice-covered Arctic waters.

Lei Dai

Lei Dai is an Associate Professor at the Department of Transportation Engineering in Shanghai Jiao Tong University (SJTU), China. He received his Ph.D. degree majoring in Ocean Engineering (2015) and his Bachelor's degree majoring in Transportation Engineering (2011) from SJTU. Before joining SJTU, he served as a management consultant at PricewaterhouseCoopers Management Consulting from 2015 to 2018, where he provided extensive consultancies to large state-owned enterprises and multi-national corporations in the public policy and supply chain sectors. After joining SJTU in 2018, he focused on the research area of green shipping management, especially on exploring the impacts of energy and environment policies in the shipping sector. He has about 30 publications in the corresponding fields. He has won one first (ranked 4) and two second prizes (both ranked 1) of progress in science and technology from the China Federation of Logistics and Purchasing.

Qing Yu

Qing Yu is the director of the Maritime Risk and Behavioral Sciences Lab, in the School of Navigation, Jimei University, China. He holds a B.Sc. in Civil Engineering (2012), an M.Sc. in Marine Operation and Management (2014), and a PhD of Transportation Information and Control Engineering (2021). For his doctoral thesis, he received the Best Thesis Award from the Wuhan University of Technology. His research focuses on maritime risk assessment, mainly on maritime traffic analysis and risk assessment theory. He has about 30 publications in the corresponding fields. He is a member of the China Communications and Transportation Association and the Institute of Marine Engineering, Science, and Technology, and was a Guest Editor of *ASCE-ASME Journal of Risk and Uncertainty in Engineering Systems, Part A: Civil Engineering* and *Journal of Marine Science and Engineering*.

Preface

The ever-evolving landscape of maritime science and engineering continually challenges researchers to explore innovative methodologies and cutting-edge technologies. In the pursuit of safer and more efficient maritime operations, scholars worldwide have embarked on comprehensive investigations, the results of which are compiled in this remarkable collection of papers.

In the first paper, "A Semantic Network Method for the Identification of Ship's Illegal Behaviors Using Knowledge Graphs: A Case Study on Fake Ship License Plates," authors Wan, Fu, Zhang, and Xiao introduce a pioneering approach that employs knowledge graphs to detect illicit activities related to ship license plates. Their method opens up new avenues for enhancing maritime security and law enforcement.

The second paper, "Leverage Bayesian Network and Fault Tree Method on Risk Assessment of LNG Maritime Transport Shipping Routes: Application to the China–Australia Route," authored by Chang, He, Fan, Guan, and He, addresses the critical issue of risk assessment in LNG maritime transport. By merging Bayesian networks and fault tree analysis, the authors present an innovative solution for assessing the safety of shipping routes, particularly the China–Australia route.

In the third contribution, "Online Estimation of Ship Dimensions by Combining Images with AIS Reports," Huang, Hu, Lu, Mei, and Yang propose a novel approach for estimating ship dimensions by combining image data with Automatic Identification System (AIS) reports. This methodology has significant implications for real-time vessel tracking and maritime surveillance.

The fourth paper, "Probabilistic Modeling of Maritime Accident Scenarios Leveraging Bayesian Network Techniques," by Liao, Weng, Zhang, Li, and Li, explores the application of Bayesian network techniques to model maritime accident scenarios. By adopting a probabilistic framework, the authors offer a valuable tool for improving maritime safety and accident prevention.

In the fifth paper, "A COLREGs-Compliant Ship Collision Avoidance Decision-Making Support Scheme Based on Improved APF and NMPC," authors Li, Wang, Wu, and Ni present a collision avoidance decision-making support scheme compliant with the International Regulations for Preventing Collisions at Sea (COLREGs). Their work showcases advancements in ship collision avoidance techniques and navigational safety.

The sixth paper, "A Deep Learning Method for Ship Detection and Traffic Monitoring in an Offshore Wind Farm Area," authored by Liu, Hu, Ji, Zhang, and Yu, introduces a deep learning approach for ship detection and traffic monitoring in offshore wind farm areas. This innovative technique enhances offshore operations and safety by providing comprehensive monitoring capabilities.

The seventh paper, "A Ship Route Planning Method under the Sailing Time Constraint," by Li, Cui, Zhang, and Yang, presents a ship route planning method that considers the constraint of sailing time. The authors' approach aids in optimizing routes for efficient and timely maritime operations.

In the eighth paper, "Parameter Prediction of the Non-Linear Nomoto Model for Different Ship Loading Conditions Using Support Vector Regression," Lan, Zheng, Chu, and Ding delve into ship modeling and parameter prediction. Their work enhances our understanding of ship behavior under varying loading conditions, contributing to the advancement of maritime engineering.

The ninth paper, "An Intelligent Algorithm for USVs Collision Avoidance Based on Deep Reinforcement Learning Approach with Navigation Characteristics," by Sun, Fan, and Wang, introduces a deep reinforcement learning approach for unmanned surface vehicles (USVs) collision avoidance. This innovation leverages navigation characteristics to enhance the safety and efficiency of USV operations.

The collection concludes with the tenth paper, "Neural Network, Nonlinear-Fitting, Sliding Mode, Event-Triggered Control under Abnormal Input for Port Artificial Intelligence Transportation Robots," authored by Zhu, Zhang, Liu, Hu, and Zhang. Their research explores the intricacies of control mechanisms for artificial intelligence-based transportation robots in port environments, highlighting the potential of these robots to revolutionize maritime logistics.

In this compendium, each paper embodies the commitment of the global maritime science and engineering community to address multifaceted challenges and drive forward the frontiers of knowledge. These contributions collectively underscore the imperative of continuous innovation and cross-disciplinary collaboration to ensure the sustainability, safety, and efficiency of maritime endeavors. We extend our gratitude to the authors for their diligent efforts and to the readers for their interest in advancing the field of marine science and engineering.

Mingyang Zhang, Xinyu Zhang, Shanshan Fu, Lei Dai, and Qing Yu
Editors

Journal of
Marine Science and Engineering

MDPI

Editorial

Recent Developments and Knowledge in Intelligent and Safe Marine Navigation

Mingyang Zhang [1], Xinyu Zhang [2,*], Shanshan Fu [3], Lei Dai [4] and Qing Yu [5]

1 Department of Mechanical Engineering, Aalto University, 02150 Espoo, Finland; mingyang.0.zhang@aalto.fi
2 Department of Navigation College, Dalian Maritime University, Dalian 116026, China
3 College of Transport and Communications, Shanghai Maritime University, Shanghai 200135, China; ssfu@shmtu.edu.cn
4 State Key Laboratory of Ocean Engineering, Department of Transportation Engineering, Shanghai Jiao Tong University, Shanghai 200232, China; dailei1989@sjtu.edu.cn
5 School of Navigation, Jimei University, Xiamen 361021, China; qing.yu@jmu.edu.cn
* Correspondence: zhangxy@dlmu.edu.cn

1. Introduction

Marine navigation is the lifeblood of international trade and the global economy, facilitating over 80% of worldwide commerce [1]. The maritime industry's pivotal role in sustaining global trade cannot be overstated. In a rapidly evolving technological landscape, intelligent and safe marine navigation has emerged as a critical domain within the maritime sector [2]. The integration of cutting-edge technologies, including artificial intelligence (AI), machine learning (ML), and big data analytics, holds the promise of significantly enhancing the intelligence and safety of ships navigating our oceans [3]. These advances are not only transforming the ways in which ships traverse vast sea expanses but also revolutionizing maritime logistics, fleet management, and environmental monitoring [4,5].

With the advent of AI and ML, predictive algorithms are now being employed to optimize routes, thereby reducing fuel consumption and minimizing the environmental impact [6]. Big data analytics enables the processing of massive amounts of data from various sources, such as satellite imagery, oceanographic data, and automatic identification system (AIS) data, providing insights that were previously unattainable. This convergence of technology and marine expertise is paving the way for autonomous ships, which promise to further revolutionize the industry by enhancing efficiency, safety, and reliability [7,8]. Furthermore, the integration of IoT (Internet of Things) devices in maritime operations is facilitating the real-time monitoring and maintenance of critical ship components, thus preventing failures and ensuring smoother operations at sea [5]. Cybersecurity, too, has become a paramount concern, with these technological integrations necessitating robust security protocols to safeguard navigational and operational data from cyber threats [9].

As we enter this new era of safe, intelligent, and sustainable navigation, it is evident that the maritime industry is on the brink of a technological revolution [10]. The potential benefits extend far beyond the shipping companies themselves, impacting global trade and environmental conservation and even reshaping international regulatory frameworks [11]. In essence, the fusion of technology with traditional maritime practices is not just a trend but also a transformative movement, setting the course for a more efficient, safe, and sustainable future in marine navigation.

Citation: Zhang, M.; Zhang, X.; Fu, S.; Dai, L.; Yu, Q. Recent Developments and Knowledge in Intelligent and Safe Marine Navigation. *J. Mar. Sci. Eng.* **2023**, *11*, 2303. https://doi.org/10.3390/jmse11122303

Received: 2 November 2023
Accepted: 21 November 2023
Published: 5 December 2023

2. An Overview of the SI and Published Articles

In response to these transformative developments, the *Journal of Marine Science and Engineering* (JMSE) proudly presents this Special Issue (SI), entitled "Recent Developments and Knowledge in Intelligent and Safe Marine Navigation". This SI serves as a platform to showcase original contributions that explore and apply emerging and frontier technologies to bolster the intelligence and safety of operational ships in real-world conditions.

This SI brings together a collection of research papers that not only contribute to academic discourse but also have substantial practical relevance to the maritime industry. The insights and innovations presented within this SI are vital to ensuring the continued advancement of intelligent and safe marine navigation. The contributions discuss recent developments on big data analytics and big data fusion for ship detection (contributions 3 and 6); ship-to-ship collision avoidance using deep reinforcement learning approach (contributions 5 and 9); navigation risk evaluation (contributions 2 and 4); ship system identification for the development of digital twins of autonomous ships (contribution 8); intelligent route planning (contribution 7); and artificial intelligence transportation robots (contribution 10). These methods draw from a confluence of expertise, merging insights from ship science, big data science, AI, and their interdisciplinary interactions. Expanding on this knowledge, this SI also delves into the realm of advanced environmental monitoring techniques, leveraging AI to predict and mitigate the impacts of maritime activities on marine ecosystems (contribution 8). It further explores the integration of IoT and sensor networks for enhanced shipboard/offshore monitoring and diagnostics (contribution 3), with potential to enhance predictive maintenance and operational efficiency. Additionally, this SI addresses the critical aspect of cybersecurity/fake ship license plates (illegal ship behaviors) in maritime operations (contribution 1), providing novel strategies that can be employed to protect sensitive navigational and operational data in an increasingly connected and digitalized maritime environment. It also includes case studies and reports on real-world applications, highlighting the practical implementation of these technologies in various maritime settings. These studies provide invaluable insights into the challenges and successes of integrating high-tech solutions in the dynamic and often unpredictable marine environment.

3. Conclusions

This SI, "Recent Developments and Knowledge in Intelligent and Safe Marine Navigation", stands as a testament to the innovative spirit and forward-thinking approach of the maritime industry. It encapsulates a diverse array of research and developments that are not only pushing the boundaries of maritime science but also exploring a potential course for safer, more efficient, and sustainable marine navigation in the future.

We are greatly appreciative of the authors who have contributed their original research to this SI. Their commitment to enhancing marine navigation technology is clearly reflected in the quality and innovation of the papers they have presented. This collection demonstrates the collective efforts being undertaken to improve safety, intelligence, and sustainability in our oceans. This SI is more than a collection of academic papers; it is a hub where innovative ideas and transformative technologies in the maritime domain meet. Each contribution enriches our understanding of maritime science and expands the possibilities of marine navigation and safety. We encourage readers and researchers from across the world to explore the articles published in this SI. Whether you are a professional in the industry, an academic, or an enthusiast in marine science and engineering, these pages offer a wealth of knowledge and inspiration. The research and practical applications presented here are likely to shape the future of maritime navigation and safety.

In essence, this SI reflects the ongoing progress in this field and serves as an indicator of the potential for a more informed, efficient, and environmentally mindful maritime future. We anticipate that the insights offered in this collection will inspire further research, collaboration, and innovation globally, contributing to the ongoing advancement of intelligent and safe marine navigation.

Author Contributions: Conceptualization, M.Z.; writing—original draft preparation, M.Z.; writing—review and editing, M.Z., X.Z., S.F., L.D. and Q.Y.; funding acquisition, X.Z., S.F., L.D. and Q.Y.; All authors have read and agreed to the published version of the manuscript.

Funding: This research was funded by the National Natural Science Foundation of China (grant numbers: 52271363, 52371359, 72061127003, and 52201412), and the Shanghai Rising-Star Program (grant numbers: 22QC1400600).

Conflicts of Interest: The authors declare no conflict of interest.

List of Contributions:

1. Wan, H.; Fu, S.; Zhang, M.; Xiao, Y. A Semantic Network Method for the Identification of Ship's Illegal Behaviors Using Knowledge Graphs: A Case Study on Fake Ship License Plates. *J. Mar. Sci. Eng.* **2023**, *11*, 1906. https://doi.org/10.3390/jmse11101906.
2. Chang, Z.; He, X.; Fan, H.; Guan, W.; He, L. Leverage Bayesian Network and Fault Tree Method on Risk Assessment of LNG Maritime Transport Shipping Routes: Application to the China–Australia Route. *J. Mar. Sci. Eng.* **2023**, *11*, 1722. https://doi.org/10.3390/jmse11091722.
3. Huang, Z.; Hu, Q.; Lu, L.; Mei, Q.; Yang, C. Online Estimation of Ship Dimensions by Combining Images with AIS Reports. *J. Mar. Sci. Eng.* **2023**, *11*, 1700. https://doi.org/10.3390/jmse11091700.
4. Liao, S.; Weng, J.; Zhang, Z.; Li, Z.; Li, F. Probabilistic Modeling of Maritime Accident Scenarios Leveraging Bayesian Network Techniques. *J. Mar. Sci. Eng.* **2023**, *11*, 1513. https://doi.org/10.3390/jmse11081513.
5. Li, H.; Wang, X.; Wu, T.; Ni, S. A COLREGs-Compliant Ship Collision Avoidance Decision-Making Support Scheme Based on Improved APF and NMPC. *J. Mar. Sci. Eng.* **2023**, *11*, 1408. https://doi.org/10.3390/jmse11071408.
6. Liu, X.; Hu, Y.; Ji, H.; Zhang, M.; Yu, Q. A Deep Learning Method for Ship Detection and Traffic Monitoring in an Offshore Wind Farm Area. *J. Mar. Sci. Eng.* **2023**, *11*, 1259. https://doi.org/10.3390/jmse11071259.
7. Li, Y.; Cui, J.; Zhang, X.; Yang, X. A Ship Route Planning Method under the Sailing Time Constraint. *J. Mar. Sci. Eng.* **2023**, *11*, 1242. https://doi.org/10.3390/jmse11061242.
8. Lan, J.; Zheng, M.; Chu, X.; Ding, S. Parameter Prediction of the Non-Linear Nomoto Model for Different Ship Loading Conditions Using Support Vector Regression. *J. Mar. Sci. Eng.* **2023**, *11*, 903. https://doi.org/10.3390/jmse11050903.
9. Sun, Z.; Fan, Y.; Wang, G. An Intelligent Algorithm for USVs Collision Avoidance Based on Deep Reinforcement Learning Approach with Navigation Characteristics. *J. Mar. Sci. Eng.* **2023**, *11*, 812. https://doi.org/10.3390/jmse11040812.
10. Zhu, Y.; Zhang, Q.; Liu, Y.; Hu, Y.; Zhang, S. Neural Network, Nonlinear-Fitting, Sliding Mode, Event-Triggered Control under Abnormal Input for Port Artificial Intelligence Transportation Robots. *J. Mar. Sci. Eng.* **2023**, *11*, 659. https://doi.org/10.3390/jmse11030659.

References

1. UNCTAD. *Review of Maritime Transport 2022*; UNCTAD: Geneva, Switzerland, 2022.
2. Guo, Y.; Liu, R.W.; Qu, J.; Lu, Y.; Zhu, F.; Lv, Y. Asynchronous Trajectory Matching-Based Multimodal Maritime Data Fusion for Vessel Traffic Surveillance in Inland Waterways. *IEEE Trans. Intell. Transp. Syst.* **2023**, *24*, 12779–12792. [CrossRef]
3. Zhang, M. Big Data Analytics Methods for Collision and Grounding Risk Analysis in Real Conditions: Framework, Evaluation, and Applications. Ph.D. Thesis, Aalto University, Espoo, Finland, 2022.
4. Liu, R.W.; Guo, Y.; Nie, J.; Hu, Q.; Xiong, Z.; Yu, H.; Guizani, M. Intelligent edge-enabled efficient multi-source data fusion for autonomous surface vehicles in maritime internet of things. *IEEE Trans. Green Commun. Netw.* **2022**, *6*, 1574–1587. [CrossRef]
5. Liu, R.W.; Guo, Y.; Lu, Y.; Chui, K.T.; Gupta, B.B. Deep network-enabled haze visibility enhancement for visual IoT-driven intelligent transportation systems. *IEEE Trans. Ind. Inform.* **2022**, *19*, 1581–1591. [CrossRef]
6. Beşikçi, E.B.; Arslan, O.; Turan, O.; Ölçer, A.I. An artificial neural network based decision support system for energy efficient ship operations. *Comput. Oper. Res.* **2016**, *66*, 393–401. [CrossRef]
7. Ghaderi, H. Autonomous technologies in short sea shipping: Trends, feasibility and implications. *Transp. Rev.* **2019**, *39*, 152–173. [CrossRef]
8. Zhang, M.; Kujala, P.; Musharraf, M.; Zhang, J.; Hirdaris, S. A machine learning method for the prediction of ship motion trajectories in real operational conditions. *Ocean Eng.* **2023**, *283*, 114905. [CrossRef]
9. Bolbot, V.; Basnet, S.; Zhao, H.; Banda, O.V.; Silverajan, B. Investigating a novel approach for cybersecurity risk analysis with application to remote pilotage operations. In *European Workshop on Maritime Systems Resilience and Security*; Konetekniikan Laitos: Espoo, Finland, 2022.

10. IMO. *Regulatory Scoping Exercise for the Use of Maritime Autonomous Surface Ships (MASS)*; MSC 99/WP.9; IMO: London, UK, 2018.
11. Kim, M.; Joung, T.H.; Jeong, B.; Park, H.S. Autonomous shipping and its impact on regulations, technologies, and industries. *J. Int. Marit. Saf. Environ. Aff. Shipp.* **2020**, *4*, 17–25. [CrossRef]

Journal of
Marine Science and Engineering

MDPI

Article

A Semantic Network Method for the Identification of Ship's Illegal Behaviors Using Knowledge Graphs: A Case Study on Fake Ship License Plates

Hui Wan [1,2], Shanshan Fu [3], Mingyang Zhang [4,*] and Yingjie Xiao [1]

1 Merchant Marine College, Shanghai Maritime University, Shanghai 201306, China; dhnthui@163.com (H.W.); xiaoyj@shmtu.edu.cn (Y.X.)
2 Shanghai Chart Center, Donghai Navigation Safety Administration of MOT, Shanghai 200090, China
3 College of Transport & Communications, Shanghai Maritime University, Shanghai 201306, China; ssfu@shmtu.edu.cn
4 Department of Mechanical Engineering, School of Engineering, Aalto University, 02150 Espoo, Finland
* Correspondence: mingyang.0.zhang@aalto.fi

Abstract: With the advancement of intelligent shipping, current traffic management systems have become inadequate to meet the requirements of intelligent supervision. In particular, with regard to ship violations, on-site boarding is still necessary for inspection. This paper presents a novel approach for enhancing ships' management and service capabilities through scientific knowledge graph technology to develop a ship knowledge graph. The proposed approach extracts key characteristics of ship violations from the ship knowledge graph, such as monitoring ships, expired ship certificates, multiple ship tracks, inconsistent ship tracks with port reports, and ships not reported to the port for a long time. Combining the characteristics of ship violations, the approach uses reasoning and identification techniques to detect specific instances of falsely licensed ships and other violations. The development of the ship knowledge graph analysis system enables the identification and verification of illegal ships using fake license plates, while also improving the effective utilization of maritime data and enhancing the ability to make informed decisions related to ship safety. By leveraging cognitive approaches and knowledge graphs, this study offers the potential to develop an intelligent decision-making system for maritime traffic management.

Keywords: ship knowledge graph; illegal behavior; fake ship license plates; decision-making; traffic management

Citation: Wan, H.; Fu, S.; Zhang, M.; Xiao, Y. A Semantic Network Method for the Identification of Ship's Illegal Behaviors Using Knowledge Graphs: A Case Study on Fake Ship License Plates. *J. Mar. Sci. Eng.* **2023**, *11*, 1906. https://doi.org/10.3390/jmse11101906

Academic Editor: Fausto Pedro García Márquez

Received: 7 September 2023
Revised: 20 September 2023
Accepted: 25 September 2023
Published: 1 October 2023

1. Introduction

1.1. Background

With the advancement of intelligent shipping, traditional maritime supervision systems such as ship automatic identification systems, ship traffic management systems, very high frequency (VHF) wireless, and ship video supervision systems have become inadequate to meet the requirements of intelligent supervision. These systems are unable to achieve the safety supervision purpose of wide area coverage, intelligent decision-making, and rapid response [1,2]. In particular, with regard to ship violations, on-site boarding is still necessary for inspection. The current practice of comparing the automatic identification system (AIS) dynamic information of the inspected ship with the AIS information on the maritime supervision platform through manual identification, to identify the ship as abnormal or violating if it does not match, is time-consuming and inefficient for timely and accurate maritime supervision [3,4]. Moreover, challenges such as irregular and renamed ship names, diverse configurations of ship AIS equipment, multiple ships with one nine-digit code or one ship with multiple nine-digit codes [5,6], and irregular encoding of ship nine-digit code information also present significant obstacles to the precise detection of ship violations [7,8].

Currently, research on identifying ship's illegal behaviors primarily focuses on ship traffic management. Liu and Tian [9] introduced a traditional method for analyzing and discriminating AIS ship data to identify illegal coding problems of ship AIS equipment, with a single recognition target and low performance. Rong et al. [10] proposed a new method for automatically identifying ship collision avoidance behavior from ship trajectories using AIS trajectory data, and the data source is relatively small and fails to consider more comprehensive factors, resulting in insufficient reliability and accuracy of identification [11]. Kaluza et al. [12] elaborated on the attribute characteristics of key ship data in maritime data according to different ship uses and stowage methods, focused on the data attributes themselves and failed to attempt to explore the association relationships between attributes. Veenstra et al. [13] proposed a framework for port sensing and computing based on maritime big data. While this method uses multi-source data, such as ship GPS tracks, ship attributes, port geographic information, and port facility parameters to construct maritime big data, it still relies on system-level data analysis. Acharya et al. [14] used spatial clustering to identify accident-prone areas, analyze the causes of accidents, and develop risk prevention and control strategies based on maritime accident data, but did not consider other dimensions to make comprehensive assessment decisions. Fu et al. [15] proposed an object-oriented Bayesian network model to quantitatively assess the risk of maritime accident scenarios in ice-covered Arctic waters concerning human and organizational factors. Similarly, Kehrer and Hauser [16] reviewed visualization for multifaceted scientific data and visual analysis techniques but did not use knowledge graphs as visual objects.

This analysis of the literature shows that the identification of violations in multi-source maritime ship data is predominantly based on the manual matching of information. Research methods for processing before data analysis are also mostly conventional, and the source data used for data analysis are relatively homogeneous, failing to analyze the behavior of ships comprehensively. Most research on maritime mass data focuses on the attribute characteristics that can be directly reflected by the data themselves, failing to explore the potential behavior of ships under big data. Although research on ship applications based on maritime data is relatively abundant, the functions are often similar and discrete, leading to issues such as the homogeneity of ship behavior analysis and inconsistency of analysis points. Currently, research on ship multi-source data processing and behavior analysis based on knowledge graph technology is lacking, especially regarding the application research on ship knowledge graph construction, ship violation analysis, and identification for maritime ship data [17,18]. Given the rapid development of shipping technology driven by new technologies such as big data, the Internet of Things, and artificial intelligence, the need for intelligent transformation and development of water traffic management and services has become more pressing. It is necessary to conduct in-depth research on the innovative application of maritime data combined with new technologies.

To improve the effectiveness of ship management and services, this study proposes an innovative approach that constructs a ship static and dynamic relationship graph based on multi-source maritime ship static and dynamic data. The study analyzes the correlation relationship between ships, extracts ship behavioral features, and comprehensively reasons and identifies ship violations using these features. The proposed method is validated through a practical case study, which demonstrates the feasibility and effectiveness of the knowledge-graph-based approach for identifying ship violations. This method integrates maritime multi-source data to achieve universal ship knowledge graph modeling. Based on violation behavior rules, graph inference technology is used to efficiently determine ship violations, improving the efficiency of ship behavior judgment, and offering the potential to develop an intelligent decision-making system for maritime traffic management.

1.2. The Definition of Ship Illegal Behavior

The term "ship violations" generally pertains to ship-related behaviors that contravene the stipulations outlined in the Maritime Traffic Safety Law of the People's Republic of China. These behaviors can include violations of ship and marine facility management orders, crew management orders, navigation, berthing, and operational management orders, dangerous goods carriage safety supervision and management orders, and marine search and rescue management orders, as well as marine traffic accident investigation and handling orders [4,19–21]. The definitions of several illegal behaviors are listed as follows:

- ***Speed violation***: a ship sailing at a speed exceeding the prescribed speed, including speeding or not complying with speed restrictions in a specific area;
- ***Improper navigation rules***: the failure of a ship to follow the prescribed navigation rules, such as failure to follow the navigation guidelines, or failure to comply with the rules for crossing ships;
- ***Unqualified seaworthiness safety inspection***: the ship's seaworthiness certificate information is incomplete or abnormal, the navigation safety equipment is incomplete or defective, and the ship fails to pass the maritime ship safety inspection, and is deemed unseaworthy;
- ***Violation of maritime management requirements***: the failure of a ship to comply with maritime safety management requirements during navigation, such as ship misregistration, AIS closure, failure to report port, failure to maintain appropriate ship spacing, failure to comply with navigation signs and signals, and failure to take appropriate collision prevention measures;
- ***Improper use of communication and navigation equipment***: the failure or incorrect use of communication and navigation equipment by ships during navigation, such as navigation radio communication and navigation equipment, satellite communication and navigation equipment, radar, or GPS.

This article, however, focuses specifically on the issue of fake ship license plates, which constitute a breach of navigation, berthing, and operational management orders [22,23]. Utilizing a fake license plate involves the unlawful use of the same ship name and number belonging to another or the use of a canceled ship name and number painted on a new, renewed, modified, or purchased ship without approval. These fake ship license plates are used under the guise of legitimacy with the intent of deceiving others. To counter this problem, the Ministry of Transport and Communications has been actively working to combat the use of fake ship license plates. In line with this, the maritime department has been exerting significant pressure to suppress illegal acts of licensing of ships.

2. Method

2.1. Framework

This research focuses on the data object associated with maritime ships, encompassing both static and dynamic data. The research process involves three key aspects: stage 1 pertains to processing ship data, while stage 2 involves the construction of a knowledge graph model related to ships. Finally, stage 3 centers on the application of the ship knowledge graph, as shown in Figure 1. The research process can be distilled into three main stages:

- ***Stage 1: Data and processing.*** Ship data processing is necessary due to the heterogeneity of data from multiple sources and issues such as redundancy, anomalies, irregularities, and other paradigmatic problems. Data pre-processing techniques including data de-duplication, data noise reduction [24], data supplementation, data fusion, and other methods are utilized to address these challenges;
- ***Stage 2: Ship knowledge graph modeling (takes center stage).*** This involves using knowledge graph technology and the maritime supervision business model to establish ship graph semantic rules and achieve the semanticization of the ship data graph [25,26]. Entity extraction and relationship extraction methods are employed to transform triadic data structures [27]. At the same time, knowledge fusion and denotation disambiguation techniques are utilized to complete the construction of the ship static graph [28]. The construction method of the ship knowledge graph in this study has significant innovative advantages in understanding actual business rules, integrating professional knowledge in the field, automating data updates, and

providing powerful query interfaces. It better meets actual business needs, improves reliability and operability. Subsequently, based on AIS data and the definition of ship violations, calculations are made for ship violation features such as monitoring key ships, expired ship certificates, inconsistent ship tracks with reported ports, multiple ship tracks, and ships that do not report ports for extended periods [29]. These calculated violation features, in combination with the ship name graph identified through on-site inspections, are used to infer ship licensing violations, ultimately leading to the creation of a complete ship knowledge graph that integrates both dynamic and static data [30,31];

- ***Stage 3: The application and validation of the ship knowledge graph.*** Using fused regional ship static and dynamic graphs, E-chart technology is used to conduct graph analysis and display ship violations [32]. The feasibility of the method is then verified by combining it with actual cases.

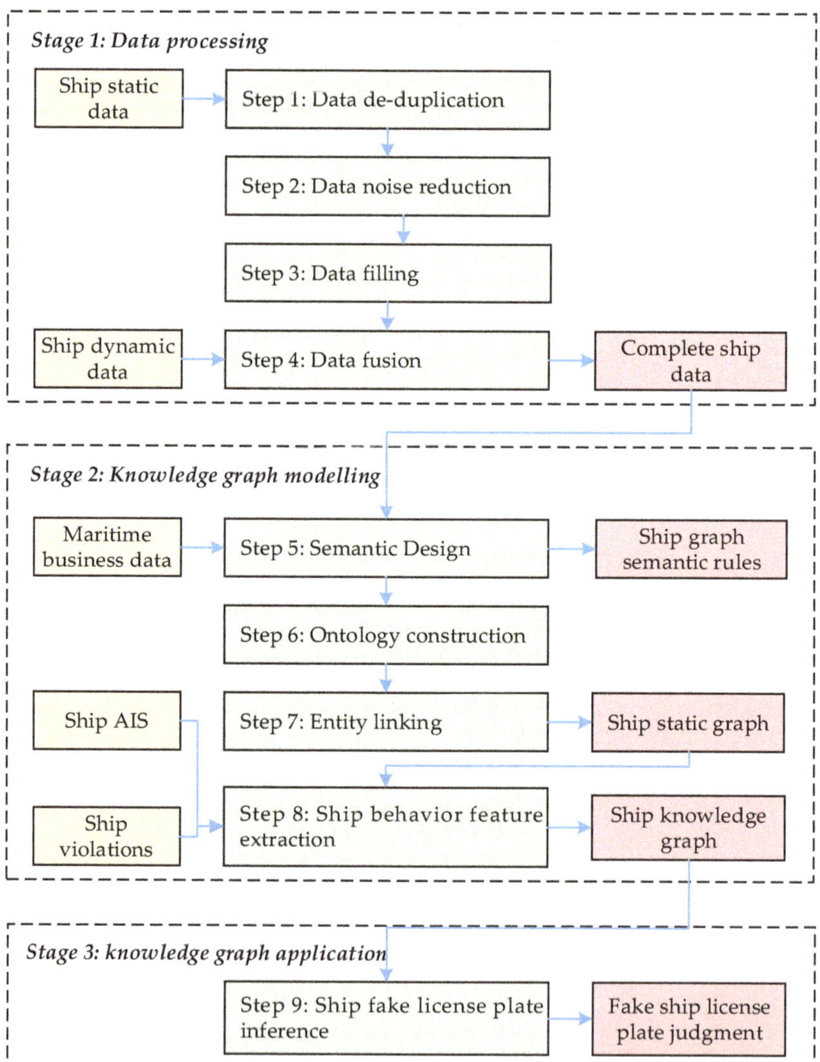

Figure 1. Research framework of the identification of ship's illegal behaviors using knowledge graph.

2.2. Data and Processing

2.2.1. Data Origin

This study area mainly focuses on the identification of illegal behaviors of fake ship license plates based on the ship's knowledge graph. The main task is to study the construction of the ship's knowledge graph and identify the fake ship license plate behaviors. The primary data involved are the static data of ships in maritime management and the dynamic business data of maritime supervision. The dataset includes basic ship information, ship company information, key tracking ship information, ship nationality certificates, ship entry and exit reports, ship violation inspection data, and ship AIS information [33–35], as shown in Table 1.

Table 1. Data source.

Dataset	Information
Basic ship information	Ship number, Chinese and English names of the ship, ship type, ship length, ship tonnage, ship owner, ship registration location, ship location, ship status.
Ship company	Serial number, Chinese and English names of the ship, legal representative, telephone number.
Key tracking ship	Tracking number, ship registration number, Chinese and English names of the ship, port of registry, MMSI, tracking reason code, tracking reason name, and date of creation, among others; these data mainly have data redundancy and missing data.
Ship nationality certificates	Ship number, Chinese name of the ship, MMSI, owner of the ship, contact number, starting validity, expiry date, registration authority, certificate printing number.
Ship entry and exit reports	Ship number, ship registration number, Chinese and English names of the ship, MMSI, ship type, gross tonnage, net tonnage, gross engine power, passenger capacity, ship length, beam, depth, ship owner, port name, port number, type of port, name of port, reporting time, berthing code, berth of call, actual cargo volume, local port volume, goods, passenger.
Ship violation inspection	Inspection number, inspection code, inspection content, inspection results, problems, description.
Ship AIS	MMSI, Chinese and English names of the ship, heading, course, speed, longitude, latitude, draft, received time.

- ***Basic ship information*** comprises essential ship data such as ship number, Chinese and English names of the ship, Maritime Mobile Service Identify (MMSI), nationality, ship type, and initial registration number [36]. The main issues with these data are data redundancy, data noise, and data being missing;
- ***Ship company information*** encompasses the shipowner's company serial number, Chinese and English names of the company, legal representative, and contact telephone number, among other details. These data mainly have data ambiguity issues;
- ***Key tracking ship information*** includes tracking number, ship registration number, Chinese and English ship names, port of registry, MMSI, tracking reason code, tracking reason name, and date of creation, among others. These data mainly have data redundancy and missing data;
- ***Ship nationality certificate data*** includes ship registration number, Chinese name of the ship, MMSI, owner of the ship, contact number, starting validity period, expiry date of the certificate, registration authority, and certificate printing number, among other key details. The main issue with these data is missing data;
- ***Ship entry and exit report data*** is comprised of ship identification number, ship registration number, Chinese and English ship names, MMSI, ship type, gross tonnage, net tonnage, gross engine power, passenger capacity, overall ship length, beam, depth, ship owner, port name, port number, type of port of entry/exit, name of next port, reporting time, berthing code, berth of call, actual cargo volume, local port unloading/loading volume, actual dangerous goods volume, local port unloading/loading dangerous goods volume, actual passenger volume, local port drop-off/pick-up volume, actual vehicle volume, local port unloading/loading vehicle volume, number

of barges, local port unloading/loading barges, actual container volume, local port unloading/loading container volume, and other relevant information. The main issue with these data is data noise;

- ***Ship violation inspection information*** comprises data such as the supervision and inspection number, inspection content code, inspection content, inspection results, violations or problems found, and description of violations or problems. The main issues with these data include missing data and data ambiguity;
- ***Ship AIS information*** includes MMSI, Chinese and English ship names, heading, course, speed, longitude, latitude, draft, received time, and other relevant data points. The main issues with these data include data noise and missing data.

Duplicate ship names and MMSI mainly characterize the problem of duplicate and redundant maritime ship data. The problem of ship data noise includes attribute data noise and spatial data noise. Attribute noise mainly refers to the problem of unclear data features caused by the complexity of field design. In contrast, spatial data noise mainly refers to the problem of data features not being displayed and data spatial anomalies caused by large amounts of data. The problem of missing ship data mainly includes features such as MMSI and trajectory being missing. The ambiguity problem of ship data is mainly caused by inconsistent information such as the ship's ownership company, ship contact person, and communication address. Based on the knowledge graph of ships, identifying illegal behavior of ship deck rigging may pose a threat to the safety of ships and may also pose a threat to the safety of crew members. The behavior judgment process requires collecting and processing a large amount of ship and crew information, which may involve personal privacy and require measures to protect it.

2.2.2. Data Processing

Ship data pre-processing is a key step for accurate and efficient data mapping transformation, analysis, and application. The principle of data processing starts with removing abnormal data and updating supplementary data as much as possible. The pre-processing methods used include data de-duplication, data noise reduction, data filling, and data fusion, as shown in Figure 2. The following are specific descriptions of these steps.

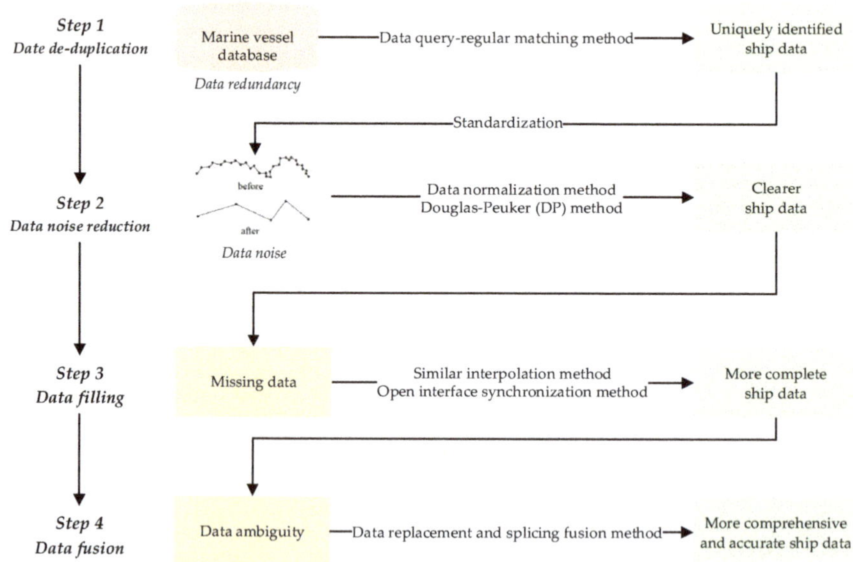

Figure 2. Data processing.

- ***Step 1: Data de-duplication.*** For the problem of duplicate redundancy of maritime ship data, data cleaning is carried out by data query and regular matching methods, including redundancy processing by using MMSI and ship name segment and redundancy processing by ship registration number and ship name segment to ensure that ship data can be uniquely identified;
- ***Step 2: Data noise reduction.*** For the ship data noise problem, the data normalization method is used for standardized field template design and field screening, and in principle, information such as primary key, Chinese name, English name, type, and content is retained, and attribute field information is simplified; for the ship spatial data noise problem, the Douglas–Peuker (DP) method is used for the abnormal ship spatial trajectory extraction thinning to simplify the ship spatial field information to achieve the purpose of clearer data. The DP algorithm is used to determine the key waypoints of the ship's trajectory [35]. These key waypoints can be used to represent the geographical locations where the ship changes course. Figure 3 and Table 2 show the steps of the DP algorithm. Step 1 can be used to generate an approximated line segment between the departure and destination points. Steps (2, 3, . . . , *n*) introduce sub-line segments using various threshold parameters ε [23].

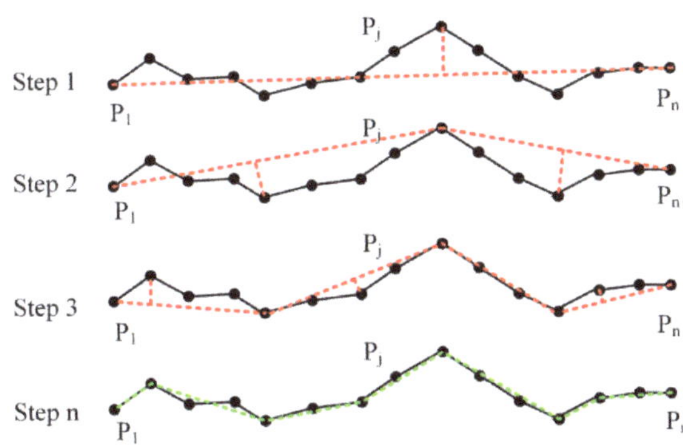

Figure 3. Douglas–Peuker algorithm.

Table 2. DP algorithm for the abnormal ship spatial trajectory extraction.

DP Algorithm

Inputs : Require :
ship trajectory (Points); ε (thresholds) Parameter d (distance), dm (maximum distance), index
Outputs : Waypoints $WP = \{p_1, p_2, \ldots, p_k\}$
Process : For $i = 2$ to $i = (n-1)2$: d = VD ($CL[i], Line[CL[1], CL[n]]$)
If d > dm
index = *i*
dm = d
End if
End if
If dm >= €
WaypointLList = *DP* (*CL* [1. . . . . .index], ε)
WaypointRList = *DP* (*CL* [index. . . . . .n], ε)
Waypoints = {WaypointLList, WaypointRList}
Else
Waypoints = {*CL* [1], *CL* [n]}
End if
Return Waypoints

The steps of the DP algorithm (the black line represents the original ship trajectory centerline, and the red line represents the simplified ship trajectory. The idea of the simplified ship trajectory centerline is to approximate the original trajectory in green).

- ***Step 3: Data filling.*** For the problem of missing ship data, the same kind of cubic spline function interpolation, K-mean clustering, and mean value method [23] are used to estimate and fill the missing data. Making full use of the third-party complete ship data, adopting the method of data interface synchronization, and completing the data filling of MMSI, ship name, ship type, etc., based on the third-party data interface, and completing the filling of spatial information such as port and ship trajectory based on the open GIS data interface make ship data more complete. Data filling effectively improves the integrity of ship data, improves the accuracy of ship data analysis and ship knowledge graph modeling, enhances the distribution characteristics of ship data, and improves the efficiency of ship data analysis and processing. The cultivation process fully considers the nature of ship data and the types of missing values, and combines the goal of building a ship knowledge graph to ensure that the filling process does not introduce noise or deviation;
- ***Step 4: Data fusion.*** In response to the problem of ship data ambiguity, we adopted data replacement and splicing fusion methods, such as splicing and fusing multiple contact fields to form more accurate contact descriptions, so as to eliminate the heterogeneity between data and improve data integrity and reliability.

2.3. *Knowledge Graph for the Identification of Ship Illegal Behaviors*

The process of constructing a knowledge graph for the domain of ships typically involves several interrelated steps. These steps typically include the semantic design of the graph, ontology construction, entity linking, graph computation, and inference. Semantic design involves the identification of domain-specific concepts and relationships, which form the backbone of the knowledge graph. This step is essential for ensuring that the knowledge graph accurately reflects the domain of interest and can provide meaningful insights. Ontology construction is the process of creating a formal specification of the concepts and relationships in the domain, typically using a standardized language such as OWL or RDF. This step is crucial for ensuring that the knowledge graph is structured in a way that can be easily queried and analyzed. Entity linking involves identifying and linking instances of concepts in the knowledge graph to external sources of information, such as databases or websites. This step can help to enrich the knowledge graph with additional information, and make it more useful for real-world applications. Graph computation is the process of analyzing the structure of the knowledge graph and extracting useful insights or patterns. This step is essential for understanding the relationships between concepts and identifying domain interest areas. Finally, inference is the process of making logical deductions or predictions based on the knowledge graph, using techniques such as rule-based reasoning or machine learning. This step can help to uncover new insights and generate hypotheses for further investigation.

A top–down construction method [37] is used. The top–down construction method is a widely used approach to knowledge graph construction, which involves starting with a high-level view of the domain and then refining the details and relationships over time. This method helps to ensure that the knowledge graph is accurate, consistent, and reflects the most important concepts and relationships in the domain.

2.3.1. Step 5: Semantic Design

To construct a ship knowledge graph business model, under the process of Section 2.2, a Mysql [38], ship relational database, is utilized for field extraction after a thorough understanding of ship maritime supervision and navigation security business, as shown in Figure 4.

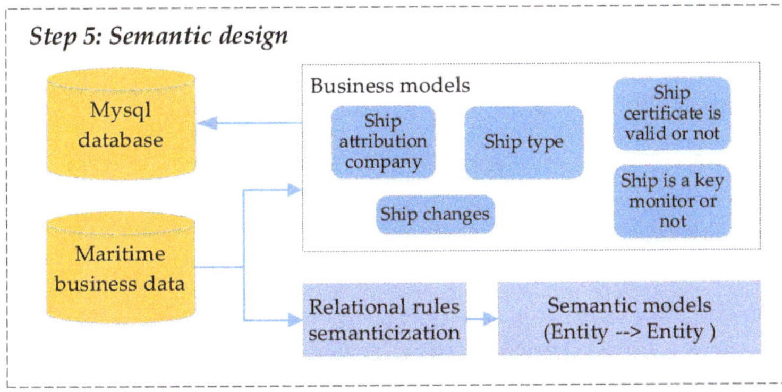

Figure 4. Semantic design for the identification of ship's illegal behaviors.

The model encompasses the ship company, ship type, ship history changes, ship certificates, ship key tracking and reasons for tracking, and ship dynamics port call. The relationships between business models are integrated with actual maritime business logic. Semanticized rule descriptions are conducted via entity-relationship-entity ternary semantic definitions to accomplish this. In this approach, nodes represent objects while concepts and edges signify relationships between nodes. The definition of association combinations is established to formulate a ship graph semantic rule that comprehensively describes the entire maritime business.

2.3.2. Step 6: Ontology Construction

In order to extract relevant information from the processed ship data, the semantic rules pertaining to entity, attribute, and relationship data are employed. The relationships and events between the entities are established using primary and foreign keys, creating a relational subject–predication–object (SPO) triad. Subsequently, the triadic data are combined and transformed into basic ship graph ontology elements, as shown in Figure 5.

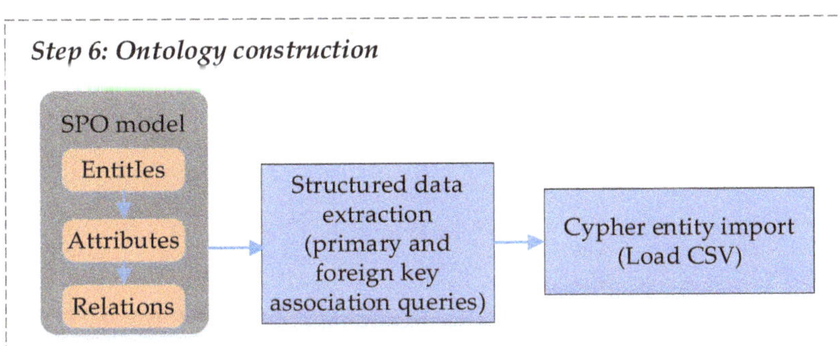

Figure 5. Ontology construction.

The primary objective of entity identification and extraction is to identify named entities of various categories, including shipping companies, ships, crew, certificates, key surveillance, ports, berths, dynamic reporting visas, etc. The extraction process entails identifying structured entity-relational data, defining entity boundaries, and determining entity types through Mysql-based primary and foreign key association queries. Following this, semantic rules defined in the previous step are used to extract two or more entities from the relational data and establish matching semantic relationships. The Cypher Load CSV function [39] is employed within the neo4j graph database to convert the relational SPO ternary data into entity graph data.

2.3.3. Step 7: Entity Linking

In the process of integrating ship data, multiple representations for a single entity often lead to a reduction in the quality of the integrated data, as shown in Figure 6. This paper proposes a pattern-matching method to address the problem of entity referent disambiguation. The method involves discovering the graph relationships between attributes in different relational data sources and calculating the similarity of related descriptions to match patterns. Entities and relational objects are deleted, replaced, and fused to solve conflicts between predicates in triples, and to ensure the unique integration of heterogeneous data sources. Furthermore, the meaning of ship entities may vary in different contexts, necessitating entity disambiguation. This paper adopts word-sense annotation to model disambiguation. The semantic features introduced in the previous section include semantic class information based on syntactic relations, such as semantic classes of subject/object central words, and semantic role annotation class information. Combining semantic information for contextual semantic discrimination classification can address the problem of diversity in the representation of ship entities.

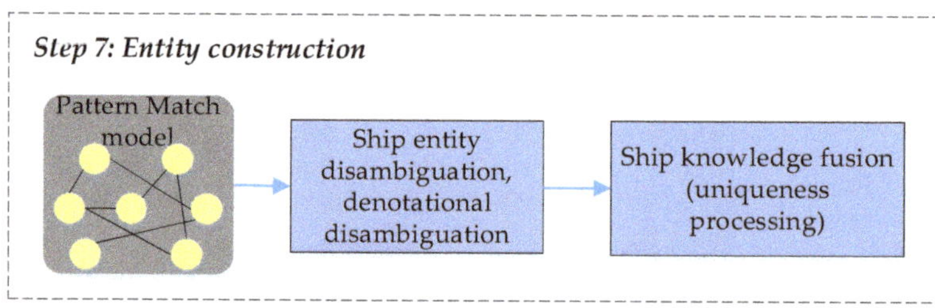

Figure 6. Entity construction.

2.3.4. Step 8: Ship Behavior Feature Extraction

Based on the developed database, a ship behavior feature extraction method is proposed to identify ship behavior feature features, as shown in Figure 7.

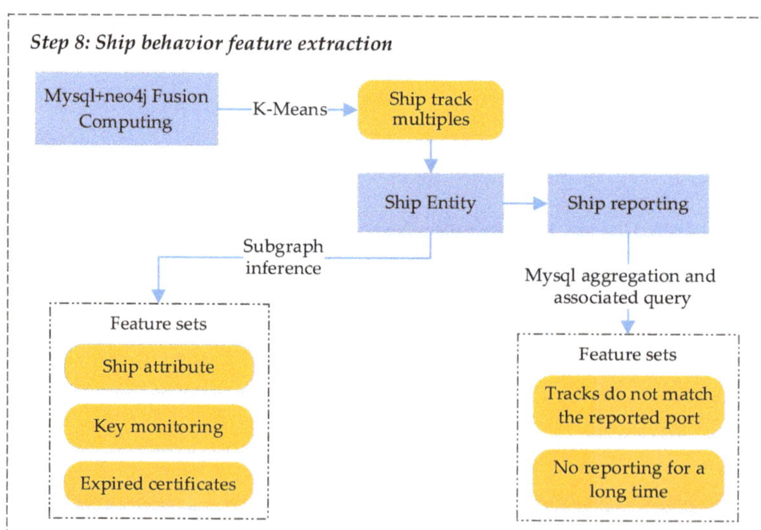

Figure 7. Ship behavior feature extraction.

Based on the features, K-means algorithm is used to cluster this feature information of ship trajectory in the case study area [35]. The mathematical background to this process is displayed in Equations (1)–(4) below and the pseudocode of the process is summarized in Table 3. For a point p_i in the way of a trajectory defined as per Equation (1), the locations of the departure and destination ports are defined as in Equation (2). Then, the length of the ship trajectory is calculated as in Equations (3) and (4).

$$p_i = \{MMSI, TIMESTAMP, LON, LAT, SOG, COG, Draft\} \tag{1}$$

$$T_{se} = \{(lon_1, lat_1), (lon_n, lat_n)\} \tag{2}$$

$$d(p_j, p_{j+1}) = (TimeStamp_{j+1} - TimeStamp_j) \times (sog_j + sog_{j+1})/2 \tag{3}$$

$$T_{length} = \sum_{j=1}^{n-1} d(p_j, p_{j+1}) \tag{4}$$

where (lon_1, lat_1) and (lon_n, lat_n) denote the longitude and latitude of the departure and destination ports, respectively. $d(p_j, p_{j+1})$ denote the distance between p_j and p_{j+1} (see Figure 8). This is example 2 of an equation:

Table 3. Ship trajectories clustering method using K-means algorithm.

K-Means Algorithm
Input: Dataset, $D = \{x_1, x_2, ..., x_m\}$, clustering number K, the maximum number of iterations N **Output:** Clustering $C = \{c_1, c_2, ..., c_k\}$ **Process:** 1. Select K trajectories as the centre trajectories $\{\mu_1, \mu_2, ..., \mu_k\}$; 2. Initially cluster division $C_t = \{c_1, c_2, ..., c_k\}$; 3. For n = 1, 2, ..., N: 4. For i = 1, 2, ..., m: 5. Calculate distance between the trajectory x_i and $\mu_j (j = 1, 2, ..., k)$ $d_{ij} = \lvert x_i - \mu_j \rvert$; 6. Mark category as j corresponding the smallest d_{ij}; 7. End for 8. For j = 1, 2, ..., K: 9. Calculate the centre trajectories based on the new clustering result $\mu_j = \sum x(x \in \mu_j)/\lvert \mu_j \rvert$ 10. End for 11. If the clustering result remains consistent: 12. Go to line 17; 13. Else: 14. Go to line 4; 15. End if 16. End for 17. Output $C = \{c_1, c_2, ..., c_k\}$.

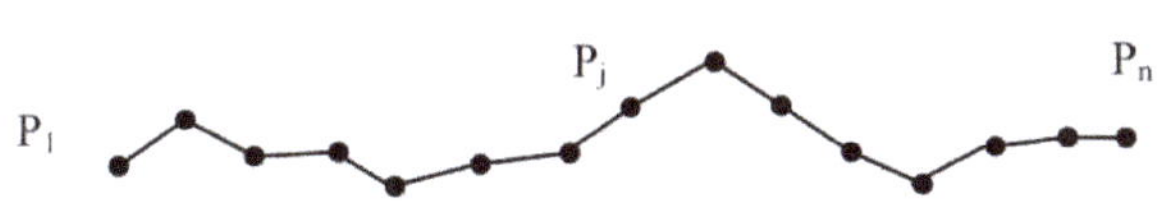

Figure 8. Defined ship trajectory using AIS data.

Finally, the ship multi-track determination method traces through Mysql relational data and neo4j graph database fusion calculation with the following rules:

(a) Define the data set $X = \{x_1, x_2, ..., x_n\}$, where each x_i represents a point in the ship trajectory;
(b) Preprocessing the dataset, e.g., sorting the trajectories in chronological order and representing each point as a triplet of (longitude, latitude, time);
(c) Choose the number of clusters k, initialize k prime centers $c_1, c_2, ..., c_k$;
(d) For each point x_i, calculate its distance d_{ij} to each center of mass c_j and classify it into the cluster s_j to which the center of mass with the closest distance belongs;
(e) Update the center of mass, and for each cluster s_j, update its center of mass c_j to be the average of all its points;
(f) Repeating the above steps (d) and (e) until the center of mass no longer changes or reaches the specified number of iterations, obtaining the number of ship multiple trajectories, and finally labeling and refining the ship multiple trajectory feature information into the ship feature graph.

Based on the ship entity (ship number, ship name), extract local subgraphs centering on ship attributes (such as ship name, MMSI, ship type, etc.), key monitoring relationship and certificate expiration relationship, obtain ship attributes, ship key monitoring and ship certificate expiration subgraph features and refine them into the new ship feature graph with specific markings; use Mysql aggregation and correlation query to retrieve ship reporting information. The ship characteristics are extracted and added to the ship characteristics graph by using Mysql aggregation and correlation query to retrieve the ship reporting information; the ship multiple trajectory features and the ship reporting time sequence information are fused, and the inconsistent trajectory and reporting features are inferred and added to the ship characteristics graph.

2.3.5. Step 9: Fake Ship License Plates Inference

As shown in Figure 9, the rules for fake ship license plates can be defined in order of whether the ship is an inland ship, MMSI or ship name information does not match, certificate expiration or invalidity, inconsistent or non-reporting of ports and trajectories, and ship focus monitoring. Using ship knowledge graph feature analysis technology, ship feature vectors can be extracted from ship information from monitoring chokepoints before combining them with feature weights for weighted assignment. The calculation process includes the following four sub-steps:

Step 9-1: Discover the ships that may be snared from the mapping of ship behavior characteristics in the study area, mainly acquiring the ships with multiple trajectories (this ship is usually a sea ship), and extract the abnormal trajectory data of the ship at sea;

Step 9-2: Retrieve the surveillance chokepoints within 50 kn of the surrounding area with this ship as the center. Assume the longitude and latitude of the ship i (i = 1, 2, 3) are lat_i and lat_j respectively; the longitude and latitude of the monitoring chokepoint j (j = 1, 2, 3) are lon_i and lat_i, respectively. The distance D_{ij} between ship i and monitoring chokepoint j can be calculated as Equations (5)–(7).

$$D_{ij} = R * 2a * tan^2(\sqrt{a}, \sqrt{1-a}), \quad (5)$$

$$a = sin^2(\Delta lat/2) + cos(lat_i) * cos(lat_j) * sin^2(\Delta lon/2)\ . \quad (6)$$

$$\Delta lon = lon_i - lon_j,\ \Delta lat = lat_i - lat_j. \quad (7)$$

where R is the radius of the Earth and takes the value of about 6371 km. After finding the distance between any point and the current position, determine whether it is less than or equal to 50 kn (1 kn is about 1.852 km) to obtain the corresponding result;

Step 9-3: Calculate the spatial distance between the multi-track ship and the surveillance chokepoint, calculate the average speed of the ship, combine the two to calculate the time difference, and collect the ship's information before the time difference of the surveillance chokepoint. The calculation formula of average ship speed is as Equation (8):

$$V_n = \sum_{i=1}^{n} v_i/n = (v_1 + v_2 + \ldots + v_n)/n \tag{8}$$

V_n denotes the average speed, V_i is the ship sailing speed at each point, and n denotes the total number of points;

Step 9-4: The collected ship information set is correlated and compared with the ship behavior characteristic graph in the current region, and combined with the fake license plate rules, according to the weight order of whether the ship is a river ship, whether the ship information does not match (ship name/MMSI), whether the ship certificate is invalid, whether the ship does not have the reported port record (if so, also judge whether the ship trajectory is inconsistent with the reported port), or whether the ship is a key monitoring, etc. The highest possible ship is finally deduced.

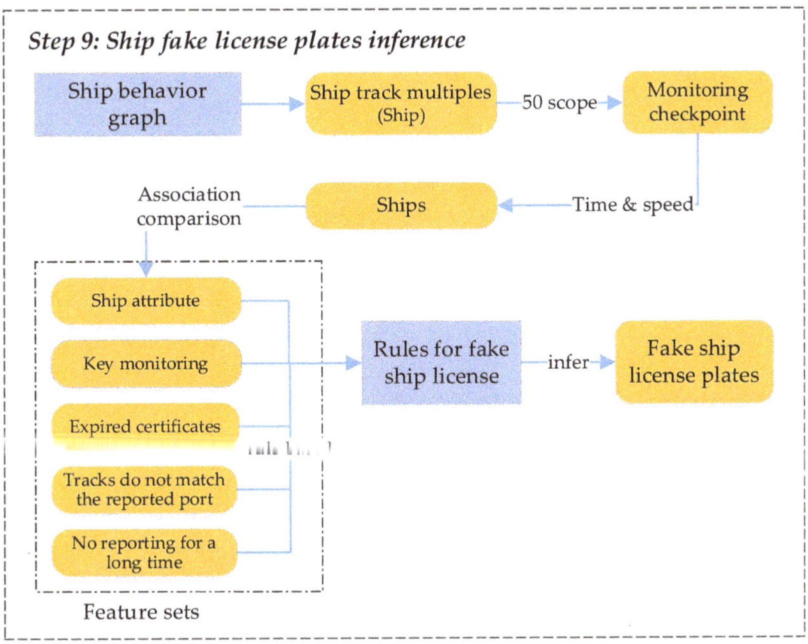

Figure 9. Fake ship license plates inference.

3. Case Study

The semantic rules for ship graph data are constructed with reference to the business rules of maritime supervision and maritime security, and the ternary processed ship ontology data are combined with the semantic rules to form a standard ship graph knowledge representation. A case study was carried out on river–sea direct ships.

3.1. Ship Knowledge Graph Construction

The ontology construction based on the graph semantic rules can form the ship base graph model, which includes the ship's owner company, ship type, key tracking or not, ship certificate validity, ship port call, etc. After the completion of the ontology construction, the initial prototype of the ship knowledge graph is formed, as shown in Figure 10.

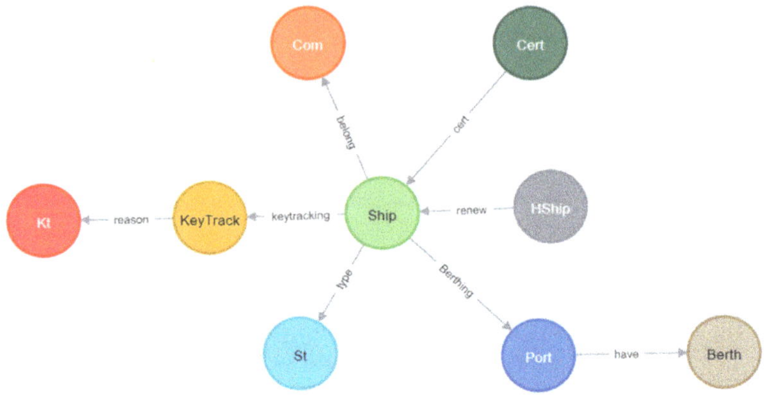

Figure 10. Ship semantic model.

Ontologies include Ship, Historyship(HShip), Shiptype(St), Company(Com), Cert, Key Track, Key Type(Kt), Port, Berth, etc. Relations include type, renew, cert, belong, keytracking, reason, have, etc.

Also included is a description of the category that a ship belongs to by establishing a type of entity link relationship; a description of the ship's change history by establishing a renewed relationship; the ship certificate information is included by establishing a cert relationship; a description of the ship company that the ship belongs to by establishing a Belong_to relationship; a description of whether a ship is a key tracking ship by establishing a key tracking relationship; a description of the specific reason why a ship is a key tracking ship by establishing a reason relationship. By establishing the relationship to describe the berth information under the port, the corresponding ship relationship graph is shown in Figure 11.

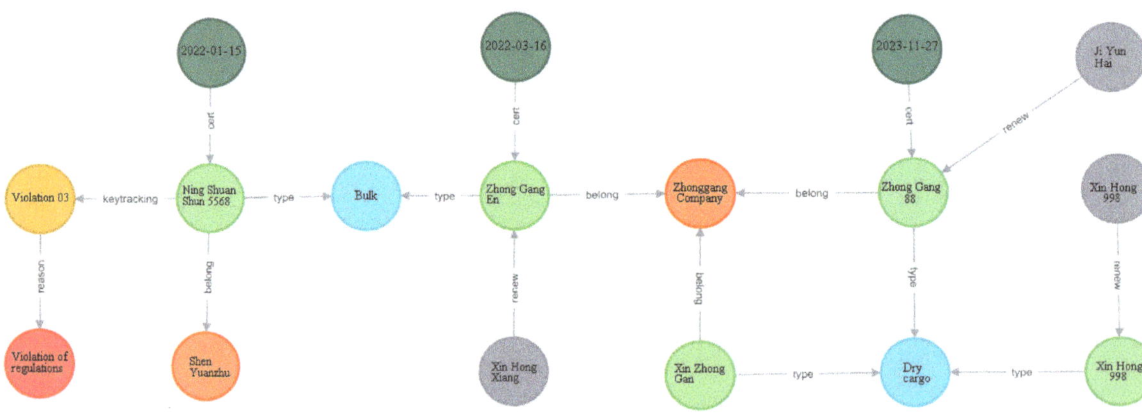

Figure 11. Ship relationship graph.

The ship knowledge graph database established by setting up ship characteristic rules [40] and combining the ship maritime supervision and navigation protection business model can realize the presentation of the overall relationship network of ships and the ship portrait of the static relationship of specific ships [34]. Take the ship named "Ning Shuangshun 5568" as an example. By drawing its portrait, we can see that it is a river ship, the type of ship is a bulk carrier, it belongs to the maritime key tracking ship, it has violated regulations in Nanjing and disobeyed the authorities or evaded them, its certificate is valid (valid until 10 January 2021), and it belongs to Shen Yuanzhu, a self-employed person.

3.2. Extraction of Ship Behavior Characteristics Graph

To extract the key waypoints, the key waypoints of ship navigation lanes were determined for each ship trajectory using the DP algorithm (see Figure 3 and Table 2 for the theoretical concept). Taking the "Xinhong 998" ship as an example, the historical trajectory of the ship for a certain period of time was first obtained from the MySQL database. Due to the dense trajectory, it is not convenient to perform data analysis. Therefore, the DP algorithm was used to optimize the ship's historical trajectory. The following figure shows the optimization effect of the ship's trajectory under different threshold conditions. As part of this process, sub-line segments were extracted using the DP algorithm with ε = 0.0025 (see Figure 12).

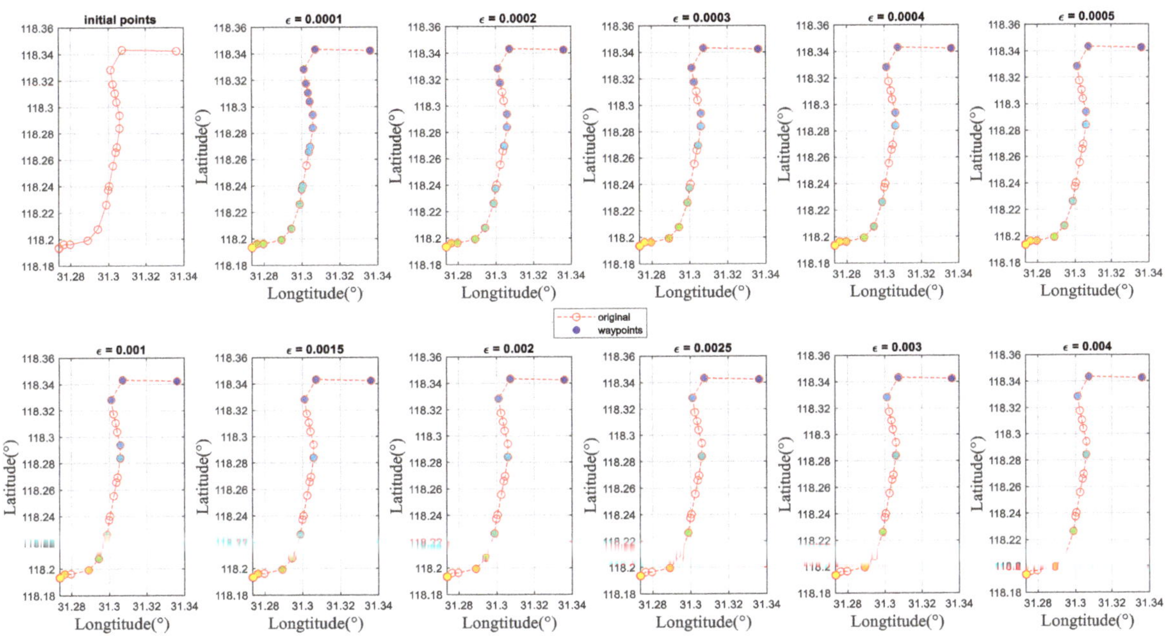

Figure 12. Key waypoints determination using DP algorithm with the various threshold parameters.

For each ship trajectory, the K-means algorithm is used to cluster and analyze whether the ship has multiple abnormal trajectories (theoretical concepts are shown in Figure 8 and Table 3). In order to determine whether the "Xinhong 998" ship has multiple trajectory features, the K-means algorithm is used to cluster and analyze the optimized ship's historical trajectory, quickly identifying the multiple trajectory features of the ship. As shown in Figure 13, two types of trajectories can be found during the same period of the ship, indicating the existence of multiple trajectory features.

Based on Mysql and neo4j fusion calculation, the K-means clustering algorithm is used to identify multiple abnormal features of ship trajectory in the study area, and the acquired features are added into the ship behavior feature graph; based on specific ship ontology, the fusion calculation (subgraph inference, Mysql aggregation, and correlation query) is used to obtain ship data with abnormal features such as key monitored ships, expired ship certificate, inconsistent ship trajectory and reported port, etc., and the feature information is improved into the ship behavior feature graph. The ship data with abnormal features, such as ship certificate expired, ship trajectory inconsistent with reported port, ship not reported port for a long time, etc., are obtained through fusion calculation (subgraph inference, Mysql aggregation, and correlation query) to form a complete ship behavior feature graph. The calculation process is shown in Table 4.

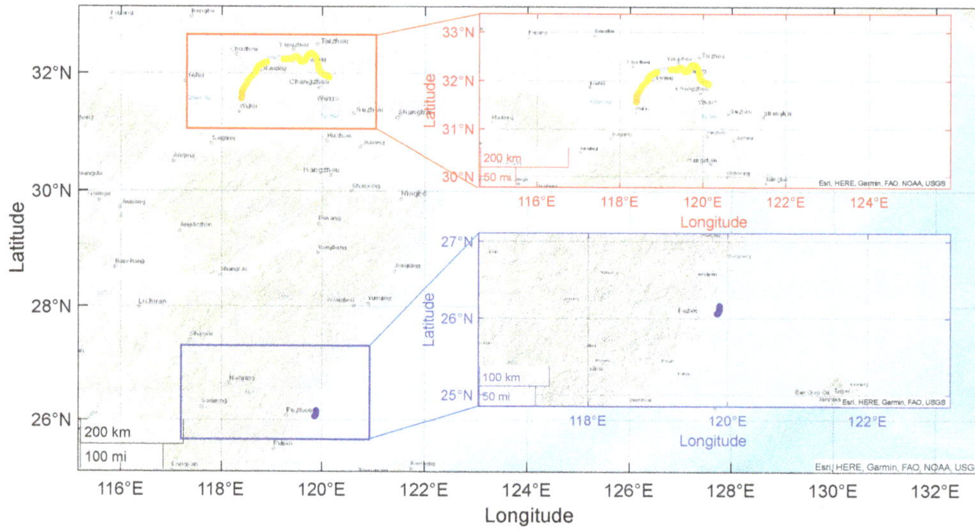

Figure 13. Ship trajectories clustering using K-means algorithm.

Table 4. Ship behavior feature extraction process.

No.	Features	Calculation Methods	Procedures
1	Multiple ship tracks	K-means multiple trajectory clustering	1. Calculate the ship history trajectory; 2. Obtain the number of ship multiple trajectories by K-means multiple trajectory clustering method.
2	Maritime key monitoring ships	Subgraph inference	1. Mysql retrieves data on ships that have been classified as maritime priority monitoring; 2. neo4j queries and identifies ship entities on the retrieved result data.
3	Ships with expired certificates	Subgraph inference	1. Mysql to ship certificate relationship data retrieval, find certificate expiry information; 2. neo4j fused results data for ship entity query and identification.
4	Inconsistent ship tracks with reported ports	Mysql aggregation and associated query	1. Calculate the historical trajectory of a ship; 2. Calculate the ship's route to report port; 3. Compare ship history trajectories with ship reporting paths based on spatial similarity; 4. Judge the consistency of the ship's trajectory with the reported port path.
5	Ships not reporting ports for a long time	Mysql aggregation and associated query	1. Mysql combines multiple ship reporting relationship tables for correlation analysis and generates results data for ships that have not reported to port for a long time; 2. neo4j fused results data for ship entity query and identification.

The subgraph search based on ship knowledge graph can quickly extract static features of ships, such as basic information of ships (including MMSI, ship name, ship type), validity of ship certificate, whether the ship is focused on tracking and the reasons for tracking, etc., and add them to the static feature graph of ship behavior; combined with dynamic port reporting information and ship AIS information, it can extract dynamic features of ships that combine motion and static, and realize the full domain mastery. By analyzing the relationship between time and space trajectory, we can find out the characteristics such as ships not reporting to port and inconsistency between the path of ships reporting to port and the actual AIS route trajectory, and add them into the dynamic feature graph of ship behavior. The combination of ship static feature graph and dynamic feature graph forms a complete ship behavior feature graph.

3.3. Fake Ship License Plate Judgment

3.3.1. Fake Ship License Plate Discovery

Based on ship behavior feature graph analysis, we focus on the abnormal behavior of double trajectories of ships in specific areas. As shown in Figure 14, taking "Xinhong 998" ship chart node in the upstream section of the Yangtze River as an example for analysis, this ship is a sea ship with static chart characteristics, and in the time range of '21 March 2021 00:00:00' to '21 March 2021 03:16:14', the ship dynamically reports ports to call at Jiangyin, Diagang, and Niutoushan ports. Checking the AIS trajectory of this ship, it is found that one is the inland trajectory of the Yangtze River, and the inland trajectory is consistent with the reported path. The other is the sea navigation trajectory, and there is no matching sea reporting information, which means that some ships have multiple abnormal trajectories. The abnormal trajectory is the sea section, so it can be inferred that this ship has the risk of being licensed by the inland ship for sea navigation.

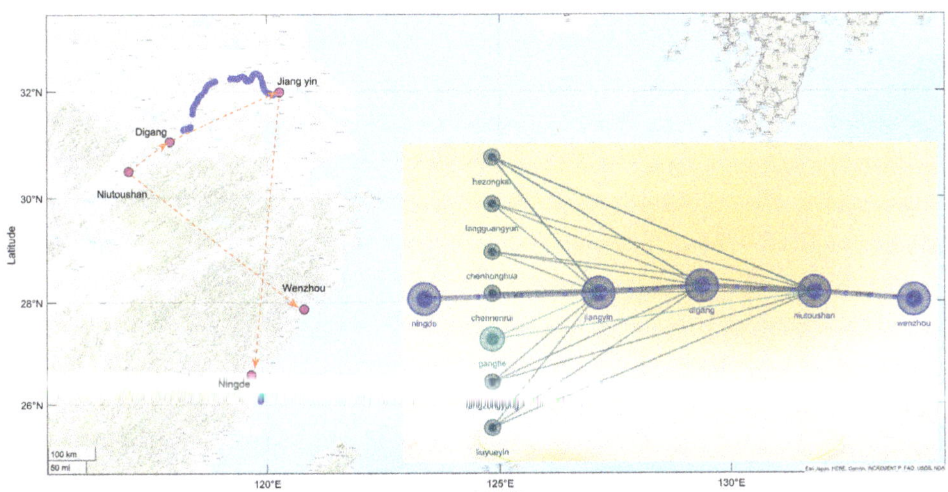

Figure 14. Ship multiple trajectory abnormalities.

3.3.2. Calculation and Reasoning of Fake Ship License Plate

For "Xinhong 998" multiple areas of sea track, we obtained the center point of the ship track, taking the center point as the circle, 50 nautical miles (see more in Figure 15) as the radius, and searching the ship monitoring chokepoint within the range (query to the monitoring chokepoint of Lianjiang Maritime Office). We calculated the distance between the center point of the ship track and the monitoring chokepoint, combined with the average speed of the ship's property characteristics. To obtain the approximate time difference between the ship sailing from the monitoring chokepoint and the track area (roughly 2 h and 15 min by calculation), we subtracted the time difference from the multiple track time period of "Xinhong 998", and obtained the time range of the monitoring chokepoint as '20 March 2021 09:45:00' to '21 March 2021 01:01:14'. We obtained the ship information of the surveillance chokepoint in this time range, identifying "Ning Shuangshun 5568" as the active ship under the current surveillance chokepoint in the current time range, comparing the ship information with the ship behavior characteristics graph in the current area through circular association, and combining with the fake ship license plate rules. The ship "Ning Shuangshun 5568" is an inland river ship, and its AIS track cannot be found, which means the ship MMSI information does not match, the ship certificate has exceeded the validity period, the ship reporting information related to this ship cannot be found, and the ship belongs to the maritime key tracking ship. Thus, it meets many characteristics of the fake ship license violation, the matching degree is high, and it can be judged that the ship has the characteristics of the fake ship license plate violation. There is a high possibility that the ship set "Xinhong 998" is sailing at sea, and engaging in illegal activities.

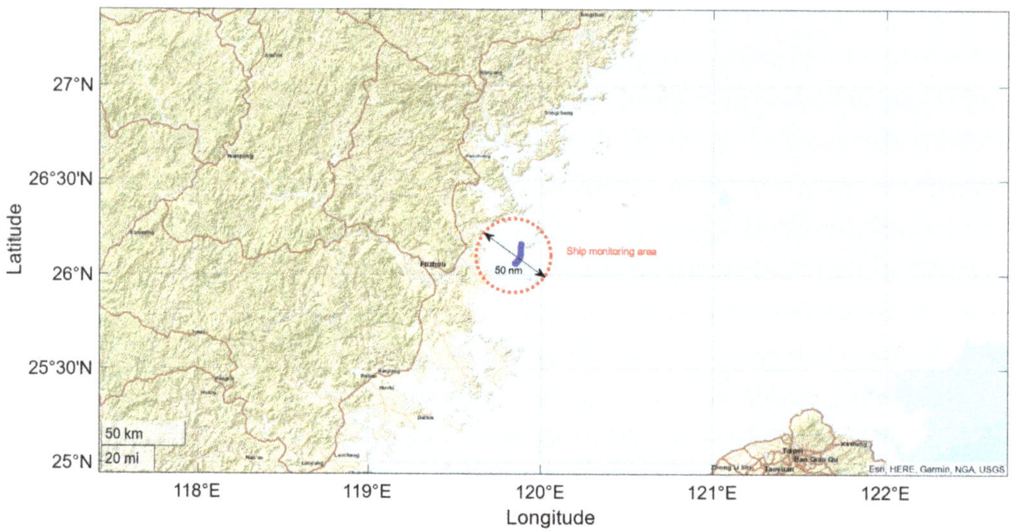

Figure 15. Monitoring checkpoint search.

This study collected questionnaire feedback from experts, and after statistical calculation, obtained the weight allocation information of ship deck rules, as shown in Table 5.

Table 5. Weighted features of fake ship license plates.

Feature	Weight
Inland waterway ship	0.2
Ship information matches	0.4
Certificate is invalid	0.1
Not reported port for a long time	0.2
The port report is inconsistent with the trajectory	0.1
Key monitoring ship	0.1

Based on the weighted of inland river-ship-licensed seagoing ships as shown in Table 6, the following algorithm is used to determine whether the ship is a licensed ship:

(a) Check if the ship is an inland waterway ship. If so, assign a weight of 0.1 to it; otherwise, no weight will be added;
(b) Check if the ship information matches. If it does not match, assign a weight of 0.4 to it; otherwise, no weight will be added;
(c) Check if the ship is using certificate information or if the certificate is invalid. If so, assign a weight of 0.1 to it; otherwise, no weight will be added;
(d) Check if the ship has not reported port for a long time or if the port report is inconsistent with the trajectory. If so, assign a weight of 0.2 to it; otherwise, no weight will be added;

(e) Check whether the ship is a key monitoring ship. If so, assign a weight of 0.1 to it; otherwise, no weight will be added. Add all weighted weight values together to obtain the total weight W of the ship;

(f) If W is greater than a certain threshold t (such as 0.6), it is determined that the ship is a licensed ship; otherwise, it is determined that it is not a licensed ship.

Table 6. Weight for fake ship license plates.

Number	Ship Number	Ship Name	Inland Ship	Ship Information Mismatch	Invalid Certificate	Abnormal Ship Reporting		Key Monitoring	Weight
						No Port Reporting	Inconsistent Trajectory		
1	60117000047	runcheng9	0.2						0.2
2	70106000226	jinfuxin28	0.2						0.2
3	80112000047	dafeng3	0.2		0.1	0.2		0.1	0.6
4	60019000021	fanzhou10	0.2		0.1	0.2		0.1	0.6
5	80617000090	jinjiangpeng29	0.2		0.1				0.3
6	60115000027	shunqiang68							
7	60016000039	ninghua417							
8	270117000007	ningshuangshun5568	0.2	0.4	0.1	0.2		0.1	1

Among them, steps (a)~(d) are to check the characteristics of the ship and give corresponding weights based on the characteristics; step (e) is to add the weight of ownership to obtain the total weight of the ship; and step (f) is to determine whether the ship is a licensed ship based on the total weight. In practical applications, it is necessary to select an appropriate threshold t based on the specific situation, and continuously optimize and adjust the features and weights to improve the accuracy and reliability of the algorithm.

Based on the above weight determination algorithm, ships that meet the definition threshold of ship fraud with a value greater than 0.6 include Ning Shuangshun 5568 (1), Dafeng 3 (0.6), Fanzhou 10 (0.6), etc. Among them, Ning Shuangshun 5568 has the highest weight value and can be determined as a ship with fraudulent behavior.

4. Discussion

Based on the analysis of the collision accident between the "Zhonggang 88" ship and the "Ningshuangshun 5568" ship in Fuzhou on 21 March 2021, the "Zhonggang 88" ship, owned by Fujian Zhonggang Shipping Co., Ltd., loaded 4950 tons of wheat from Nantong to Quanzhou, collided with the "Ningshuangshun 5568" ship owned by Shen XX in the waters about 1.5 nautical miles west of Mazu Nangan Island (approximate position: 26°10.4′ N/119°53.2′ E). The accident caused the sinking of the "Ningshuangshun 5568" ship, and the stowage information of the "Zhonggang 88" ship was basically consistent with the portrait of the "Zhonggang 88" ship. No relevant records were found for "Ning Shuangshun 5568".

A visualization analysis system for ship knowledge graphs was developed based on the ship knowledge graph technology studied in this study and combined with 3D-force-graph technology. The portrait analysis of the "Ningshuangshun 5568" ship shows that it belongs to an inland navigation ship, and its appearance at sea violates the rules of cross-inland and ocean navigation. Due to its lack of dynamic port reporting records and historical trajectory, it can be theoretically inferred that there is an illegal act of ship licensing. Combining the above analysis with the risk of "Xinhong 998" being falsely licensed for sea navigation by inland ships during the same period, and combining the abnormal trajectory of "Xinhong 998" in the current period analyzed earlier with the consistency of the monitoring checkpoint distance and speed calculation of the "Ningshuangshun 5568" falsely licensed ship, the combination of the two can indirectly infer that "Ningshuangshun 5568" falsely licensed "Xinhong 998" has a high probability of crossing inland rivers and sea navigation.

The final accident investigation report also confirmed that the "Ningshuangshun 5568" was falsely licensed as the "Xinhong 998"; "Ningshuangshun 5568" evaded normal maritime supervision and did not comply with the maritime regulations on the exclusive supervision of AIS for ships. The validity period of the certificate expired on 10 January 2021. When the accident occurred, the validity period of the certificate had expired. In addition, "Ningshuangshun 5568" had not yet fulfilled the declaration business of ship reporting operations as required, and it should bear the primary responsibility for this ship collision accident.

The inference method based on the graph of illegal behavior of ship deck rigging can also be applied to the field of ship collision risk assessment, and this method has a certain degree of universality. The inference technology for the graph of illegal behavior of ship cheating mainly includes extracting illegal behavior features of ship cheating, the design of ship cheating rules, the allocation of feature weights of ship cheating behavior, graph calculation, and graph inference process, etc. It is also applicable to the determination of ship collision risk. A ship collision risk assessment based on graph calculation and graph reasoning can be achieved by identifying and extracting key risk factors for ship collisions, designing ship collision rules, and assigning weights to key risk factors for ship collisions.

5. Conclusions

This paper presents the construction of a ship knowledge graph using science knowledge graph technology, which allows for extracting ship violation features and identifying specific fake ship license plates through inference and case analysis verification. The results of this research can significantly enhance the decision-making ability of maritime for ship safety supervision. The ship static and dynamic knowledge graph facilitates ship portrait and ship hidden relationship mining and analysis, allowing for the rapid identification of maritime key tracking ships, ships with expired certificates, and the consistency of ship trajectories and reported ports. It also enables the identification of multiple trajectory anomalies of ships and the retrieval of ships that have not reported ports for a long time. Through the analysis of the ship graph, the existence of abnormal ships navigating across inland rivers and seas, as well as the violation of the ship's license plate, can be determined. Furthermore, in conjunction with the identification of ship names under on-site supervision, ship graph comparison analysis can be conducted to enable rapid screening and key monitoring of abnormal risks similar to "Ning Shuangshun 5568", thereby enhancing the ability to prevent accidents beforehand. The ship graphical portrait provides full information on the ship belonging to the company, certificate information, the last port from where it departed, the next port where it will arrive, local cargo, and the ship's crew, as well as violation behavior characteristics. This information allows for timely response to ship monitoring and tracking business requirements and enhances after-accident investigation, tracing, and control disposal capability through the use of graph path query and source tracing methods.

In summary, this research enhances the maritime industry's pre-accident prevention capability and post-accident handling capability through the construction and analysis of a ship knowledge graph, providing valuable insights for ship safety supervision decision-making.

Author Contributions: Conceptualization, validation, writing, original draft, methodology, and formal analysis—H.W., S.F. and M.Z.; writing, review and editing, supervision, funding acquisition—H.W., S.F. and Y.X. All authors have read and agreed to the published version of the manuscript.

Funding: This study is supported by the National Natural Science Foundation of China under grants 52271363 and Shanghai Rising-Star Program 22QC1400600.

Institutional Review Board Statement: Not applicable.

Informed Consent Statement: Not applicable.

Data Availability Statement: Not applicable.

Conflicts of Interest: We declare that we have no financial and personal relationships with other people or organizations that can inappropriately influence our work, there is no professional or other personal interest of any nature or kind in any product, service, and/or company that could be construed as influencing the position presented in, or the review of, the manuscript entitled, 'A semantic network method for the identification of ship illegal behaviors using knowledge graph: A case study on fake ship license plates'.

References

1. Ma, X.; Shen, J.; Liu, Y.; Qiao, W. A Methodology to Evaluate the Effectiveness of Intelligent Ship Navigational Information Monitoring System. *IEEE Access* **2020**, *8*, 193544–193559. [CrossRef]
2. Su, P.; Abera, T.; Guan, Y.; Pellikka, P. Image-to-Image Training for Spatially Seamless Air Temperature Estimation with Satellite Images and Station Data. *IEEE J. Sel. Top. Appl. Earth Obs. Remote Sens.* **2023**, *16*, 3353–3363. [CrossRef]
3. Shi, Y.; Long, C.; Yang, X.; Deng, M. Abnormal Ship Behavior Detection Based on AIS Data. *Appl. Sci.* **2022**, *12*, 4635. [CrossRef]
4. Wang, H.; Liu, Z.; Wang, X.; Graham, T.; Wang, J. An analysis of factors affecting the severity of marine accidents. *Reliab. Eng. Syst. Saf.* **2021**, *210*, 107513. [CrossRef]
5. Tan, Z.; Liu, H.; Shao, S.; Liu, J.; Chen, J. Efficiency of Chinese ECA policy on the coastal emission with evasion behavior of ships. *Ocean. Coast. Manag.* **2021**, *208*, 105635. [CrossRef]
6. Zhang, M.; Taimuri, G.; Zhang, J.; Hirdaris, S. A deep learning method for the prediction of 6-DoF ship motions in real conditions. *Proc. Inst. Mech. Eng. Part M J. Eng. Marit. Environ.* **2023**, 14750902231157852. [CrossRef]
7. Duan, H.; Ma, F.; Miao, L.; Zhang, C. A semi-supervised deep learning approach for vessel trajectory classification based on AIS data. *Ocean. Coast. Manag.* **2022**, *218*, 106015. [CrossRef]
8. Liu, H.; Liu, Y.; Li, B.; Qi, Z. Ship Abnormal Behavior Detection Method Based on Optimized GRU Network. *J. Mar. Sci. Eng.* **2022**, *10*, 249. [CrossRef]
9. Liu, T.; Ma, J. Ship Navigation Behavior Prediction Based on AIS Data. *IEEE Access* **2022**, *10*, 47997–48008. [CrossRef]
10. Rong, H.; Teixeira, A.P.; Soares, C.G. Ship collision avoidance behaviour recognition and analysis based on AIS data. *Ocean. Eng.* **2022**, *245*, 110479. [CrossRef]
11. Liu, Z.; Zhang, B.; Zhang, M.; Wang, H.; Fu, X. A quantitative method for the analysis of ship collision risk using AIS data. *Ocean. Eng.* **2023**, *272*, 113906. [CrossRef]
12. Kaluza, P.; Kölzsch, A.; Gastner, M.T.; Blasius, B. The complex network of global cargo ship movements. *J. R. Soc. Interface* **2010**, *7*, 1093–1103. [CrossRef] [PubMed]
13. Veenstra, A.W.; Harmelink, R.L. Process mining ship arrivals in port: The case of the Port of Antwerp. *Marit. Econ. Logist.* **2022**, *24*, 584–601. [CrossRef]
14. Acharya, T.D.; Yoo, K.W.; Lee, D.H. GIS-based spatio-temporal analysis of marine accidents database in the coastal zone of Korea. *J. Coast. Res.* **2017**, *79*, 114–118. [CrossRef]
15. Fu, S.; Zhang, Y.; Zhang, M.; Han, B.; Wu, Z. An object-oriented Bayesian network model for the quantitative risk assessment of navigational accidents in ice-covered Arctic waters. *Reliab. Eng. Syst. Saf.* **2023**, *238*, 109459. [CrossRef]
16. Kehrer, J.; Hauser, H. Visualization and visual analysis of multifaceted scientific data: A survey. *IEEE Trans. Vis. Comput. Graph.* **2012**, *19*, 495–513. [CrossRef]
17. Hogan, A.; Blomqvist, E.; Cochez, M.; d'Amato, C.; Melo, G.D.; Gutierrez, C.; Kirrane, S.; Gayo, J.E.L.; Navigli, R.; Neumaier, S.; et al. Knowledge graphs. *ACM Comput. Surv. (CSUR)* **2021**, *54*, 1–37. [CrossRef]
18. Singhal, A. Official Google Blog: Introducing the Knowledge Graph: Things, Not Strings [EB/OL]. Available online: https://blog.google/products/search/introducing-knowledge-graph-things-not/ (accessed on 27 October 2022).
19. Fu, S.; Yu, Y.; Chen, J.; Xi, Y.; Zhang, M. A framework for quantitative analysis of the causation of grounding accidents in arctic shipping. *Reliab. Eng. Syst. Saf.* **2022**, *226*, 108706. [CrossRef]
20. Kim, H.; Kim, G.; Na, H.; Lee, S. An Implementation of Detecting Abnormal Ship Navigation and Ship Safety Navigation Guidance. *J. Digit. Contents Soc.* **2021**, *22*, 1903–1912. [CrossRef]
21. Wang, X.; Xia, G.; Zhao, J.; Wang, J.; Yang, Z.; Loughney, S.; Fang, S.; Zhang, S.; Xing, Y.; Liu, Z. A novel method for the risk assessment of human evacuation from cruise ships in maritime transportation. *Reliab. Eng. Syst. Saf.* **2023**, *230*, 108887. [CrossRef]
22. Pan, W.; Zhou, X.; Zhou, T.; Chen, Y. Fake license plate recognition in surveillance videos. *Signal Image Video Process.* **2022**, *17*, 937–945. [CrossRef]
23. Zhang, M.; Zhang, D.; Fu, S.; Kujala, P.; Hirdaris, S. A predictive analytics method for maritime traffic flow complexity estimation in inland waterways. *Reliab. Eng. Syst. Saf.* **2022**, *220*, 108317. [CrossRef]
24. Wei, Z.; Xie, X.; Zhang, X. AIS trajectory simplification algorithm considering ship behaviours. *Ocean. Eng.* **2020**, *216*, 108086. [CrossRef]
25. Song, R.; Wen, Y.; Tao, W.; Zhang, Q.; Papadimitriou, E.; van Gelder, P. Semantic Modeling of Ship Behavior in Cognitive Space. *J. Mar. Sci. Eng.* **2022**, *10*, 1347. [CrossRef]
26. Wen, Y.; Zhang, Y.; Huang, L.; Zhou, C.; Xiao, C.; Zhang, F.; Peng, X.; Zhan, W.; Sui, Z. Semantic modelling of ship behavior in harbor based on ontology and dynamic bayesian network. *ISPRS Int. J. Geo-Inf.* **2019**, *8*, 107. [CrossRef]
27. Li, Q.; Li, F.; Li, S.; Li, X.; Liu, K.; Liu, Q.; Dong, P. Improving Entity Linking by Introducing Knowledge Graph Structure Information. *Appl. Sci.* **2022**, *12*, 2702. [CrossRef]
28. Hu, L.; Ding, J.; Shi, C.; Shao, C.; Li, S. Graph neural entity disambiguation. *Knowl. Based Syst.* **2020**, *195*, 105620. [CrossRef]

29. Dong, J.; Jing, X.; Lu, X.; Liu, J.; Li, H.; Cao, X.; Du, C.; Li, J.; Li, L. Process knowledge graph modeling techniques and application methods for ship heterogeneous models. *Sci. Rep.* **2022**, *12*, 2911. [CrossRef]
30. Yan, J.; Wang, C.; Cheng, W.; Gao, M.; Zhou, A. A retrospective of knowledge graphs. *Front. Comput. Sci.* **2018**, *12*, 55–74. [CrossRef]
31. Zhong, S.; Wen, Y.; Huang, Y.; Cheng, X.; Huang, L. Ontological Ship Behavior Modeling Based on COLREGs for Knowledge Reasoning. *J. Mar. Sci. Eng.* **2022**, *10*, 203. [CrossRef]
32. Li, D.; Mei, H.; Shen, Y.; Su, S.; Zhang, W.; Wang, J.; Zu, M.; Chen, W. ECharts: A declarative framework for rapid construction of web-based visualization. *Vis. Inform.* **2018**, *2*, 136–146. [CrossRef]
33. Chen, J.; Ye, J.; Zhuang, C.; Qin, Q.; Shu, Y. Liner shipping alliance management: Overview and future research directions. *Ocean. Coast. Manag.* **2022**, *219*, 106039. [CrossRef]
34. Wang, X.; Liu, Z.; Yan, R.; Wang, H.; Zhang, M. Quantitative analysis of the impact of COVID-19 on ship visiting behaviors to ports-A framework and a case study. *Ocean. Coast. Manag.* **2022**, *230*, 106377. [CrossRef]
35. Zhang, M.; Montewka, J.; Manderbacka, T.; Kujala, P.; Hirdaris, S. A big data analytics method for the evaluation of ship-ship collision risk reflecting hydrometeorological conditions. *Reliab. Eng. Syst. Saf.* **2021**, *213*, 107674. [CrossRef]
36. Xu, L.; Di, Z.; Chen, J.; Shi, J.; Yang, C. Evolutionary game analysis on behavior strategies of multiple stakeholders in maritime shore power system. *Ocean. Coast. Manag.* **2021**, *202*, 105508. [CrossRef]
37. Li, M.G.; Chen, J.J.; Xu, A.J.; Xia, X.H.; Wang, J.H. Case study of innovative top-down construction method with channel-type excavation. *J. Constr. Eng. Manag.* **2014**, *140*, 05014003. [CrossRef]
38. Widenius, M.; Axmark, D. *MySQL Reference Manual: Documentation from the Source*; O'Reilly Media, Inc.: Sebastopol, CA, USA, 2002.
39. Gong, Z.; Yu, X.; Fu, W.; Che, X.; Mao, Q.; Zheng, X. The Construction of Knowledge Graph for Personalized Online Teaching. In Proceedings of the Data Mining and Big Data: 6th International Conference, DMBD 2021, Guangzhou, China, 20–22 October 2021; Proceedings, Part II 6. Springer: Singapore, 2021; pp. 98–107.
40. Webber, J. A programmatic introduction to neo4j. In Proceedings of the 3rd Annual Conference on Systems, Programming, and Applications: Software for Humanity, Tucson, AZ, USA, 21–25 October 2012; Volume 10, pp. 217–218.

Journal of
Marine Science and Engineering

MDPI

Article

Leverage Bayesian Network and Fault Tree Method on Risk Assessment of LNG Maritime Transport Shipping Routes: Application to the China–Australia Route

Zheng Chang [1], Xuzhuo He [1], Hanwen Fan [1,*], Wei Guan [2] and Linsheng He [3]

1 College of Transportation Engineering, Dalian Maritime University, Dalian 116026, China; chang_zheng@dlmu.edu.cn (Z.C.); hexvzhuo@163.com (X.H.)
2 Navigation College, Dalian Maritime University, Dalian 116026, China; gwwtxdy@163.com
3 Shandong Shipping Tanker Co., Qingdao 266000, China; tanker-safety@sdshipping.cn
* Correspondence: hwfan@dlmu.edu.cn

Abstract: The China–Australia Route, which serves as the southern economic corridor of the '21st Century Maritime Silk Road', bears great importance in safeguarding maritime transportation operations. This route plays a crucial role in ensuring the security and efficiency of such activities. To pre-assess the risks of this route, this paper presents a two-stage analytical framework that combines fault tree analysis and Bayesian network for evaluating the occurrence likelihood of risk of transporting liquefied natural gas (LNG) on the China–Australia Route. In the first stage, our study involved the identification of 22 risk influencing factors drawn from a comprehensive review of pertinent literature and an in-depth analysis of accident reports. These identified factors were then utilized as basic events to construct a fault tree. Later, we applied an expert comprehensive evaluation method and fuzzy set theory, and by introducing voting mechanism into expert opinions, the prior probability of basic events was calculated. In the second stage, a fault tree was transformed into a Bayesian network, which overcame the deficiency that the structure and conditional probability table of the Bayesian network find difficult to determine. Consequently, the employment of the Bayesian network architecture was applied to forecast the likelihood of LNG maritime transport along the China–Australia shipping pathway. The probability importance and critical importance of each basic event was calculated through an importance analysis. The development of a risk matrix was achieved by considering the two primary dimensions of frequency and impact, which were subsequently utilized to categorize all relevant risk factors into high, moderate, or low risk categories. This allowed for effective risk mitigation and prevention strategies to be implemented. Finally, assuming that the final risk occurs, we calculated the posterior probability of the basic event to diagnose the risk. The research findings indicate that the primary reasons for the risk of transporting LNG on the China–Australia Route are the impact of natural forces and epidemics, piracy and terrorist attacks, and the risk of LNG explosions. In the final section, we provide suggestions and risk control measures based on the research results to reduce the occurrence of risks.

Keywords: risk assessment; maritime transport; fault tree analysis; Bayesian network; Liquid Natural Gas (LNG)

Citation: Chang, Z.; He, X.; Fan, H.; Guan, W.; He, L. Leverage Bayesian Network and Fault Tree Method on Risk Assessment of LNG Maritime Transport Shipping Routes: Application to the China–Australia Route. *J. Mar. Sci. Eng.* **2023**, *11*, 1722. https://doi.org/10.3390/jmse11091722

Academic Editor: Claudio Ferrari

Received: 7 August 2023
Revised: 21 August 2023
Accepted: 29 August 2023
Published: 1 September 2023

1. Introduction

The rapid development of the shipping industry has significantly contributed to the growth of global trade, but it has also posed challenges to maritime safety [1]. The safety of routes has emerged as a critical domain of emphasis for the shipping sector and its associated enterprises. As the southward economic channel of the '21st Century Maritime Silk Road', the China–Australia and New Zealand route has gained significant attention for its transportation function and significance. However, according to a report by the International Maritime Organization, the safety situation of maritime transportation along this route is not optimistic. Nevertheless, as per a report published by the International Maritime Organization, the safety status of maritime transportation along this course is not sanguine. How to scientifically and reasonably build the risk assessment model of the route is an important premise and calls for mastering the risk status of the route and identifying the influencing factors in the risk scenario.

From 2010 to 2023, very serious levels of maritime casualties and accidents accounted for 60–70% of the total number of accidents along the China–Australia and New Zealand route, highlighting the seriousness of safety risks along the route. For example, on 17 May 2023, a fishing vessel from China to Australia capsized in the Indian Ocean. The vessel with 39 crewmembers onboard is still missing. To ensure the sustainable development of the shipping industry, preventing maritime accidents and ensuring safety at sea should be the primary goal of the International Maritime Organization. Therefore, describing various risks and developing a quantitative evaluation model for the risk of LNG maritime transportation on this route to accurately predict and prevent risks that have not yet occurred can guide the safety of maritime transportation activities to a certain extent [2].

Previous studies have demonstrated that due to the numerous influencing factors involved in the indicator system, the diverse types and complex structures of indicators, as well as the participation of experts, maritime safety risk assessment problems have uncertainties brought about by the diverse types of indicators, structural uncertainties brought about by complex system structures, and cognitive uncertainties brought about by human cognitive limitations. Taking into account various uncertainties, risk factors including cargo characteristics, ship conditions, environmental conditions, human error, and management issues ensure that the safety of maritime transport is a complex task [3]. Addressing these challenges requires the collection and investigation of the latest data from recent maritime accidents, analysis of the causes of such accidents, identification of key risk influential factors (RIFs) under different scenarios, and the prediction of associated risks.

While classical risk analysis methods, including Failure Mode and Effects Analysis, Analytic Hierarchy Process, Markov Model, Human Factors Analysis and Classification System, and Fault Tree Analysis (FTA), have been widely used to identify critical factors for enhancing maritime safety, they are not practical for uncertain risk analysis with changing environmental conditions [4]. Among these models, Failure Mode and Effects Analysis with transparent and simple features [5] and Analytic Hierarchy Process with the ability to assign different weights to indicators [6] are widely used. However, both methods require experts to make evaluations in highly complex assessment environments [7]. Professional level, personality traits, and subjective judgment can all affect the ability to make accurate evaluations, leading to ambiguity and uncertainty in the final results, which cannot be used for risk assessment under multiple indicators. Human Factors Analysis and Classification System is designed to analyze different degrees of human factors in accidents. Celik M [8] introduced the Human Factors Analysis and Classification System model for water traffic accidents and identified the lowest level human factors in the model through a fuzzy analytic hierarchy process. However, its shortcomings are obvious. This method is only applicable to risk assessment caused by human factors in smaller models and cannot express other relevant factors. These limitations have necessitated the development of advanced risk analysis methods, such as Fuzzy Logic (FL) and the Bayesian Network.

Among the advanced methods used for maritime risk analysis, BN has gained significant attention due to its ability to explain the relationships among multiple variables under uncertainty, based on probabilistic information for risk assessment. Risk analysis using BN has become a promising technique in complex and uncertain shipping scenarios [9]. For example, Fan et al. [10] used the Naïve Bayesian Network (NBN) to model maritime accident risk analysis and identified 16 RIFs based on the analysis of 161 accident reports collected from 2012 to 2017. Similarly, Jiang and Lu [11] proposed a dynamic Bayesian network (DBN) model to assess dynamic contingencies in the Indian Ocean sea lanes based on incident data from 2007 to 2018.

However, the Bayesian network faces some challenges when capturing the conditional probability tables (CPTs) among the influencing factors and make it hard to determine a rations structure among the nodes: (1) when determining the CPT, the traditional method is expert scoring or data-driven, but these two methods require a high number of data samples, consuming time and energy; and (2) regarding structural learning, expert judgment is usually introduced, which can lead to strong subjectivity and bias, especially when multiple nodes are involved. Zhao [8] has shown that FTA can cleverly solve this problem, and combining FTA and BN can be used for maritime risk analysis under complex environmental impacts, which can be beneficial in compensating for their respective shortcomings. They use fuzzy fault tree analysis and noise or gate Bayesian network to estimate the probability of navigation accidents. The fault tree analysis is constructed from the navigation accident investigation report, and then the fault tree analysis is transformed into a Bayesian network using Noisy-OR gate. Finally, the model was applied to Qinzhou Port and reasonable conclusions were drawn by comparing it with the calculation results of other waterways.

This paper endeavors to rectify the aforementioned shortcomings by presenting the following measures. The primary objective of this paper is to propose a two-stage model framework for evaluating the risk of LNG maritime transport on the China–Australia route. The first stage involves constructing a fault tree based on the influencing factors identified from the related literature and accident reports, followed by calculating the prior probability of basic events using the expert comprehensive evaluation method and fuzzy set theory. In the second stage, the fault tree is transformed to a BN model, and the results of FTA in the first stage are input into the BN model in the second stage as initial values. After completing the BN model validation, the next step is to predict, prevent, and diagnose the risk of LNG maritime transport, and introduce a risk matrix to analyze the risks from the perspectives of importance and frequency. This paper offers three significant contributions: (1) the introduction of fuzzy set theory and expert voting mechanism addresses the challenges associated with handling conflicts in expert opinions and the inherent fuzziness in the expert scoring process and enriches the application of expert scoring methods in the field of risk assessment; (2) the probability importance degree and key importance degree obtained from the fault tree analysis are regarded as two inputs of the risk matrix. This enables decision-makers to clearly perceive the frequency and severity of risks, as well as their interrelationships, and assigns them priority levels, expanding the application and development of traditional risk management theory where risk is directly multiplied by frequency and importance; and (3) the richness of influencing factors in the LNG maritime transportation process leads to the complexity and uncertainty of the indicator system structure. Through the transformation of FTA to BN, a rational and scientific Bayesian risk assessment model has been constructed, expanding the application of LNG risk transportation under uncertain factors.

The remainder of this paper is organized as follows. Section 2 reviews the literature related to maritime accident research and the application of BN and FTA in maritime risk analysis and explores relevant research gaps. Section 3 proposes a risk assessment framework for LNG maritime transportation. This includes an illustration of the basic theory, identification of influencing factors in the risk assessment model, and explanation of the voting mechanism and risk matrix used in the subsequent analysis. In Section 4, the methodology is applied to evaluate the risk of LNG maritime transport on the China–Australia route. Section 5 proposes improvement measures for risks with a high impact on the model output.

2. Literature Review

2.1. Risk Assessment of LNG Maritime Transport

LNG maritime transportation belongs to high-risk cargo transportation, and there is rich research in the academic community on the risks of LNG maritime transportation. Vanem et al. [12] conducted a high-risk assessment of the global navigation of LNG ships. The analysis collects and combines information from multiple sources and available information from different sources has been structured in the form of event trees for different generic accident categories. Five different types of LNG-related risks have been identified, namely collision, grounding, contact, fire and explosion, and accidents that occur during loading and unloading at the dock. The results showed that the highest risk is collision. On the basis of traditional evaluation models, Martins [13] proposed a complete quantitative risk analysis method (QRA) for potential risk accidents that may occur during the offshore terminal loading and unloading processes of LNG ships. By comparing it to traditional models, the advantages and limitations of the new model are pointed out. Marroni et al. [14] developed a simplified method for the risk assessment of LNG ships in port areas. Based on the standard characteristics of the ship, a set of reference accident scenarios that need to be considered in risk assessment has been determined, providing specific guidance for determining hazards, estimating frequency of occurrence, and consequences. Finally, a customized risk matrix was adopted to support decisions on prevention and mitigation measures. Abdussamie et al. [15] proposed a fuzzy set method to deal with the uncertainty in expert opinions used in qualitative risk assessment research (such as a risk matrix). The risk parameters are modeled using fuzzy set, and the fuzzy risk values of several dangerous scenarios at different stages of the ship berthing operation are calculated.

Among the existing methods for quantitative risk analysis, fault tree analysis (FTA) and Bayesian network (BN) are conventional tools. For example, Zhou [16] took the loading and unloading process of a ship as an example, constructed a modified FTA for ship accident leakage, and introduced human reliability analysis (HRA) to predict human errors in the loading and unloading processes of LNG ships. Finally, the results of FTA and human reliability analysis are combined, and a Monte Carlo simulation (MCS) is used to evaluate the risk. Additionally, Zhao et al. [17] used Bayesian network to identify potential risks, calculate accident probability, and evaluate the severity of consequences for the safe anchoring system of LNG ships. Yeo [18] analyzed and identified potential hazardous events that may occur during the unloading process of LNG transport vessels at floating terminals. They use Bayesian networks to dynamically analyze the safety of LNG ships during loading and unloading to identify the most likely types of accidents. The result is similar to Vanem: collision is the most probable accident to occur during the offloading process of an LNG carrier at berth, which may have catastrophic consequences. Li et al. [19] proposed a process risk-based decision-making method for LNG ships colliding with Arctic Sea ice or obstacles based on the dynamic Bayesian network (DBN) risk assessment model, indicating that the decision-making process of ship navigation is dynamically related to time. Additionally, Melani et al. [20] combined the two methods, using FTA to analyze the failure of the unloading equipment of LNG ships. They combined it with pre-hazard analysis and causality diagram to calculate the probability of various accidents through Bayesian probability. Finally, they used the risk matrix for risk analysis and provided corresponding improvement measures and suggestions.

2.2. FTA and BN in Maritime Risk Analysis

In this section, the advantages of FTA and BN in risk modelling are further demonstrated by a systematic review of its applications in maritime accident/risk analysis. FTA and BN models have been extensively applied in the field of maritime transport risk assessment and have yielded several notable results. FTA aims to determine the root cause by using a top-down method to build the accident chain and evaluate its impact on the accident. Fu et al. [21] proposed a fuzzy event tree method for Frank copula, which evaluated the risk of major ship accidents in Arctic waters under the consideration of uncertainty. Ugurlu [22] used FTA for qualitative and quantitative analysis to determine the root causes of ship-to-ship collisions statistically. Results show that the violation of the COLREG Rules is the most important and effective factor for collision accidents.

A Bayesian network model is used for quantitative assessment of risks under certain conditions. For instance, Wan et al. [23] develop a novel model to assess the risk factors of maritime supply chains by incorporating a fuzzy belief rule approach with Bayesian networks. The new model, compared to traditional risk analysis methods, has the capability of improving result accuracy under a high uncertainty in risk data. In another study, Baksh et al. [24] employed BN to assess the transport risks during navigation in the Arctic Sea. The researchers first discussed the causes of maritime accidents and calculated the prior probability of Bayesian network nodes based on historical data and expert judgment. Subsequently, the possibility of accidents was determined through a sensitivity analysis of the model. The results revealed that sea ice was the main influencing factor of the accident, and appropriate management measures were proposed accordingly. Additionally, Chen et al. [25] proposed an evidence-based fuzzy Bayesian network method to build a maritime accident Statistical model. Using maritime accident reports, the Bayesian network was constructed from a systematic perspective and its reliability was verified by three axioms.

FTA can establish a linear or sequential relationship between events leading to an accident and provide a known conditional probability table and a clear model structure for the Bayesian network. For example, Sakar et al. [26] mapped FTA to BN to analyze the causes of grounding accidents and found that navigation factors had the most significant impact on grounding accidents. Sokukcu et al. [27] considered the limitations of Fault tree analysis in terms of conditional dependence and stationarity, proposed a Bayesian network mapping method based on Fault tree analysis to overcome this limitation, and conducted a probabilistic risk analysis on collision events. Kaushik et al. [28] proposed a comprehensive method based on intuitionistic fuzzy fault tree and Bayesian network to evaluate the fault probability of a system in cases of imprecise and insufficient fault data. The results indicate that when the statistical failure data of components are inaccurate, this method can be used as an alternative method for reliability probability assessment.

Based on the literature review presented above, two research gaps have been identified: (1) regarding structural learning, most studies use data-driven TAN or traditional BN models, which require expert guidance or data-driven development, leading to high energy and time consumption. Additionally, when multiple evaluation indicators are involved, determining the causal relationship between nodes and CPT in the network can be challenging. To address these issues, a FTA transformation method can be applied to remedy the structural defects of a traditional BN model and (2) when obtaining the quantitative value of the prior probability of the root nodes, most studies use the method of expert questionnaire. However, the possible deviations in the questionnaire results are not handled, resulting in a large deviation in the prior probability of some root nodes. By introducing an expert voting mechanism, we consider screening and retaining the expert opinions with significant deviations. Thus, the accurate prior probability value can be obtained.

In order to address the aforementioned gaps in the research, this study employs a comprehensive, multi-step framework. First, we identify risk factors that may lead to hazards in LNG maritime transport by reviewing relevant literature and develop a fault tree model accordingly. Next, the probability of failure of basic events is calculated according to the fuzzy set theory and expert scoring method. W also introduce a voting mechanism into expert scoring results to handle results with significant differences in opinions. For example, when experts have similar ratings for the frequency of events, their opinions are taken into consideration for subsequent analysis. However, when most experts give a relatively unified opinion on the frequency of an event, while very few experts have opposite opinions, the opinions of the very few experts are discarded, and the unified opinions of other experts are retained.

Then, the fault tree is transformed into a Bayesian network, and the risk prediction, prevention, and diagnosis are carried out in turn. The novelty of this research lies in integrating the fault tree with the Bayesian network, where FTA analyzes the causal relationship between risk factors, compensating for the challenge of determining the causal relationship of nodes in the Bayesian network model. After establishing the Bayesian network model, we can perform forward prediction and backward diagnosis, which overcome the limitation of the fault tree's inability to carry out probabilistic quantitative analysis.

3. Model Construction

3.1. Establish a Maritime Risk Assessment Framework

FTA and BN are two widely used methods for risk assessment. However, using FTA for risk assessment requires calculating the top event state based on the probability of basic events, resulting in generating a large number of calculations when reasoning in the forward direction, and backward reasoning cannot be carried out in this model. These problems can be overcome in the BN model, which allows for both forward and backward reasoning. Additionally, fault tree transformation is the primary approach for constructing a Bayesian network.

Maritime risk analysis studies that use FTA-BN are typically conducted through several established steps, including data collection, variable identification, FTA structure transformation, BN model validation, and sensitivity analysis [29]. The methodology in this paper is no exception, and it consists of five parts: (1) identification of variables, (2) construction of the fault tree, (3) transformation of the fault tree into a Bayesian network, (4) calculation of the probability of root nodes, and (5) risk prediction, prevention, and diagnosis, as depicted in Figure 1.

3.2. Risk Factor Identification

RIFs, which stands for Risk Influencing Factors, are the variables that impact the security and safety of maritime transportation. Identifying risk influencing factors can provide support for subsequent survey questionnaires, concretizing and digitizing indicators. This can identify potential problems in the entire process of LNG maritime transportation and provide complete and scientific information for risk control measures. The widely used identification method currently is to identify RIFs based on the construction of a maritime accident database, utilizing relevant literature and the existing maritime accident record guidance from the IMO. After searching on the Web of Science, 16 typical journal papers were selected, which described the risk factors and were further analyzed against each of the retrieved results, as shown in Appendix A. Next, 22 RIFs were identified and are listed in Figure 2, with the frequency of occurrence of each risk factor. It is evident that the safety performance of LNG ships, high waves, heavy fog, strong sea breeze, unsafe behavior of personnel, piracy, and terrorist attacks are the top six RIFs identified in previous research. Once the RIFs were identified, previous studies in the field usually simplified the definition of their states to reduce the high data demand in quantifying their interdependencies, such as CPTs in BN. In this paper, all RIFs were set to two states: normal and fault. Experts scored the occurrence frequency of each RIF, and the fault probability was obtained through data processing, which was then input into the BN as prior probability.

3.3. Fault Tree Construction

FTA is a well-structured and widely used tool for the risk assessment and root cause analysis (RCA) of complex systems. In a fault tree there are top events, intermediate events, and basic events. The fault tree analysis begins with a final result, which is the top event, and decomposes layer by layer from top to bottom according to the causal relationship of logic gates until it cannot be decomposed any further. Finally, the basic events that caused the final failure are identified. This method can intuitively analyze various ways of system fault occurrence, effectively finding the fault source of the system. FTA produces graphical displays that show the logical connections between failure and the path toward the failure of a system. Due to the visualization and predictability of FTA, it has been widely used in the maritime field [21].

The event structure and relationship of the risks associated with the maritime transportation of liquefied natural gas (LNG) are illustrated in Table 1. This table outlines the specific factors that are represented by each event, and their relationship with intermediate events.

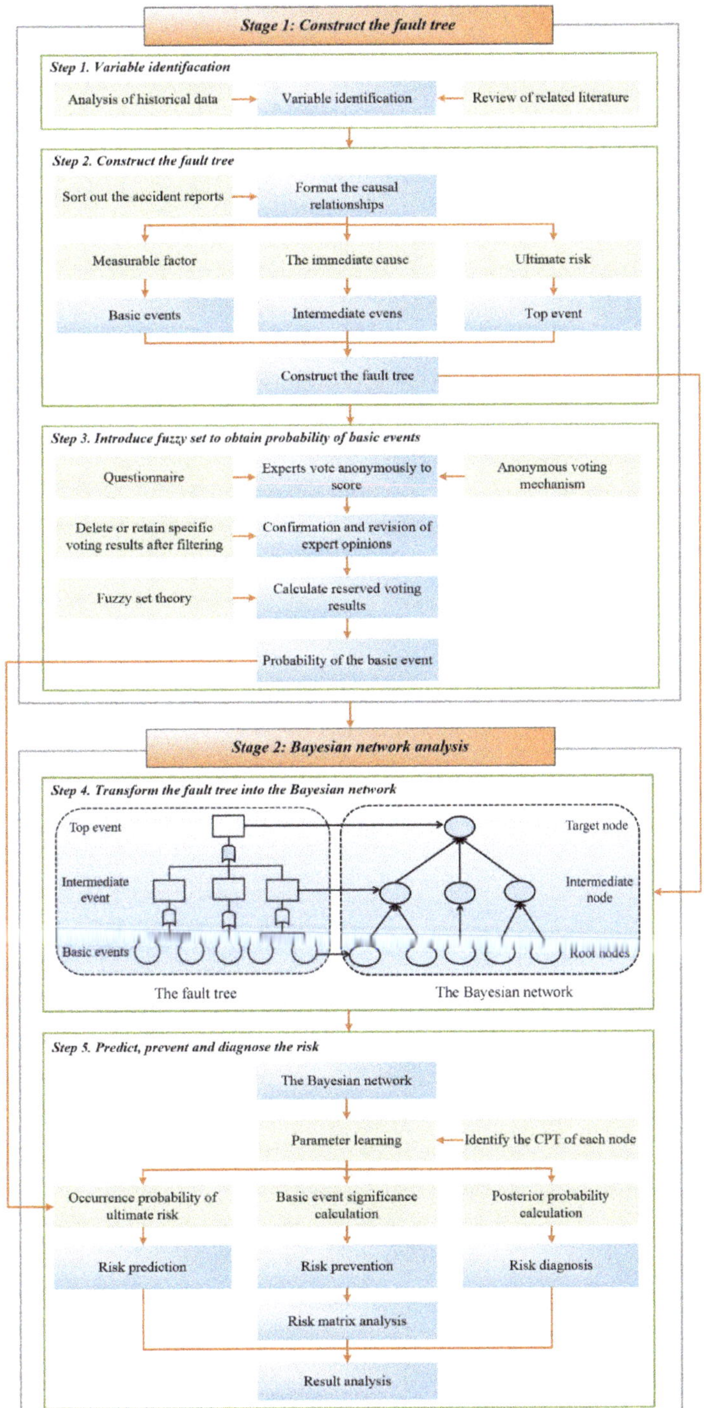

Figure 1. Developed framework for maritime transport risk assessment.

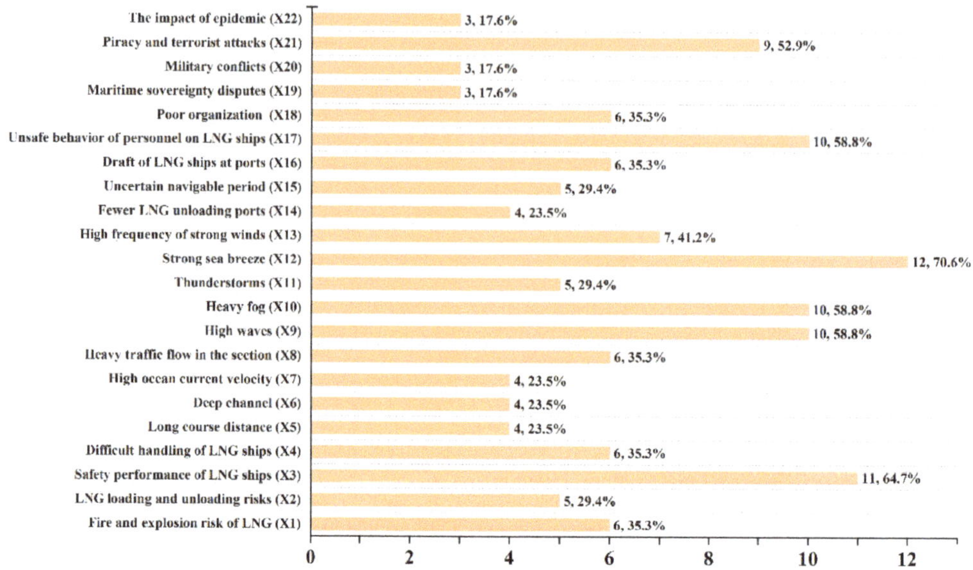

Figure 2. Frequency of RIFs in the retrieved literature and accident reports.

In view of the event design for the risk fault tree of LNG maritime transport in Table 1, this paper obtained the fault tree structure by drawing with Visio, as shown in Figure 3.

3.4. Introduce Fuzzy Set to Obtain Probability of Basic Events

The fuzzy set theory (FST) was introduced by Zadeh [30] as a means of dealing with imprecision and vagueness. A fuzzy set on a given domain U means that, for any $x \in U$, there is a number $u(x) \in [0,1]$ corresponding to it. The membership function $u(x)$ represents the membership value of x in U. The triangular fuzzy number is a simple and widely used method for representing the membership function, where a and b are the lower and upper limits of the fuzzy number, respectively, and m is the value with the highest possibility. In this paper, a triangular fuzzy number $U = (a, m, b)$ is utilized to represent the fuzzy failure probability of the nodes, and the membership function is given by Equation (1).

$$u(x) = \begin{cases} \frac{x-a}{m-a}, a < x \leq m \\ \frac{b-x}{b-m}, m < x \leq b \\ 0, otherwise \end{cases} \tag{1}$$

The expert survey method has been widely used in risk research as an effective and feasible approach. However, this method solely focuses on the level of the expert judgment's ability, while ignoring the uncertainty in expert judgment. This can result in a certain degree of deviation in data reliability, which greatly impacts the subsequent evaluation. To address this issue, this paper proposes an expert survey method based on the confidence index, which takes into account the subjective reliability of the experts.

Table 1. Event structure relationship for risk of LNG maritime transport.

Top Event	Intermediate Events			Basic Events
Risk of LNG maritime transport (T)	LNG's own transport risks (I_1)			Fire and explosion risk of LNG (X_1)
				LNG loading and unloading risks (X_2)
	Vessels and equipment risks (I_2)			Safety performance of LNG ships (X_3)
				Difficult handling of LNG ships (X_4)
	Safety of shipping routes (I_3)	Inherent risks of the route itself (I_4)		Long course distance (X_5)
				Deep channel (X_6)
				High ocean current velocity (X_7)
				Heavy traffic flow in the section (X_8)
		Influence of weather and sea state (I_5)		High waves (X_9)
			Low visibility (I_8)	Heavy fog (X_{10})
				Thunderstorms (X_{11})
			Sea breeze effect (I_9)	Strong sea breeze (X_{12})
				High frequency of strong winds (X_{13})
		Coastal ports risk (I_6)	Objective factors (I_{10})	Fewer LNG unloading ports (X_{14})
				Uncertain navigable period (X_{15})
				Draft of LNG ships at ports (X_{16})
			Subjective factors (I_{11})	Unsafe behavior of personnel on LNG ships (X_{17})
				Poor organization (X_{18})
		Maritime security environment (I_7)	Influence of political game (I_{12})	Maritime sovereignty disputes (X_{19})
				Military conflicts (X_{20})
			Non-traditional threat to security (I_{13})	Piracy and terrorist attacks (X_{21})
				The impact of epidemic (X_{22})

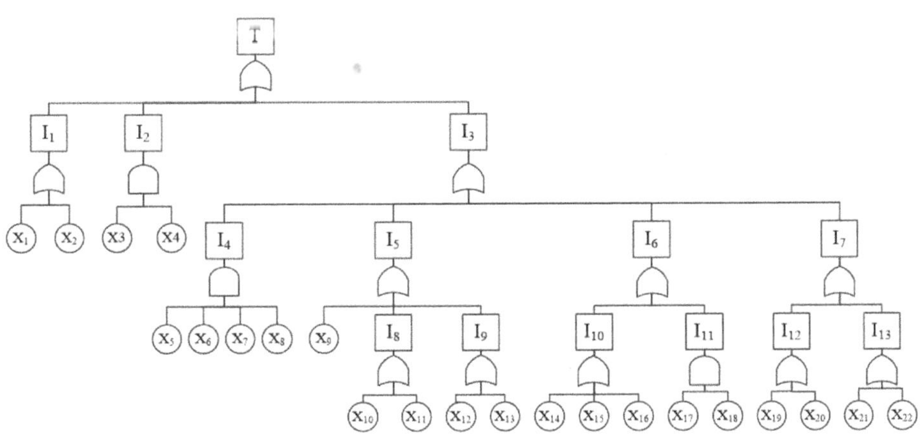

Figure 3. Fault tree diagram of LNG maritime transport risk assessment.

Firstly, the expert judgment ability, denoted as ξ, was categorized into five levels, represented by 'I, II, III, IV, and V' and corresponding to the values '0.6, 0.7, 0.8, 0.9, and 1.0', respectively. A smaller ξ value indicates less reliable expert judgment ability. The subjective reliability, represented by ψ, measures the degree of reliability of the expert in their judgment and is divided into five levels, denoted as '0.6, 0.7, 0.8, 0.9, and 1.0', respectively. A higher level of subjective reliability ψ indicates a more reliable judgment. Assuming that m experts participate in the survey, the confidence index δ_n of the nth expert can be calculated using Equation (2).

$$\delta_n = \xi_n \times \psi_n \tag{2}$$

Secondly, in this paper, we adopt five fuzzy languages, namely 'very low (VL)', 'low (L)', 'medium (M)', 'high (H)', and 'very high (VH)', to describe the failure probability [31] of basic events. Table 2 illustrates the corresponding relationship between the linguistic variables and the failure probability intervals. The i_{th} interval is defined by its lower and upper bounds $[a_i, a_{i+1}]$, and its average value $c_i (1 \leq i \leq 5)$.

Table 2. Triangular fuzzy number assignment.

	Fuzzy Languages	Lower Bound (a_i)	Mean (c_i)	Upper (a_{i+1})
1	VL	0.00	0.10	0.20
2	L	0.20	0.30	0.40
3	M	0.40	0.50	0.60
4	H	0.60	0.70	0.80
5	VH	0.80	0.90	1.00

Multiple experts are selected to score the probability of failure of the basic event, usually the expert confidence level δ is less than 1, which means that the remaining probability of the root node is $1 - \delta$, distributed among other intervals. According to the Gaussian distribution pattern of random variables, the probability of failure tends to fluctuate around its expectation and gradually decreases as it moves away from the expectation. Therefore, a simplified formula for the distribution of residual probability $1 - \delta$ in other intervals is proposed, as shown in Equations (3)–(5), where a_i is the lower bound of the triangular fuzzy number in the i_{th} interval of failure probabilities.

$$P_{xn}^{k} = \begin{cases} \delta, k = i(i = 1) \\ \frac{(a_{5+2-k} - a_1)}{\sum\limits_{n=2}^{5} (a_n - a_1)} \times (1 - \delta), 2 \leq k \leq 5 \end{cases} \tag{3}$$

$$P_{xn}^{k} = \begin{cases} \frac{(a_i - a_{i-k})}{\sum\limits_{n=1}^{i-1} (a_i - a_n)} \times \frac{1-\delta}{2}, 1 \leq k \leq i - 1 \\ \delta, k = i \\ \frac{(a_{6+i-k} - a_i)}{\sum\limits_{n=i+1}^{5} (a_n - a_i)} \times \frac{1-\delta}{2}, i + 1 \leq k \leq 5 \end{cases} \tag{4}$$

$$P_{xn}^{k} = \begin{cases} \frac{(a_5 - a_{5-k})}{\sum\limits_{n=1}^{4} (a_5 - a_n)} \times (1 - \delta), 1 \leq k \leq 4 \\ \delta, k = i(i = 5) \end{cases} \tag{5}$$

Thirdly, the failure probability evaluated by each expert on the root node X_n can be obtained through Equation (6), where c_k is the average value of the K_{th} failure probability interval, as shown in Table 2.

$$P_{X_n} = \sum_{k=1}^{5} \left(c_k \times P_{xn}^{k} \right) \tag{6}$$

Fourthly, the fuzzy number P^* of the failure probability of each basic event is obtained by calculating the mean value. Then, the failure probability P of the basic event is calculated by solving the fuzzy, as shown in Equations (7) and (8) [25].

$$z = 2.301\left(\frac{1 - P^*}{P^*}\right)^{\frac{1}{3}} \tag{7}$$

$$P = \begin{cases} \frac{1}{10^z}, P^* \neq 0 \\ 0, P^* = 0 \end{cases} \tag{8}$$

3.5. Transforming the Fault Tree into Bayesian Network

A Bayesian network is a directed acyclic graph (DAG) that encodes the joint probability distribution of a set of random variables [32]. As a tool for prediction, diagnosis, and reasoning, a BN can calculate the probability of risk occurrence. The Bayesian network has two types of reasoning: causal reasoning and diagnostic reasoning. Causal reasoning refers to predicting the probability of the failure of the target node based on the probability of the root node failure. Diagnostic reasoning involves assuming that the target node has a fault. Based on the degree of correlation between nodes, the possibility and importance of each root node can be obtained. This information can then be used to determine the specific reason for the occurrence of the target node.

The network diagram of BN can be observed as the qualitative part of the model, while the quantitative part of the model is composed of probability parameters. The joint probability of a set of random variables $(A_1, A_2, A_3, \ldots, A_n)$ based on the conditional independence and the chain rule can be obtained as follows:

$$P(A_1, A_2, A_3, \ldots, A_n) = P(A_1|A_2, A_3, \ldots, A_n)P(A_2|A_3, \ldots, A_n)\ldots P(A_{n-1}|A_n)P(A_n) \tag{9}$$

Bayes theorem [24] is used in the BN to update the failure probability (prior) of basic events given new observations to yield the consequence probability (posterior) using the following equation:

$$P(A \mid B) = \frac{P(B|A)P(A)}{P(B)} \tag{10}$$

where $P(A|B)$ is the posterior probability of A if B is true.

In addition to the posterior probability, the probability importance degree and the critical importance degree are also used to study the probability of the risk caused by the basic event.

Probability importance refers to the degree to which the change of probability of failure of basic events causes the change of probability of failure of top events. The specific calculation formula is as follows:

$$I_i^{pr} = p(T = 1|a_i = 1) - p(T = 1|a_i = 0) \tag{11}$$

Critical importance refers to the change rate of top event failure probability caused by the change rate of basic event failure probability, which essentially reflects the importance of basic events in the fault tree. The specific calculation formula is as follows:

$$I_i^{cr} = p(A_i = 1)[p(T = 1|a_i = 1) - p(T = 1|a_i = 0)]/p(T = 1) \tag{12}$$

where, T is the top event; $P(T = 1|\cdot)$ is the conditional probability of the top event; $A_i = 1, 0$ indicates the status of occurrence or non-occurrence of basic event i; and $P(A_i) = 1$ is the prior probability of the basic event i.

After establishing the relationships between events, logical relations, and symbols of the Bayesian network model in the fault tree, the fault tree was transformed into a Bayesian network, as shown in Figure 4. The top event in the fault tree is denoted by T, and the basic event is represented by $N = (x_1, x_2, \ldots\ldots, x_n)$. After transformation, the graphical structure of the developed BN is shown in Figure 5. The colors of nodes represent the target node, intermediate node, and root node in order of depth to light.

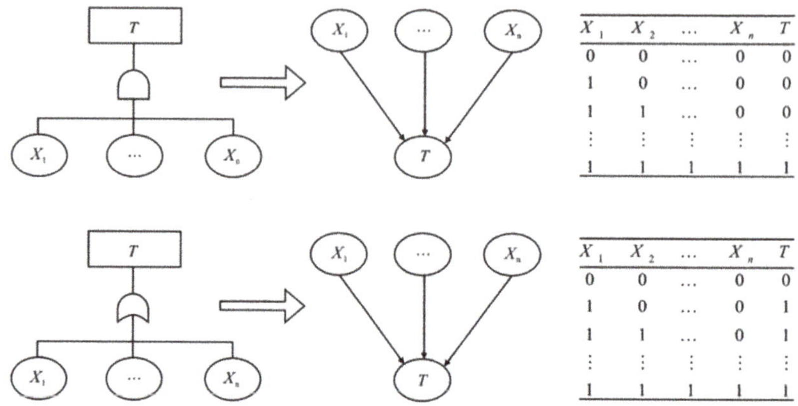

Figure 4. Transformation diagram of AND-Gate and OR-Gate.

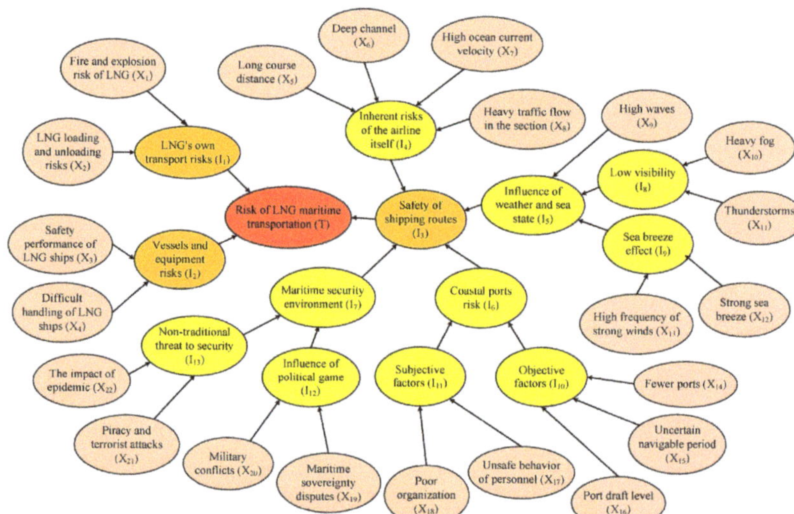

Figure 5. Bayesian network for risk assessment of LNG maritime transport.

4. Case Study

4.1. Description of the China-Australia Route

The China–Australia route, starting from China, passes through Southeast Asian countries such as the Philippines and Indonesia, as well as the South China Sea, Banda Sea and other sea areas, and finally reaches Australia and South Pacific island countries. As China's offshore route, this route is a route channel with intensive and busy economic and trade exchanges with Australia and other countries, and it is also an important maritime transport channel for national strategic energy. China imports a large amount of energy and goods such as LNG and iron ore from Australia, which is the lifeline of China–Oceania import and export trade. Therefore, the shipping activities such as container transportation, LNG, iron ore, asphalt, and other bulk cargo transportation on the China–Australia route are active and frequent.

Due to the uncertainty of risks along the China–Australia route, there are certain impacts and losses along the route due to risks, providing a certain warning for the safe operation of the route. According to the statistics on maritime casualties and accidents released by the IMO, the statistics on maritime casualties and accidents along the route can be divided into three levels: very serious, serious, and not very serious. During 2006–2020 (the global shipping industry was impacted and affected by the global COVID-19 from 2020 to 2023, so there were fewer casualties and accidents at sea), the proportion of the very serious marine casualties and accidents along the line was basically 60–70%. The proportion of serious maritime casualties and accidents is second, ranging from 20% to 30%. The proportion of less serious maritime casualties and accidents is the smallest, around 10%. This indicates that the consequences of maritime safety accidents caused by safety risks along the route are relatively serious. Therefore, the risk issues of the China–Australia route need to be taken seriously and security risk control along the route should be strengthened. In addition, according to a series of reports on Safety and Transportation Review issued by IMO, among the top ten loss areas in 2012–2019, southern China, Indonesia, and the Philippines have always been the key areas and focuses of ship losses. Among these key areas, the areas along the China–Australia route account for the vast majority, and the proportion of ship losses along the route remains around 20–30% globally. The significant safety risk losses indicate that the results caused by safety risks along the route are not optimistic and urgently need to be taken seriously by relevant parties.

4.2. Risk Prediction

In the realm of risk analysis, the utilization of expert evaluation via questionnaires is widely regarded as a reliable approach to supplement the inadequacies inherent in incomplete data, thereby affording the opportunity to procure prior probabilities of greater precision. When selecting multiple experts to score the probability of failure of basic events, the judgment ability of each expert (ξ) may be distinguished according to their working years.

The flowchart of the refinement stage of the voting mechanism is shown in step 3 of Figure 1. Firstly, each expert needs to choose a probability based on their own experience and research objectives. The five probabilities given by experts are all located in "very low (VL), low (L), medium (M), high (H), and very high (VH)". In the second step, expert opinions will be processed to further identify voting results with scores far below the average score for each factor. After discussion, it is believed that this situation was an accidental situation encountered by a certain expert and could not be symbolically summarized. Therefore, the excessive biased opinions in this situation are removed and the common opinions of the majority of experts are retained. The third step is to process and calculate the retained opinions into fuzzy set theory, converting them into probability values within the range of (0, 1). The prior probability of each basic event is obtained.

In this study, six experts were invited to conduct the expert judgements. Table 3 shows the allocation of experts according to their years of work experience, and the specific questionnaire survey results can be found in Appendix B.

Table 3. Expert assignment situation.

Expert No.	Expert Judgment (ξ)	Subjective Reliability Level (ψ)	Degree of Confidence (δ)
1	0.9	1	0.9
2	0.8	0.9	0.72
3	0.6	0.9	0.54
4	0.8	0.8	0.64
5	1	0.8	0.8
6	0.8	0.8	0.64

It can be observed from the questionnaire results that the scoring results of most basic events are relatively average. The Mean c_i corresponding to each fuzzy language is averaged, and the error between the scoring results of six experts under 22 basic events and the average value is calculated, respectively. Five experts are involved, so a ratio of 0.2 is used for screening. We believe that scores exceeding the average value by 20% conflict with the opinions of most experts, so we have manually removed this situation to ensure consistency in expert judgment. If the error is within 0.2, it is considered reasonable, and the scoring result error table is shown in Table 4. If the score error of four Basic events is higher than 0.2, the expert opinion with a larger error in this case will be discarded and the other average expert opinions will be retained. In this way, the fairness of expert scoring opinions has been reasonably addressed. Meanwhile, the deviation of unequal information in the subjective judgment of experts is overcome, and the accuracy of the prior probability value of basic events is guaranteed.

Table 4. The scoring result deviation.

RIFs	Average Value	No. 1	No. 2	No. 3	No. 4	No. 5	No. 6
X_1	0.43	0.13	0.07	0.07	0.07	0.13	0.07
X_2	0.30	0.00	0.20	0.00	0.20	0.20	0.20
X_3	0.30	0.20	0.00	0.00	0.00	0.00	0.20
X_4	0.23	0.07	0.07	0.07	0.07	0.13	0.13
X_5	0.20	0.10	0.10	0.10	0.10	0.10	0.10
X_6	0.23	0.07	0.13	0.07	0.27	0.13	0.13
X7	0.50	0.20	0.20	0.20	0.20	0.00	0.40
X_8	0.53	0.03	0.17	0.17	0.03	0.23	0.03
X_9	0.57	0.07	0.13	0.07	0.07	0.07	0.13
X_{10}	0.77	0.07	0.07	0.13	0.07	0.07	0.13
X_{11}	0.40	0.10	0.10	0.10	0.10	0.10	0.10
X_{12}	0.40	0.10	0.10	0.10	0.10	0.10	0.10
X_{13}	0.70	0.00	0.00	0.20	0.00	0.20	0.00
X_{14}	0.70	0.00	0.00	0.20	0.20	0.00	0.00
X_{15}	0.30	0.00	0.00	0.20	0.00	0.00	0.20
X_{16}	0.30	0.00	0.00	0.20	0.20	0.00	0.00
X_{17}	0.17	0.07	0.07	0.13	0.07	0.13	0.07
X_{18}	0.23	0.07	0.13	0.07	0.07	0.07	0.13
X_{19}	0.27	0.17	0.03	0.03	0.23	0.03	0.17
X_{20}	0.43	0.27	0.13	0.07	0.07	0.13	0.13
X_{21}	0.73	0.03	0.03	0.17	0.03	0.03	0.03
X_{22}	0.80	0.10	0.10	0.10	0.10	0.10	0.10

The prior probabilities of each basic event in LNG maritime transport risk can be calculated using Equations (3)–(8), and the calculation results are shown in Table 5.

Table 5. Probability transformation of expert language.

Basic Event	Prior Probability	Basic Event	Prior Probability	Basic Event	Prior Probability
X_1	3.13×10^{-3}	X_9	7.51×10^{-3}	X_{17}	4.63×10^{-4}
X_2	1.40×10^{-3}	X_{10}	1.86×10^{-2}	X_{18}	7.78×10^{-4}
X_3	1.33×10^{-3}	X_{11}	2.49×10^{-3}	X_{19}	9.66×10^{-4}
X_4	7.38×10^{-4}	X_{12}	2.47×10^{-3}	X_{20}	3.25×10^{-3}
X_5	5.53×10^{-4}	X_{13}	1.40×10^{-2}	X_{21}	1.71×10^{-2}
X_6	8.44×10^{-4}	X_{14}	1.42×10^{-2}	X_{22}	2.37×10^{-2}
X_7	5.78×10^{-3}	X_{15}	1.28×10^{-3}		
X_8	6.05×10^{-3}	X_{16}	1.28×10^{-3}		

The probabilities of top event and intermediate events can be obtained after importing the prior probabilities of various basic events into the Bayesian network model. This step becomes 'risk prediction'. By performing calculations using NETICA, the probability of the occurrence of risks during LNG maritime transport on the China–Australia route is 0.106, which is the predicted result. Due to the direct connection between the three intermediate events I_1, I_2, and I_3 and the top event, they are considered as the direct cause of the risk occurrence. The direct causes of accidents are also evaluated, with a probability of 0.102 assigned to 'Shipping routes safety risks (I_3)', 4.53×10^{-3} to 'LNG's own transport risks (I_1)', and 9.82×10^{-7} to 'Vessels and equipment risks (I_2)'. As a result, measures to reduce the risk of shipping routes should be prioritized over other risk-reduction measures.

4.3. Model Validation

The validity of the BN model can be evaluated using the two axioms. The reliability of partial nodes in the network is verified by applying these axioms sequentially.

Axiom 1: The change of the failure probability of the target node is observed in the BN model by changing the prior probability value of the relevant intermediate node. This test determines whether the model meets the requirements of Axiom 1.

Axiom 2: The total impact of the combination of probability changes from 'evidence nodes' on the target value should always be greater than the combination of probability changes from 'secondary evidence nodes'.

The results of Axiom 1, depicted in Figure 6, show that the prior probability value of the target node and the intermediate node exhibit similar fluctuation trends. When the probability of the intermediate nodes being in the 'normal' state is 0%, the probability of the target node being in the 'normal' state is also 0%. This is because, in CPT, any abnormality in the three intermediate nodes leads to the occurrence of the final accident.

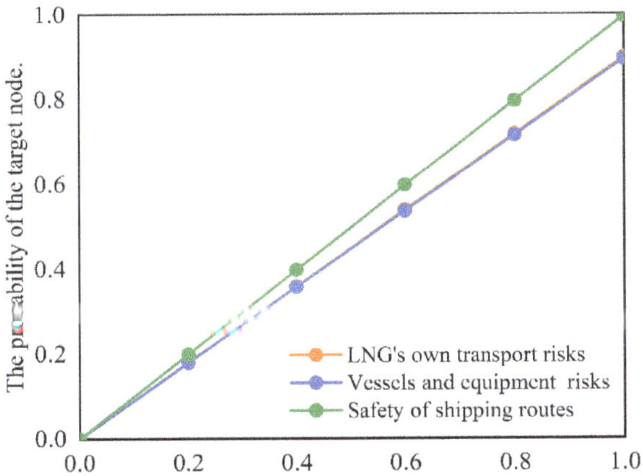

Figure 6. Test result of Axiom 1 under various prior probabilities.

To verify Axiom 2, it is necessary to first determine 'evidence nodes' and 'secondary evidence nodes'. The state of 'Safety of shipping routes (I_3)' is related to 'Inherent risks of the airline itself (I_4)', 'Influence of weather and sea state (I_5)', 'Coastal ports risk (I_6)', and 'Maritime security environment (I_7)'. Therefore, in this paper, 'Safety of shipping routes (I_3)' is considered an 'evidence node', while 'inherent risks of the airline itself (I_4)', 'Influence of weather and sea state (I_5)', 'Coastal ports risk (I_6)', and 'Maritime security environment (I_7)' are regarded as 'secondary evidence nodes'.

When new evidence is introduced into the Bayesian network and the probability of four nodes being in the 'normal' state is 100%, the probability of 'Risk of LNG maritime transport (T)' being in the 'normal' state is 89.4%, 93.5%, 90.9%, and 93.5%, respectively. When the probability of the four nodes being in the 'normal' state simultaneously is 100%, the probability of 'Risk of LNG maritime transport (T)' being in the 'normal' state is 99.5%. This value is greater than the probability value of 'Risk of LNG maritime transport (T)' caused by the individual change probabilities of the four nodes, thereby satisfying the validation conditions of Axiom 2. In addition, tests were conducted on other corresponding secondary evidence nodes, which also met the validation criteria of Axiom 2.

4.4. Risk Prevention

By calculating and ranking the critical importance and probability importance of basic events, it is possible to clarify the degree of influence of each event on the occurrence risk. Taking control measures for events with a high degree of impact and a relatively easy reduction in the failure probability can effectively prevent the occurrence of LNG maritime transport risk.

4.4.1. Calculation of Importance

The probability importance and critical importance of basic events can be calculated, and the results are shown in Table 6.

Table 6. Basic event significance calculation results.

Basic Event	Probability Importance	Critical Importance	Basic Event	Probability Importance	Critical Importance
X_1	0.897	0.026	X_{12}	0.896	0.021
X_2	0.895	0.012	X_{13}	0.907	0.12
X_3	0.001	8.26×10^{-6}	X_{14}	0.907	0.121
X_4	0.001	8.26×10^{-6}	X_{15}	0.895	0.011
X_5	0	0	X_{16}	0.895	0.011
X_6	0	0	X_{17}	0.001	0
X_7	0	0	X_{18}	0	0
X_8	0	0	X_{19}	0.895	0.008
X_9	0.901	0.064	X_{20}	0.897	0.027
X_{10}	0.911	0.16	X_{21}	0.909	0.146
X_{11}	0.896	0.021	X_{22}	0.916	0.204

4.4.2. Rank of Importance

The risk matrix is a qualitative analysis tool used to rank the likelihood and consequences and specify the level of risk. It mainly analyzes and evaluates risks from two dimensions: the likelihood of risk factors occurring, and the severity of damage caused. This evaluation method is a combination of qualitative and quantitative methods. The form of the risk matrix is represented by a two-dimensional table, and the basic risk matrix coordinate diagram is shown in Figure 7. By drawing a risk matrix diagram, multiple risks in the system can be more intuitively compared, and the corresponding order and methods of risk factors can be further determined based on the comparison results.

The basic risk matrix mainly divides the risk level into three regions: A, B, and C. If a risk factor is located in Region A, it is considered a high-level risk factor. Preventive measures should be taken well, and rules and regulations should be established to avoid such situations. If a risk factor is located in Region B, it is considered a moderate risk factor and reasonable control methods and solutions need to be developed. If a risk factor is in region C, it is considered a very low-level risk factor. Under the existing security management system, there is no need for additional control.

In order to comprehensively evaluate the risk of LNG maritime transport, the risk matrix analysis is introduced on the basis of the Bayesian network. Risk matrix is a qualitative analysis tool used to grade the possibility and consequence and specify the risk level. It can comprehensively represent the frequency and severity of risk accidents. According to the calculation results of probability importance and critical importance, they are standardized and presented in the table in the form of quadrants. By classifying the analyzed risks, we can obtain three different risk categories: 'high risk factor (HR)', 'medium risk factor (MR)', and 'low risk factor (LR)'. The specific division is shown in Figure 8.

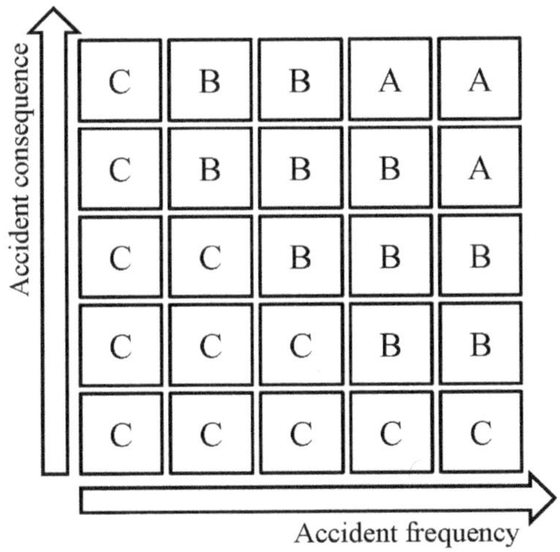

Figure 7. Basic risk matrix.

Very serious		X_{14}	X_{10}	X_{22}
		X_{15}	X_{13}	X_{21}
Serious	X_{16}	X_{11}	X_{12}	X_{20}
Commonly	X_{3}	X_{19}	X_{2}	X_{9}
Slight	X_{7}	X_{8}	X_{17}	X_{1}
	X_{5}	X_{6}	X_{18}	X_{4}
	Rarely	Occasionally	Likely	Frequently

Figure 8. Risk matrix.

4.4.3. Analysis of Importance Ranking Results

The high-risk factors are located at the intersection of the peaks of probability importance and critical importance, which means the events with a high risk of influencing top events and easily reduce the priori probability [33]. Taking measures to reduce the possibility of failure of these events will quickly and effectively reduce the risk of LNG maritime transport.

From the calculation results of probability importance, we find that there is no significant difference between the importance values of the top 14 basic events, which is much higher than the bottom 8. The order of critical importance of the top 14 is the same as probability importance, but the critical importance of the top 5 is far higher than that of others. Consequently, the fundamental occurrences associated with the uppermost five were scrutinized as perilous elements. Specifically, these comprise 'Non-traditional threat to security (I_{13})', 'Heavy fog (X_{10})', 'High frequency of strong winds (X_{13})', and 'Fewer LNG unloading ports (X_{14})'. First of all, non-traditional security threats along the China–Australia route are increasingly prominent. Piracy and terrorist attacks have international characteristics, increasing the probability of transport risks. Moreover, the impact of the global epidemic has made maritime public health safety one of the important factors affecting transport risks. Therefore, we should include this factor in the assessment and response to transport risks and strengthen joint prevention and control measures to build a solid maritime security defense line. Secondly, 'Heavy fog (X_{10})' and 'High frequency of strong winds (X_{13})' fall under the category of 'Influence of weather and sea state (I_5)'. In case of severe weather, navigation ships need to take evasive measures such as entering the port or avoiding navigation. Finally, the small number of seaports on the China–Australia route has a high impact on LNG maritime transport risks, mainly because some ports along the route do not have facilities and equipment for LNG storage, loading, and unloading, resulting in poor connectivity with other ports, thereby increasing the risks in LNG transport.

Meanwhile, eight basic events with probability importance and critical importance close to zero are considered low-risk factors, namely 'Safety performance of LNG ships (X_3)', 'Difficult handling of LNG ships (X_4)', 'Long course distance (X_5)', 'Deep channel (X_6)', 'High ocean current velocity (X_7)', 'Heavy traffic flow in the section (X_8)', 'Unsafe behavior of personnel on LNG ships(X_{17})', and 'Poor organization (X_{18})'. Reducing the occurrence probability of these events has less effect on improving the safety of LNG maritime transport and is more difficult to reduce [33]. We can summarize this as two intermediate events: 'Vessels and equipment risks (I_2)' and 'Inherent risks of the route itself (I_4)'. In the case of limited resources, we can ignore these factors.

Moreover, medium-risk factors are events with high probability importance but low critical importance. Although such factors have a high risk of influencing the occurrence of top events, their own occurrence frequency is very low. Therefore, daily inspection and management should be strengthened for such factors, and potential symptoms should be found and handled in a timely manner.

4.5. Risk Diagnosis

The cause of risk can be diagnosed by sorting the posterior probability of basic events, with the support of the ability of BN's binary risk diagnosis and risk prediction. The posterior probability of each basic event is calculated through NETICA, as shown in Table 7.

Table 7. Basic event posterior probability.

Basic Event	Posterior Probability	Basic Event	Posterior Probability	Basic Event	Posterior Probability
X_1	2.95×10^{-2}	X_9	7.07×10^{-2}	X_{17}	4.70×10^{-4}
X_2	1.32×10^{-2}	X_{10}	1.75×10^{-1}	X_{18}	7.80×10^{-4}
X_3	1.30×10^{-3}	X_{11}	2.35×10^{-2}	X_{19}	9.10×10^{-3}
X_4	7.50×10^{-4}	X_{12}	2.33×10^{-2}	X_{20}	3.06×10^{-2}
X_5	5.50×10^{-4}	X_{13}	1.32×10^{-1}	X_{21}	1.61×10^{-1}
X_6	8.40×10^{-4}	X_{14}	1.34×10^{-1}	X_{22}	2.23×10^{-1}
X_7	5.80×10^{-3}	X_{15}	1.21×10^{-2}		
X_8	6.03×10^{-3}	X_{16}	1.21×10^{-2}		

The posterior probability ranking results of basic events are as follows.

$$\begin{aligned} &P_{x22} > P_{x10} > P_{x21} > P_{x14} > P_{x13} > P_{x9} > P_{x20} > P_{x1} > P_{x11} > P_{x12} > P_{x2} \\ &> P_{x15} = P_{x16} > P_{x19} > P_{x8} > P_{x7} > P_{x3} > P_{x6} > P_{x18} > P_{x4} > P_{x5} > P_{x17} \end{aligned}$$

If there are risks in LNG maritime transport on the China–Australia route, the most likely reasons are 'The impact of epidemic (X_{22})', 'Piracy and terrorist attacks (X_{21})', 'Fewer LNG unloading ports (X_{14})', and 'Influence of weather and sea state (I_5)'. When accidents occur in maritime transport, priority can be given to checking whether these events occur, so as to save time and cost.

5. Suggestions and Discussion

Based on the diagnosis of maritime transport risks on the China–Australia route, this paper puts forward suggestions and measures to reduce risks by reducing the probability of events with high posterior probability in Table 7.

As shown in Table 7, when accidents occur in LNG maritime transport, the posterior probability ranking of the top six basic events is significantly higher than others. To mitigate the risks associated with these events, the following suggestions and measures are proposed. From the perspective of risk control, these six events are divided into two categories for consideration: preventable and uncontrollable. Among them, the posterior probability ranking first and third are 'The impact of epidemic (X_{22})' and 'Piracy and terrorist attacks (X_{21})', which are considered as preventable events due to their low probability of failure. In response to the impact of the epidemic, shipping companies are predominantly affected, so it is crucial to establish a response mechanism for epidemic prevention and control for them; for piracy and terrorist attacks, the country should strengthen its domestic naval capacity building for defense. The rest are uncontrollable risks. The fourth-ranked event is 'Fewer LNG unloading ports (X_{14})', and as this situation cannot be improved in a short period of time, the establishment of alternative route mechanisms at ports can alleviate this impact to some extent. Finally, the second, fifth, and sixth-ranked events, namely 'Heavy fog (X_{10})', 'High frequency of strong winds (X_{13})', and 'High waves (X_9)' can be summarized as the intermediate event 'Influence of weather and sea state (I_5)'. Given the unchangeable nature of these objective conditions, the only viable approach is to avoid LNG loading and unloading operations during adverse weather conditions.

The following are specific further discussions based on the above suggestions. For shipping enterprises, they should establish epidemic prevention and control response mechanisms. The outbreak of the global COVID-19 pandemic has increased the risk of shipping routes due to port suspensions in some relevant countries. To prevent similar situations from causing greater risks in the future, shipping enterprises should take targeted actions. For example, they should formulate comprehensive and feasible emergency plans for epidemic prevention and control under the leadership and guidance of the national government. This would help establish and improve relevant emergency response mechanisms to reduce the harm caused by public health emergencies at sea and prevent the spread of the epidemic [34]. Furthermore, strengthening the marine defense line for epidemic prevention and control is essential. Only by maintaining the normal order of maritime activities on the China–Australia route, and reducing the risk of the route, can we promote the recovery and normalization of shipping. Thus, it is crucial to take proactive measures to ensure the safety and security of shipping activities in the face of similar global accidents.

For the country, it is necessary to strengthen the capacity building of the domestic navy. The waters surrounding the South China Sea, Indonesia, and the Philippines are known for their high levels of pirate attacks and maritime terrorism, which create multiple uncertainties for shipping routes in the region. Therefore, it is crucial to enhance the construction of the domestic navy to improve route security. Specifically, China should focus on improving the navy's long-distance combat capability and defense capability, as well as enhancing the protection mechanism for LNG ships. Additionally, the ability to patrol dangerous waters and monitor the marine environment must be strengthened, and a well-established emergency plan system for piracy and terrorist attacks must be put in place. Increasing the frequency of naval convoys on the China–Australia route and cracking down on piracy are also important measures to consider.

For ports, a route substitution mechanism could be established to address the challenges posed by the restrictions of many islands, such as the Indonesian archipelago and the South Pacific islands, and the different passing capacities of key nodes, such as the Straits. Alternative routes should be actively sought to reduce the threat of emergencies to route safety. In the face of security threats, the route substitution mechanism could be implemented to reduce losses caused by port blockade and ensure the safety of the route.

Furthermore, loading and unloading operations should be avoided as far as possible during bad weather conditions. If environmental factors such as bad weather, strong winds, or waves affect the loading and unloading operation, it should be stopped immediately, and the corresponding connecting equipment should be disconnected to ensure the safety of the ship and the wharf.

6. Conclusions

The main contribution of this paper is to build a fault tree analysis and Bayesian network model for the risk assessment of LNG maritime transport on the China–Australia routes. The proposed model combines the two methods to overcome the problem that conditional probabilities in the Bayesian network find difficult to determine. Specifically, due to the complexity of the maritime scenario, it is inappropriate for analysis to attribute accidents to a single cause or a few causes. This paper constructs accident causation networks from various perspectives of cargo, ships, route, and environment. The fault tree is established by investigating relevant literature and accident investigation reports, and expert opinions and fuzzy set are used to derive the prior probability, the fault tree is transformed into Bayesian network, and the conditional probability table in the Bayesian network is obtained through a relationship gate in fault tree analysis. From further analysis, the key influencing factors and sensitive factors can also be identified in this developed model.

In conclusion, the China–Australia route is an important maritime transport route for the trade activities between China and the Oceania region. However, the route is subject to multiple uncertainties, including piracy, terrorism, epidemic, port restrictions, and inclement weather conditions. These uncertainties pose significant risks to the safety of maritime transport activities and may result in economic losses and environmental damages. To reduce the risks and ensure the safety of maritime transport activities on this route, various measures should be taken, such as strengthening the construction of the domestic navy, establishing safety early warning systems for LNG storage and transportation, setting up appropriate meteorological monitoring departments, and avoiding LNG loading and unloading operations in bad weather conditions. By implementing these measures, we can promote the recovery and normalization of shipping on the China–Australia route and facilitate economic and trade cooperation between China and the Oceania region.

This paper analyzed the effect of risk factors from a systematic perspective based on real-world accidents. Although this paper takes LNG and the China–Australia route as an example, the proposed model can also be applied to other route to predict the probability of maritime accidents if the proposed route data have similar characteristics. Moreover, quantitative information assessed by experts due to limited data may be a biased representation of the exact real-world situation. Therefore, future work can model and analyze large amounts of data to provide additional and practical insights into enhancing marine safety.

Author Contributions: Conceptualization, X.H. and H.F.; methodology, H.F.; software, X.H.; validation, X.H., H.F. and Z.C.; formal analysis, H.F.; investigation, W.G.; resources, L.H.; data curation, X.H.; writing—original draft preparation, X.H.; writing—review and editing, H.F. and Z.C.; visualization, X.H.; supervision, W.G.; project administration, Z.C.; funding acquisition, Z.C. All authors have read and agreed to the published version of the manuscript.

Funding: This research was funded by Dalian Maritime University grant number 2023JXA03.

Data Availability Statement: The data presented in this study are available on request from the corresponding author. The data are not publicly available due to privacy.

Conflicts of Interest: The authors declare no conflict of interest.

Appendix A. The Sources of RIFs Based on the Retrieved Results

Refs.	[35]	[36]	[37]	[38]	[39]	[40]	[41]	[42]	[24]	[43]	[44]	[45]	[34]	[46]	[47]	[40]	Reports
X_1	A	A		A					A		A						A
X_2	A	A		A					A								A
X_3		A	A	A	A	A		A	A		A		A			A	A
X_4		A		A			A		A							A	A
X_5					A		A						A		A		
X_6					A								A	A	A		
X_7						A		A						A		A	
X_8							A		A	A			A	A	A		
X_9		A	A		A	A	A		A	A			A		A	A	
X_{10}			A		A	A		A		A	A		A	A	A	A	
X_{11}			A							A			A	A	A		
X_{12}	A		A	A		A		A	A	A		A	A	A	A	A	
X_{13}			A		A	A		A	A						A	A	
X_{14}										A				A	A		A
X_{15}				A			A			A		A		A			
X_{16}			A			A	A							A	A		A
X_{17}			A	A		A	A	A	A	A	A					A	A
X_{18}						A		A	A	A	A					A	
X_{19}												A	A	A			
X_{20}											A	A	A				
X_{21}		A		A	A		A		A	A	A		A		A		
X_{22}								A			A		A				

Note: 'A' means this RIF is applied in the related reference.

Appendix B. Questionnaire Results

Basic Events \ Expert	1	2	3	4	5	6
Fire and explosion risk of LNG (X_1)	L	M	M	M	L	M
LNG loading and unloading risks (X_2)	L	M	L	M	VL	VL
Safety performance of LNG ships (X_3)	M	L	L	L	L	VL
Difficult handling of LNG ships (X_4)	L	L	L	L	VL	VL
Long course distance (X_5)	VL	VL	L	VL	L	L
Deep channel (X_6)	L	VL	L	M	VL	VL
High ocean current velocity (X_7)	H	H	L	H	M	VL
Heavy traffic flow in the section (X_8)	M	H	H	M	L	M
High waves (X_9)	M	H	M	M	M	H
Heavy fog (X_{10})	H	H	VH	H	H	VH
Thunderstorms (X_{11})	L	L	M	L	M	M
Strong sea breeze (X_{12})	L	M	M	L	L	M
High frequency of strong winds (X_{13})	H	H	VH	H	M	H
Fewer LNG unloading ports (X_{14})	H	H	VH	M	H	H
Uncertain navigable period (X_{15})	L	L	M	L	L	VL
Draft of LNG ships at ports (X_{16})	L	L	M	VL	L	L
Unsafe behavior of personnel on LNG ships (X_{17})	VL	VL	L	VL	L	VL
Poor organization (X_{18})	L	VL	L	L	L	VL
Maritime sovereignty disputes (X_{19})	VL	L	L	M	L	VL
Military conflicts (X_{20})	H	L	M	M	L	L
Piracy and terrorist attacks (X_{21})	H	H	VH	H	H	H
The impact of epidemics (X_{22})	VH	VH	H	VH	H	H

Appendix C. Abbreviation and Full Name

Abbreviation	Full Name
LNG	Liquefied Natural Gas
FTA	Fault Tree Analysis
BN	Bayesian Network
IMO	International Maritime Organization
RIFs	Risk Influential Factors
FMEA	Failure Mode and Effects Analysis
HFACS	Human Factors Analysis and Classification System
FL	Fuzzy Logic
NBN	Naïve Bayesian Network
DBN	Dynamic Bayesian Network
CPTs	Conditional Probability Tables
QRA	Quantitative Risk Analysis
HRA	Human Reliability Analysis
MCS	Monte Carlo Simulation
RCA	Root Cause Analysis
FST	Fuzzy Set Theory
DAG	Directed Acyclic Graph

References

1. Li, H.; Lam, J.S.L.; Yang, Z.; Liu, J.; Liu, R.W.; Liang, M.; Li, Y. Unsupervised hierarchical methodology of maritime traffic pattern extraction for knowledge discovery. *Transp. Res. Part C Emerg. Technol.* **2022**, *143*, 103856. [CrossRef]
2. Fan, H.; Lu, J.; Chang, Z. A risk-based game theory model of navy and pirate behaviors. *Ocean Coast. Manag.* **2022**, *225*, 106200. [CrossRef]
3. Khan, R.U.; Yin, J.; Mustafa, F.S.; Liu, H. Risk Assessment and Decision Support for Sustainable Traffic Safety in Hong Kong Waters. *IEEE Access* **2020**, *8*, 72893–72909. [CrossRef]
4. Kalantarnia, M.; Khan, F.; Hawboldt, K. Dynamic risk assessment using failure assessment and Bayesian theory. *J. Loss Prev. Process Ind.* **2009**, *22*, 600–606. [CrossRef]
5. Braglia, M.; Frosolini, M.; Montanari, R. Fuzzy criticality assessment model for failure modes and effects analysis. *Int. J. Qual. Reliab. Manag.* **2003**, *20*, 503–524. [CrossRef]
6. Levary, R.W. K An analytic hierarchy process based simulation model for entry mode decision regarding foreign direct investment. *Int. J. Manag. Sci.* **1999**, *21*, 661–677. [CrossRef]
7. Zaili, Y.; Bonsall, S.; Jin, W. Fuzzy Rule-Based Bayesian Reasoning Approach for Prioritization of Failures in FMEA. *IEEE Trans. Reliab.* **2008**, *57*, 517–528. [CrossRef]
8. Celik, M.; Cebi, S. Analytical HFACS for investigating human errors in shipping accidents. *Accid. Anal. Prev.* **2009**, *41*, 66–75. [CrossRef]
9. Khan, R.U.; Yin, J.; Mustafa, F.S.; Anning, N. Risk assessment for berthing of hazardous cargo vessels using Bayesian networks. *Ocean Coast. Manag.* **2021**, *210*, 105673. [CrossRef]
10. Fan, S.; Yang, Z.; Blanco-Davis, E.; Zhang, J.; Yan, X. Analysis of maritime transport accidents using Bayesian networks. *J. Risk Reliab.* **2020**, *234*, 439–454. [CrossRef]
11. Jiang, M.L.J. Maritime accident risk estimation for sea lanes based on a dynamic Bayesian network. *Marit. Policy Manag.* **2020**, *47*, 649–664. [CrossRef]
12. Vanem, E.; Antão, P.; Østvik, I.; Comas, F.D.C.d. Analysing the risk of LNG carrier operations. *Reliab. Eng. Syst. Saf.* **2008**, *93*, 1328–1344. [CrossRef]
13. Martins, M.R.; Pestana, M.; Souza, G.; Schleder, A. Quantitative risk analysis of loading and offloading liquefied natural gas (LNG) on a floating storage and regasification unit (FSRU). *J. Loss Prev. Process Ind.* **2016**, *43*, 629–653. [CrossRef]
14. Marroni, G.; Moreno, V.C.; Ovidi, F.; Chiavistelli, T.; Landucci, G. A methodology for risk assessment of LNG carriers accessing vulnerable port areas. *Ocean Eng.* **2023**, *273*, 114019. [CrossRef]
15. Abdussamie, N.; Daboos, M.; Elferjani, I.; Shuhong, C.; Alaktiwi, A. Risk assessment of LNG and FLNG vessels during manoeuvring in open sea. *J. Ocean Eng. Sci.* **2018**, *3*, 56–66. [CrossRef]
16. Zhou, T.; Wu, C.; Zhang, J.; Zhang, D. Incorporating CREAM and MCS into fault tree analysis of LNG carrier spill accidents. *Saf. Sci.* **2017**, *96*, 183–191. [CrossRef]
17. Zhao, S.; Zhu, H. A Bayesian Network Modelling and Risk Analysis on LNG Carrier Anchoring System. In Proceedings of the 3rd International Conference on Transportation Information and Safety, Wuhan, China, 25–28 June 2015.
18. Yeo, C.; Bhandari, J.; Abbassi, R.; Garaniya, V.; Chai, S.; Shomali, B. Dynamic risk analysis of offloading process in floating liquefied natural gas (FLNG) platform using Bayesian Network. *J. Loss Prev. Process Ind.* **2016**, *41*, 259–269. [CrossRef]

19. Li, Z.; Hu, S.; Gao, G.; Yao, C.; Fu, S.; Xi, Y. Decision-making on process risk of Arctic route for LNG carrier via dynamic Bayesian network modeling. *J. Loss Prev. Process Ind.* **2021**, *71*, 104473. [CrossRef]
20. Melani, A.H.A.; Souza, G.F.M.; Silva, D.W.R. Gilberto Francisco Martha Souzaa, Use of Bayesian Network to Support Risk-Based Analysis of LNG Carrier Loading Operation. In Proceedings of the Probabilistic Safety Assessment and Management PSAM 12, Honolulu, HI, USA, 22–27 June 2014.
21. Fu, S.; Zhang, D.; Montewka, J.; Zio, E.; Yan, X. A quantitative approach for risk assessment of a ship stuck in ice in Arctic waters. *Saf. Sci.* **2018**, *107*, 145–154. [CrossRef]
22. Ugurlu, H.; Cicek, I. Analysis and assessment of ship collision accidents using Fault Tree and Multiple Correspondence Analysis. *Ocean Eng.* **2022**, *245*, 110514. [CrossRef]
23. Wan, C.; Yan, X.; Zhang, D.; Qu, Z.; Yang, Z. An advanced fuzzy Bayesian-based FMEA approach for assessing maritime supply chain risks. *Transp. Res. Part E Logist. Transp. Rev.* **2019**, *125*, 222–240. [CrossRef]
24. Baksh, A.-A.; Abbassi, R.; Garaniya, V.; Khan, F. Marine transportation risk assessment using Bayesian Network: Application to Arctic waters. *Ocean Eng.* **2018**, *159*, 422–436. [CrossRef]
25. Chen, P.; Zhang, Z.; Huang, Y.; Dai, L.; Hu, H. Risk assessment of marine accidents with Fuzzy Bayesian Networks and causal analysis. *Ocean Coast. Manag.* **2022**, *228*, 106323. [CrossRef]
26. Sakar, C.; Toz, A.C.; Buber, M.; Koseoglu, B. Risk Analysis of Grounding Accidents by Mapping a Fault Tree into a Bayesian Network. *Appl. Ocean Res.* **2021**, *113*, 102764. [CrossRef]
27. Sokukcu, M.; Sakar, C. Risk analysis of collision accidents during underway STS berthing maneuver through integrating fault tree analysis (FTA) into Bayesian network (BN). *Appl. Ocean Res.* **2022**, *126*, 103290. [CrossRef]
28. Kaushik, M.; Kumar, M. An integrated approach of intuitionistic fuzzy fault tree and Bayesian network analysis applicable to risk analysis of ship mooring operations. *Ocean Eng.* **2023**, *269*, 113411. [CrossRef]
29. Li, H.; Ren, X.; Yang, Z. Data-driven Bayesian network for risk analysis of global maritime accidents. *Reliab. Eng. Syst. Saf.* **2023**, *230*, 108938. [CrossRef]
30. Zadeh, L.A. Fuzzy Sets*. *Inf. Control* **1965**, *8*, 338–353. [CrossRef]
31. Zhang, G.-H.; Chen, W.; Jiao, Y.-Y.; Wang, H.; Wang, C.-T. A failure probability evaluation method for collapse of drill-and-blast tunnels based on multistate fuzzy Bayesian network. *Eng. Geol.* **2020**, *276*, 105752. [CrossRef]
32. Fan, S.; Blanco-Davis, E.; Yang, Z.; Zhang, J.; Yan, X. Incorporation of human factors into maritime accident analysis using a data-driven Bayesian network. *Reliab. Eng. Syst. Saf.* **2020**, *203*, 107070. [CrossRef]
33. Hao, D.; Jie, Y.; Jia, W. The Brittle Risk of China's Importing Oil from the Middle East by Maritime Transportation under "The Belt and Road". *Stat. Inf. Forum* **2018**, *33*, 107–115.
34. Xu, L. Research on Risk Assessment of China-Australia and New Zealand Route Based on Bayesian Network. 2021. Available online: https://kns.cnki.net/kcms2/article/abstract?v=3uoqIhG8C475KOm_zrgu4sq25HxUBNNTmIbFx6y0bOQ0cH_CuEtpsOE-ngEml1FeqmrZZpTGLx8mi3KrZmrxIbwIkJ3kNjwu&uniplatform=NZKPT (accessed on 6 August 2023).
35. Luketa-Hanlin, A. A review of large-scale LNG spills: Experiments and modeling. *J. Hazard. Mater.* **2006**, *132*, 119–140. [CrossRef] [PubMed]
36. Animah, I.; Shafiee, M. Application of risk analysis in the liquefied natural gas (LNG) sector: An overview. *J. Loss Prev. Process Ind.* **2020**, *63*, 103980. [CrossRef]
37. Fu, S.; Zhang, D.; Montewka, J.; Yna, X.; Zio, E. Towards a probabilistic model for predicting ship besetting in ice in Arctic waters. *Reliab. Eng. Syst. Saf.* **2016**, *155*, 124–136. [CrossRef]
38. Fu, S.; Zhang, D.; Montewka, J.; Yan, X.; Zio, E. Research on quantitative risk assessment of fuel leak of LNG-fuelled ship during lock transition process. *Reliab. Eng. Syst. Saf.* **2022**, *221*, 108368.
39. Xie, C.; Huang, L.; Wang, R.; Deng, J.; Shu, Y.; Jiang, D. Safety assessment of shipping routes in the South China Sea based on the fuzzy analytic hierarchy process. *Saf. Sci.* **2014**, *62*, 46–57.
40. Qian, H.; Zhang, R.; Zhang, Y.-J. Dynamic risk assessment of natural environment based on Dynamic Bayesian Network for key nodes of the arctic Northwest Passage. *Ocean Eng.* **2020**, *203*, 107205. [CrossRef]
41. Xia, H. Navigational risk analysis based on GIS spatiotemporal trajectory mining: A case study in Nanjing Yangtze River Bridge waters. *Arab. J. Geosci.* **2021**, *14*, 229. [CrossRef]
42. Khan, B.; Khan, F.; Veitch, B.; Yang, M. An operational risk analysis tool to analyze marine transportation in Arctic waters. *Reliab. Eng. Syst. Saf.* **2018**, *169*, 485–502. [CrossRef]
43. Mazaheri, A.; Montewka, J.; Kujala, P. Towards an evidence-based probabilistic risk model for ship-grounding accidents. *Saf. Sci.* **2016**, *86*, 195–210. [CrossRef]
44. Yang, Z.L.; Wang, J.; Li, K.X. Maritime safety analysis in retrospect. *Marit. Policy Manag.* **2013**, *40*, 261–277. [CrossRef]
45. Marciel, S.; Maritime Issues and Sovereignty Disputes in East Asia. Testimony of Deputy Assistant Secretary Scot Marciel, United States Senate, 15 July 2009. Available online: https://2009-2017.state.gov/p/eap/rls/rm/2009/07/126076.htm (accessed on 6 August 2023).

46. He, P.; Ma, X.; Zhang, J.; Qiao, W. Risk Evaluation of different legs of northeast arctic route based on fuzzy analytic hierarchy process–multilevel extension. *Chin. J. Polar Res.* **2021**, *33*, 279–293.
47. Zhou, X.; Cheng, L.; Li, M. Assessing and mapping maritime transportation risk based on spatial fuzzy multi-criteria decision making: A case study in the South China sea. *Ocean Eng.* **2020**, *208*, 107403. [CrossRef]

Journal of
Marine Science and Engineering

MDPI

Article

Online Estimation of Ship Dimensions by Combining Images with AIS Reports

Zishuo Huang [1], Qinyou Hu [1,*], Lan Lu [2], Qiang Mei [3] and Chun Yang [1]

1 Merchant Marine College, Shanghai Maritime University, Shanghai 200135, China; 202040110005@stu.shmtu.edu.cn (Z.H.); chunyang@shmtu.edu.cn (C.Y.)
2 Hainan Research Institute, Nankai University, Haikou 570100, China; nklulan@163.com
3 Navigation College, Jimei University, Xiamen 361021, China; meiqiang@jmu.edu.cn
* Correspondence: qyhu@shmtu.edu.cn; Tel.: +86-21-38282915

Abstract: Ship dimensions are an important component of static AIS information, and are a key factor in identifying the risks of ship collisions. We describe a method of extracting and correcting ship contour information using inland waterway surveillance video combined with AIS information that does not depend on ship dimension data. A lightweight object detection model was used to determine the ship's position in an image. Dynamic AIS information was included to produce multigroup control points, solve the optimal homography matrix, and create a transformation model to map image coordinates onto water surface coordinates. A semantic segmentation DeepLabV3+ model was used to determine ship contours from the images, and the actual dimensions of the ship contours were calculated using homography matrix transformation. The mAP of the proposed object detection model and the MIoU of the semantic segmentation model were 86.73% and 91.07%, respectively. The calculation error of the ship length and width were 5.8% and 7.4%, respectively. These statistics indicate that the proposed method rapidly and accurately detected target ships in images, and that the model estimated ship dimensions within a reasonable range.

Keywords: contour extraction; object detection; semantic segmentation; coordinate mapping

Citation: Huang, Z.; Hu, Q.; Lu, L.; Mei, Q.; Yang, C. Online Estimation of Ship Dimensions by Combining Images with AIS Reports. *J. Mar. Sci. Eng.* **2023**, *11*, 1700. https://doi.org/10.3390/jmse11091700

Academic Editor: Tieshan LI

Received: 24 July 2023
Revised: 16 August 2023
Accepted: 26 August 2023
Published: 29 August 2023

1. Introduction

In water traffic scenarios, ship collision avoidance needs to use an automatic identification system (AIS) or radar and other navigational equipment to obtain the movement information of ships, and the two must complement each other. Using AIS reports, it is often difficult to accurately calculate the distance of the ship from its surrounding objects without considering the shape of the ship. The AIS position is determined by the position of the GPS antenna, and the distance between the GPS antenna and the ship periphery can range from tens to hundreds of meters. For ship–pier collisions, it is necessary to consider the transverse distribution of ships in the river, and use the ship width data to calculate the collision probability. The ship collision risk index (CRI) is calculated through the distance closest point of approach (DCPA), the time closest point of approach (TCPA), and other indexes to represent the urgency of the ship collision at a micro level. In the process of calculating the CRI between two ships, an AIS-equipped ship is often shown as a triangle or rectangle with the transponder at the center [1]. There are dimensional attributes in the AIS static information to represent the length and width of the ship, but in practice, these AIS data are often unavailable. Table 1 shows the dimensions and numbers of ships passing through two inland waterway channels in China within a given time period. It is clear from the table that ship dimension information is unavailable for more than 60% of the vessels. Given the possible consequences of ship collisions, the problem of unavailable ship size data needs urgent resolution.

Table 1. Available information for ship dimensions.

	Bulk Carrier			Tanker		
	No. Observed	No. with Available Information	Ratio	No. Observed	No. with Available Information	Ratio
Jingzhou waterway	100	38	0.38	50	19	0.38
Yichang waterway	100	33	0.33	50	26	0.52

Images have become more widely used for information extraction by computer vision technologies. An augmented reality system based on a fusion of AIS and advanced image processing technology can provide auxiliary information for early warning of navigation risks for autonomous surface vehicles (ASVs) [2]. Such system can also be used for traffic supervision that enables vessels to conform to navigation regulations in key navigable waters [3].

Combining visual data with AIS information enables the estimation of the size of specific ships in the image. Remote sensing or visible light images have frequently been used to extract the contours of target ships, create matching external rectangles or ellipses based on the contour shapes, and derive longitudinal and transverse ship dimension information [4]. Therefore, contour extraction is the basis of ship size estimation, which has been widely used in the transportation sector.

There have been many studies of ship contour extraction. In conventional contour extraction methods, edge detection based on image characteristics has been used to determine contours. Yan et al. [5] improved the Canny edge detection algorithm using a two-dimensional wavelet Gaussian function to calculate the partial derivative of the structural filter gradient amplitude, and adopt maximum inhibition and threshold filters for edge detection and connection for ships. Gu et al. [6] used a binary image gradient calculation for edge detection, and determined the minimum enclosing rectangle for ship contours. Zhu et al. [7] demonstrated a ship recognition method that used a predicted shape template to determine ship contours using the Otsu method, with peak density detection and column scanning as well as a conventional area averaging algorithm. Nie et al. [8] used a binarized normed gradient (BING) algorithm to predict the location of a ship in SAR images, and used an active contour algorithm to predict ship contours iteratively. Standard ship contour extraction methods are simple but are often unable to extract deep image information, and are only suitable for simple scenes.

Convolutional neural networks (CNN) are widely used for image feature extraction in deep learning applications in networks such as VGGNet [9], GoogleNet [10], Inception [11], and ResNet [12]. CNNs have important applications in semantic segmentation; multicategory target contours can be accurately segmented using pixel-level classification of images, and they have been used in many ship contour extraction applications. The fully convolutional network (FCN) was commonly used to extract ship contours by categorizing each pixel in a remote sensing image into the bow, hull, land, and sea [13]. Bovcon et al. [14] developed a deep encoder–decoder framework (the water obstacle separation and refinement network) for autonomous crewless ship navigation that extracted the contours of several ship targets. Ust et al. [15] introduced a scaffolding learning regime (SLR) that trained an obstacle detection segmentation network under weak supervision for individual ship contour extraction. Kelm et al. [16] trained a CNN to identify central pixels; the network recognized a part of an input image and calculated a rotation angle, and used the central pixel to describe the upcoming directional change in the contour. Deep learning methods that rely on training data are more accurate than other established methods, and are highly adaptable to different scenarios.

Remote sensing data of a particular area are not frequently updated, and vessels are densely distributed on inland waterways, making it difficult to accurately extract contours from remote sensing images at any given point in time. Visible light images are generally made from a horizontal perspective, and this perspective is not particularly suitable for accurate ship size estimation. Targeting the problem of missing information of some ship dimensions in a waterway, this study innovatively proposes an intelligent identification method of ship dimensions based on a fusion of inland waterway monitoring overhead image and AIS information. The research scenario is shown in Figure 1, the important stretches of the upper reaches of the Yangtze River that have high marine traffic flow and density. The main contributions of this study are as follows:

Figure 1. The traffic flow of Jingzhou Bridge in one month: white and yellow bands are used to represent the traffic flow in two different directions.

- A deep learning ship object detection model was developed based on a lightweight object detection model, and using the SENet attention mechanism to improve the network structure and increase the effectiveness of detection;
- An optimal homography matrix solution algorithm using several AIS control points was developed to determine the mapping relationship between image coordinates and water surface coordinates;
- Ship contours were extracted using the deep learning DeepLabV3+ semantic segmentation model, in conjunction with the homology matrix transformations to determine the real size of the vessel.

2. Related Studies

2.1. Ship Object Detection

Ship object detection technology has a research history of more than twenty years, and forms the basis for combining video and AIS information in our study [17]. The established methods include those based on the water–sky boundary, saliency detection, and moving object detection. Kim et al. [18] used a background algorithm to detect ships, and combined it with AIS to match ships with ship-related information. Fefilatyev et al. [19] developed a method using optimal water–sky boundary extraction combined with Gaussian distribution and the Hough transform. Yang et al. [20] designed a ship motion tracking system based on the FPGA that differed from traditional inter-frame difference methods, which had fixed frame intervals. Its inter-frame difference method was based on the adaptive extraction of key frames, and was used to adaptively detect ships moving at different speeds.

The CNN is used mainly to solve object detection problems, either in two-stage algorithms such as Faster RCNN [21] and Mask RCNN [22], or single-stage algorithms such as YOLO [23] and SSD [24]. Object detection algorithms that use deep learning overcome the shortcomings of target detection algorithms, such as lack of targeted region selection, sliding window redundancy, and time complexity. Using a deep learning algorithm for ship object detection significantly improves detection. The rotational CNN algorithm [25] was used for text detection because of its excellent rotation detection capability that was introduced to ship target detection, and this produced good results. He et al. [26] combined the Gabor filter with the Faster RCNN to increase ship object detection accuracy from satellite images. Zhang et al. [27] preprocessed images with a support vector machine, and then processed the RoI images with a ship detection algorithm that used a regional CNN. This technique improved the recall and precision of small ship detection and the overall performance of the algorithm. Guo et al. [28] added rotation angle information to feature extraction, which increased the detection rate of ship objects at different scales, and greatly reduced the quantity of redundant information in the detection frame.

2.2. *Video Ranging Technology*

Video ranging technology is important in determining the true locations of imaged objects. The two types of video ranging are monocular ranging and binocular ranging. Monocular ranging has the benefits of a simple structure, rapid operation, and low cost; it is the main field of research at present, and a commonly used method is the Kalman filter (KF). Einhorn et al. [29] devised a feature-based extended Kalman filter (EKF) monocular visual ranging measurement algorithm that captured images with a single camera and used a depth estimation method to calculate a reliable initial estimate; the 3D positions were later reconstructed via an EKF. Chen et al. [30] introduced a monocular vision ranging measurement method based on pixel area and aspect ratio that predicted and optimized the pixel position in the subsequent frame using KF processing.

Another widely used technique was to calculate the distance to the object using object detection and camera projection. Raza et al. [31] used marker points to establish a line in the image, and used a linear equation to calculate the real-world distance between pixels based on the length of the line. Huang et al. [32] developed a monocular vision distance measurement method using object detection and segmentation; they developed a two-dimensional geometric vector model and used camera projection to calculate the distance. Zhe et al. [33] developed a monocular vision distance measurement method based on 3D detection, and created a regional distance geometric model to calculate the distance based on 3D detection and camera projection that produced good results when image detail was obscured.

The geometric principle of camera projection is shown in Figure 2, where C indicates the fixed position of the camera; A and B are points on the target; A' and B' are the respective projections of points A and B on the plane of the camera sensor, which are recorded in the image; the projection line AA' and camera optical axis plane belong to the surface method; h is the height of the center of the camera sensor above the surface; d and d' are the surface distances from the target to the vertical axis of the center of the sensor; θ is the angle between the lines OA and OB on the horizontal surface; f is the focal length of the camera. When the conditions $x = \frac{X}{2}, y < \frac{Y}{2}$, and $y = y'$ are satisfied, the distances between target points and the camera are calculated with the following:

$$d = \frac{f \cdot h}{Y/2 - y} \tag{1}$$

$$d' = \frac{d}{\cos\theta} \tag{2}$$

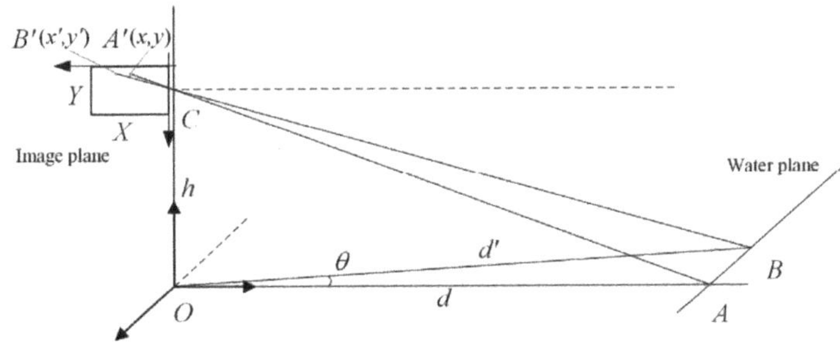

Figure 2. Geometric principle of camera projection.

3. Methods

There are four distinct stages of ship contour extraction: object detection, coordinate mapping, semantic segmentation, and image correction. Figure 3 is the technical roadmap of this research, and shows how data is transmitted between the various stages. In this section, we describe the key steps in detail.

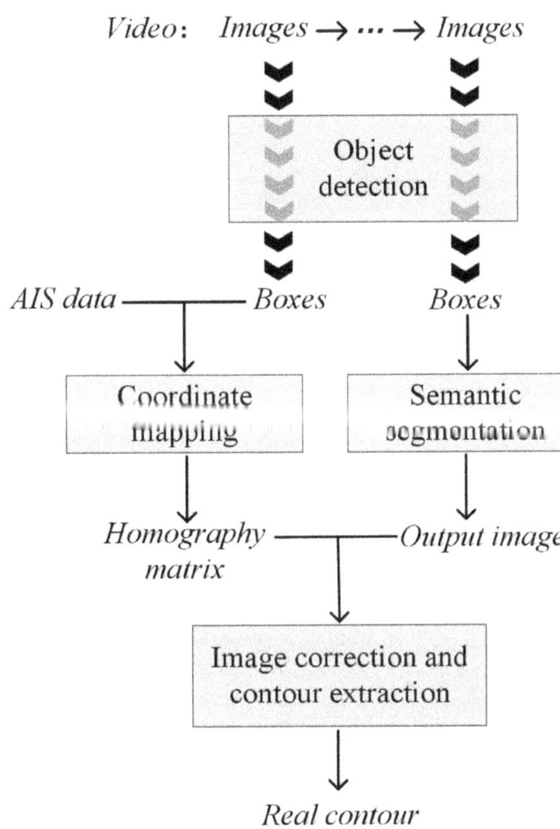

Figure 3. Technology roadmap of ship contour extraction and correction: video and AIS are input data, and the optimal homography matrix is first calculated, which is used to map the segmented image; after image correction and contour extraction, the final actual contour is output.

3.1. Object Detection Model

In general, object detection models often have a deep network structure, and require a large number of convolution layers with many parameters. Real-time inferences can only be performed if the devices they run on have adequate computing power. NanoDet is an excellent lightweight object detection model introduced in 2020 that dramatically reduces the number of parameters using a series of optimization methods. NanoDet can be quickly trained and ported to most embedded modules.

NanoDet uses several lightweight methods in the backbone, neck, and head, which enable it to balance accuracy, speed, and processing volume. The backbone is ShuffleNet V2, which removes the last layer of the convolution from the network and extracts 8, 16, and 32 downsampled features as the next inputs. ShuffleNet V2 is a CNN architecture that uses pointwise group convolutions to simplify the calculation of 1×1 convolutions, and uses channel shuffle to resist negative influences. The network greatly reduces computation, but maintains accuracy. The PAN module is a feature pyramid structure that performs upsampling and downsampling successively, which can fully integrate high-level features with low-level features. The neck is an optimized PAN that deletes all convolutions in the PAN, and only uses 1×1 convolutions extracted by the backbone for channel dimension alignment. An interpolation algorithm is used for upsampling and downsampling, and the multiscale feature map is added for feature fusion, which enables the network to learn the characteristics of multiscale targets. The FCOS is a typical anchor-free object detection algorithm with head detection through the neck of the output feature map pixel classification and bounding box regression to obtain the detection box. The optimized FCOS model was used as the detection head with abandoned weight sharing; it uses different convolutions to extract features at each layer and uses batch normalization, which uses deep separable convolution instead of group normalization. The number of convolution kernels and convolution channels also decreases, and the generalized focal loss function is used to resolve problems of convergence in training. In all, these methods greatly reduce redundant convolution and the number of parameters in the model, thus decreasing computation time.

SENet [34] is a spatial attention mechanism that increases the depth of a CNN and improves feature extraction. It consists of squeeze, excitation, and reweight functions. In the squeeze stage, the feature space with dimensions $c \times h \times w$ is compressed to $c \times 1 \times 1$ by global pooling, and the feature maps of a single channel are compressed into a weight factor. Two fully connected devices are used in the excitation stage. The first compresses the global information obtained from the global pooling; the feature dimension $c \times 1 \times 1$ is reduced to $c/r \times 1 \times 1$. The second fully connected device is used to map the feature back to $c \times 1 \times 1$ after ReLU activation. A sigmoid function is used to determine the normalized weight in a range of 0–1 of each channel to multiply the original feature map in the reweight stage.

The network structure is shown in Figure 4. SENet was added between ShuffleNetV2 and the optimized PAN with 40×40, 20×20, and 10×10 feature maps are input to emphasize useful features and suppress irrelevant features. Figure 4 shows the two feature maps before and after the SENet mechanism for comparison, and displays the SENet-enhanced features. After the calculations for PAN feature fusion and the FCOS detection head, the final output represents the locations of candidate boxes and their scores for different categories.

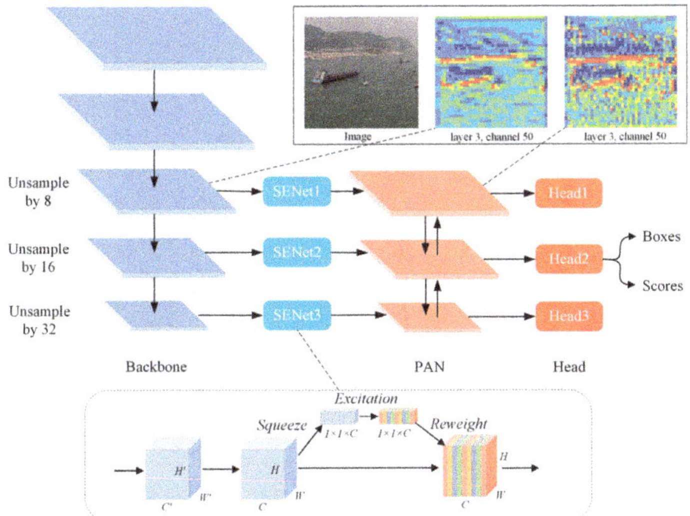

Figure 4. NanoDet–SENet network structure.

3.2. Coordinate Mapping

Mapping sensor image coordinates onto the water surface is key to matching image information with real-world information. When the camera attitude is constantly shifting, it is often necessary to combine optical ranging methods to ensure the camera view is parallel to the water surface. When parameters for camera height above the water surface, focal length, and pitch angle are combined with the projection equations, the depth map of the image can be calculated to predict the distance to each pixel on the water surface in the image. However, deploying surveillance cameras in inland river navigation areas is complex, and parameters such as height, focal length, and pitch angle are difficult to obtain in a timely manner. The use of data from fixed monitoring locations often requires using a homography transformation matrix to convert between the sensor image coordinate system and the water surface coordinate system, depending on the control points, and then mapping the pixel coordinates of the ship contour image to the water surface coordinates. The equation for the homography matrix transformation is as follows:

$$[x' \; y' \; w'] = [u \; v \; 1]\begin{bmatrix} a_{11} \, a_{12} \, a_{13} \\ a_{21} \, a_{22} \, a_{23} \\ a_{31} \, a_{32} \, 1 \end{bmatrix} = [u \; v \; 1]\boldsymbol{H} \tag{3}$$

where u and v are the pixel coordinates of the control points; the transformed coordinates are represented as (u', v'), where $u' = \frac{x'}{w'}$ and $v' = \frac{y'}{w'}$. $\boldsymbol{H}$ is the homography transformation matrix. At least four control points are required for the eight independent parameters in the solution of $\boldsymbol{H}$.

We used the real-time AIS position data and the corresponding observed ship positions in the image to create several control point coordinates. This necessitates that the camera be raised above the water's surface so that the hull takes up as much space as possible in the overhead image; on at least one channel, the camera can see the ship's side and back, as shown in Figure 5. The specific calculation steps of the optimal homographic transformation matrix are as follows.

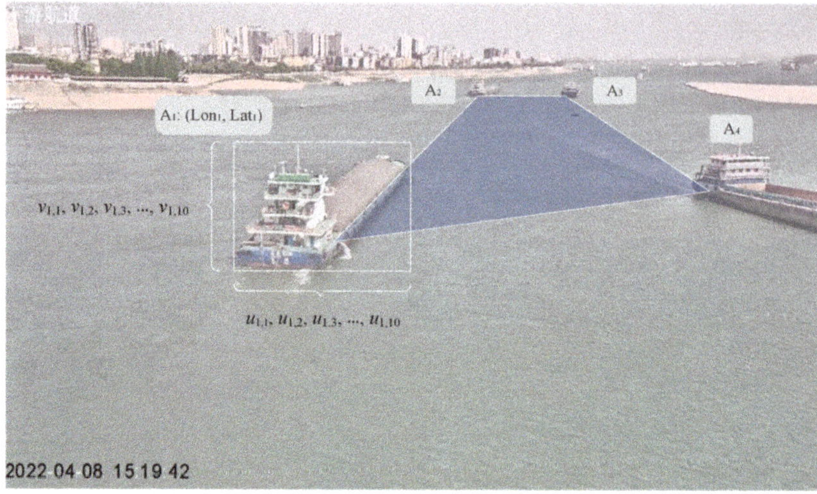

Figure 5. Multi-group control points constructed based on AIS information.

1. Obtain the video and AIS data for two ships on different courses to form four groups of matching coordinates and AIS positions. One is $A_i = \{(Lon_i,\ Lat_i),\ Box_i\}$, where Lon_i and Lat_i are the latitude and longitude coordinates of the control point i, and $Box_i = \{(u_{i1}, v_{i1}), (u_{i2}, v_{i2}) \ldots, (u_{in}, v_{in})\}$ is the set of pixel coordinates of control points in group i, which consists of coordinates of n equally spaced pixels. The control points of two ships traveling in the forward and backward directions are combined in pairs to produce several homography transformation matrices;
2. Detect the key point $\boldsymbol{p}_i$ of Box$_i$, where $\boldsymbol{p}_i$ is the intersection of the extended side in the lateral and inferior directions from the saliency detection image produced by the LC model that obtains the saliency value of a pixel by calculating the sum of the distance in color between the pixel and all other pixels in the image. The two waterlines are determined with linear fitting;
3. Select the two sets of coordinates for one ship, Box$_1$ and Box$_2$. Two key point pixels $\boldsymbol{p}_1$ and $\boldsymbol{p}_2$ are calculated, and Equation (3) is used to calculate the corresponding n^2 groups of water surface coordinates for $\boldsymbol{p}_1$ and $\boldsymbol{p}_2$. The error calculation of the homography matrix $\boldsymbol{H}_j$ is as follows:

$$\vec{\boldsymbol{P}_j} = \boldsymbol{p}'_{2j} - \boldsymbol{p}'_{1j}\ , j = 1,2,3,\ldots,n^2 \tag{4}$$

$$lon'_2 = lon_2 \times \cos\theta - lat_2 \times \sin\theta \tag{5}$$

$$lat'_2 = lat_2 \times \cos\theta + lon_2 \times \sin\theta \tag{6}$$

$$\vec{K} = (lon'_2, lat'_2) - (lon_1, lat_1) \tag{7}$$

$$\beta_j = \cos^{-1}\left(\frac{\vec{P_j}\cdot\vec{K}}{\left|\vec{P_j}\right| \times \left|\vec{K}\right|}\right), j = 1,2,3,\ldots,n^2 \tag{8}$$

where $\boldsymbol{p}'_{1j}$ and $\boldsymbol{p}'_{2j}$ are the mapped coordinates of $\boldsymbol{p}_1$ and $\boldsymbol{p}_2$, respectively, and θ, the steering angle of the ship, is determined from the AIS information. When the smallest

β_α has been obtained, the matrix $\boldsymbol{H}_\alpha$ calculated by β_α is considered to be the optimal homography matrix.

3.3. Semantic Segmentation Model

Ship contour extraction requires segmentation of the area covered by the ship surface when viewed from above, but the area is often obscured by the superstructure of the vessel. Commonly used image segmentation algorithms are often greatly affected by noise and lack robustness, and it is difficult to determine the target area when it is obscured. However, the deep learning DeepLabV3+ semantic segmentation model [35] is highly accurate, robust, and not very susceptible to noise. Therefore, it is suitable for use in the segmentation of specific targets in a complex environment.

The network structure of DeepLabV3+ is shown in Figure 6. It consists of an encoder and a decoder. The main body of the encoder is a deep CNN with dilated convolution that controls the size of the receptive field by a rate (r) without changing the size of the feature graph. A greater value of r produces a larger receptive field. The dilated convolution in the encoder is combined with a spatial pyramidal pooling module to produce multiscale information. The main constituents of the encoder are the following:

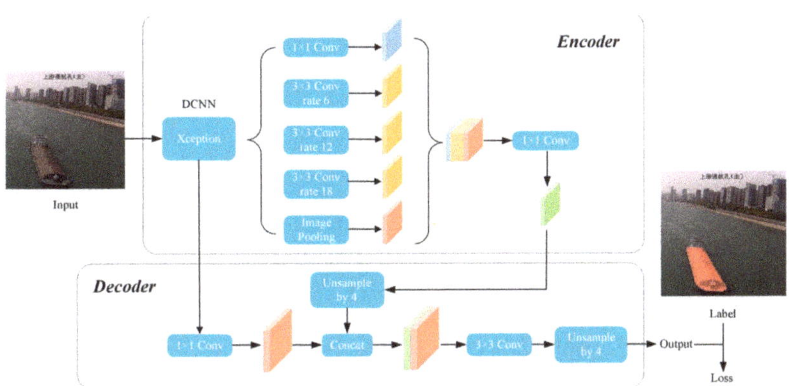

Figure 6. DeepLabV3+ network structure (The Non-English characters in the illustrations are meaningless).

a. One 1×1 convolution layer and three 3×3 empty convolution layers. The rate r is (6, 12, 18) when the output step size is 16, and is doubled when the output step size is 8.

b. A global average pooling layer is used to produce image-level features that are then input into the 1×1 convolution layer and bilinearly interpolated to the original size.

c. Five features of different scales are combined in the dimension channel, and then input into the 1×1 convolution layer to be combined to produce 256 channels of new features.

The decoder can also combine low-level features with high-level features to increase segmentation accuracy. The main steps of feature fusion are as follows. The multiscale feature information is bilinearly interpolated and upsampled. The encoder then combines it with the original features that were extracted by the CNN. The combined feature information is then convoluted for simple feature combination. Finally, the combined features are bilinearly interpolated and upsampled to produce the segmentation results.

Xception [36] was used as the backbone network for feature extraction (Figure 6). Xception, which is an improved version of Inception V3, introduces depthwise convolution derived from Inception V3 to reduce model complexity and improve segmentation. The label in Figure 6 is from the annotation of the input images. The spatial pyramid module combined with dilated convolution combines multiscale image information. A larger value of r will extract features from different regions of the image into a larger receptive field, thus reducing obscuration by the superstructure. The marked area is the area where the ship is vertically mapped onto the water surface.

4. Results

4.1. Ship Object Detection

The experimental platform was a desktop computer with Windows 10, a GTX1050Ti GPU, and the PyTorch 1.8.0 framework. We created a coco dataset for training with 2535 images collected from surveillance videos from waterways upstream and downstream of the Yichang Yangtze River Bridge and the Jingzhou Yangtze River Bridge in China. The images included ships from different angles of different sizes, and in various lighting conditions. The image count was increased to 5070 using data enhancement methods such as noise processing, random angle rotation, random brightness adjustment, and simulated rain and fog weather conditions. Labelme software was used to annotate the images, which contained 12,376 ship objects altogether. The dataset was divided for training, validation, and testing in a ratio of 8:2:1. The input size for detection was 320 × 320. Stochastic gradient descent (SGD) was used for optimization. The initial learning rate was set to 0.01 because it is a common value suitable for most deep learning models.

The object detection models were trained and tested before and after validation, and their precision and recall were calculated. Precision represents the proportion of correctly predicted targets in total predictions, and recall represents the proportion of all target predictions that were correct. In general, precision decreases as recall increases.

To assess detection improvement attributed to SENet, P–R curves were plotted using precision and recall values for the two models before and after validation. In addition, the mainstream object detection algorithms Faster RCNN and YOLOv4, and the lightweight algorithm YOLOv4-Tiny using the test set, were selected for comparison to further assess the accuracy and efficiency of the NanoDet–SENet model detection. Table 2 shows the number and size of parameters for these three models. It can be seen from the table that NanoDet–SENet is an excellent lightweight model because its network complexity is much less than that of other models.

Table 2. Params number and size of different models.

	Total Params	**Params Size**
Faster RCNN	137.08 M	522.91 MB
YOLOv4	64.36 M	245.53 MB
YOLOv4-Tiny	6.06 M	23.10 MB
NanoDet–SENet	**0.95** M	**3.62** MB

The P–R curves for the experiment are shown in Figure 7. Detection by several algorithms was assessed using four indicators: mean average precision (mAP), frames/s (FPS), precision, and recall, which are shown in Table 3. It can be seen from the table that the NanoDet–SENet model outperformed NanoDet, YOLOv4-Tiny, and Faster RCNN in terms of precision and the mAP, but did not perform as well as YOLOv4. The recall was greater than for NanoDet and YOLOv4-Tiny, but less than that for Faster RCNN and YOLOv4. The FPS was significantly greater than for all the other models except NanoDet. These results indicate that the SENet attention mechanism significantly influenced the detection effectiveness of the model, and that NanoDet–SENet detected objects almost as well as YOLOv4, although it is a simpler model.

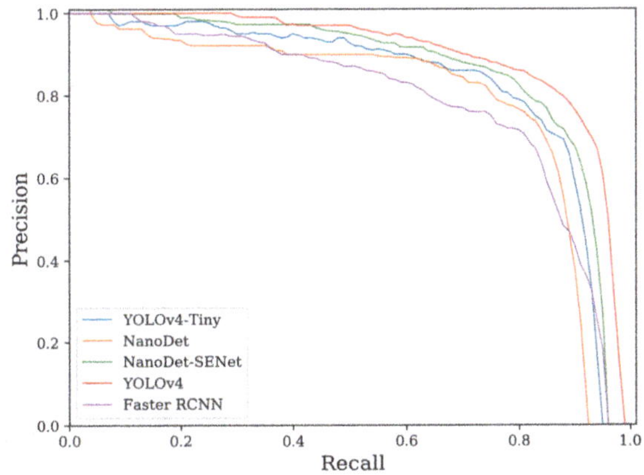

Figure 7. P–R curves for tested models.

Table 3. Comparison test results of target detection.

	Precision (%)	Recall (%)	mAP (%)	FPS
Faster RCNN	69.68	**81.51**	78.84	6.35
YOLOv4	86.23	79.64	**90.20**	9.10
YOLOv4-Tiny	84.20	74.55	83.68	30.71
NanoDet	78.96	74.72	76.55	38.62
NanoDet–SENet	**85.09**	76.83	86.73	**37.47**

4.2. Establishment of the Homography Transformation Model

We combined AIS information with video data to create the optimal homography transformation matrix between the sensor pixel coordinate system and the water surface coordinate system. The critical aspect of this algorithm is the synchronization of the time of the image sequence with the time of AIS information acquisition to ensure the accuracy of key point detection.

A Hikvision zoom network camera remotely captured video stream was transmitted using a real-time streaming protocol (RTSP), and a message queue telemetry transmission (MQTT) server was used to create [illegible] platform. The video transmission rate was 3.2 Mbps with a 3.5 s delay, and the AIS signal delay was 6 s. As described in Section 3.2, the delays were eliminated, and the video and AIS information was used to obtain the coordinates of 250 sets of control points.

The key points were then tested. The accuracy of key point detection decreases for small-target ships, so we needed to ensure the ship detection boxes we selected as the control points were of adequate size and had well-defined contours, as shown in Figure 8. Figure 8a,b show ship objects detected at two distinct time points. The ship moving away from the sensor on the left side of the images was used to validate the optimal homography matrix. The red pixels in Figure 8c,d show the LC significant image pixels of change points, and the green pixels show key points after linear fitting. Figure 8d shows that the model predicted the exact position of the key point when the ship was further away from the sensor.

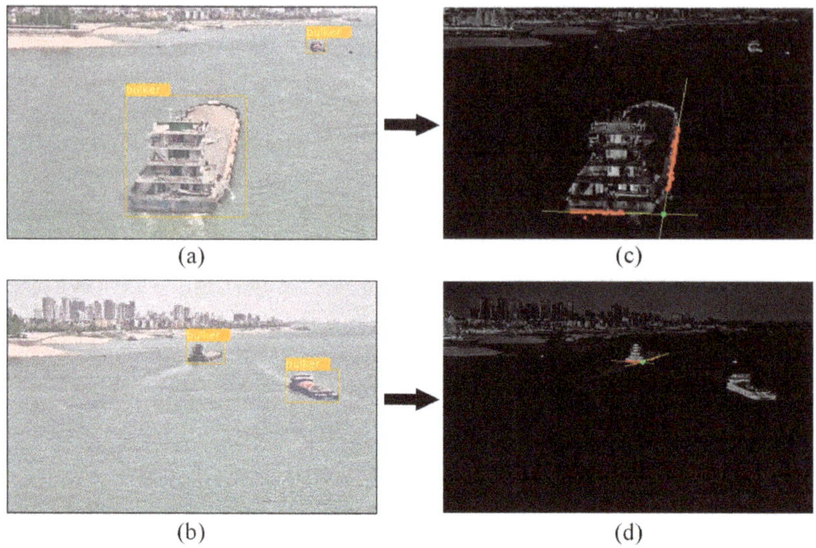

Figure 8. Key point detection progress: (**a**) ship object detection at first time, (**b**) ship object detection at second time, (**c**) key point detection at first time, and (**d**) key point detection at second time.

After we derived the optimal homography matrix, as described in Section 3.2, we used sensor data for the upstream and downstream directions to test the homography transformation model. Consecutive AIS coordinates were obtained, and the pixel coordinates of key points in the corresponding images were obtained and mapped onto the water coordinate system to predict the trajectory. Figure 9 shows that the trajectory of key points after coordinate mapping was very close to the AIS trajectory. The mean values of the distances between the corresponding points in the two sets of trajectories, 11.5 m and 77.8 m, were consistent with relative positions and distances between the real-world GPS antenna and the ship waterline inflection point. This result shows the accuracy of coordinate mapping using the homography transformation model.

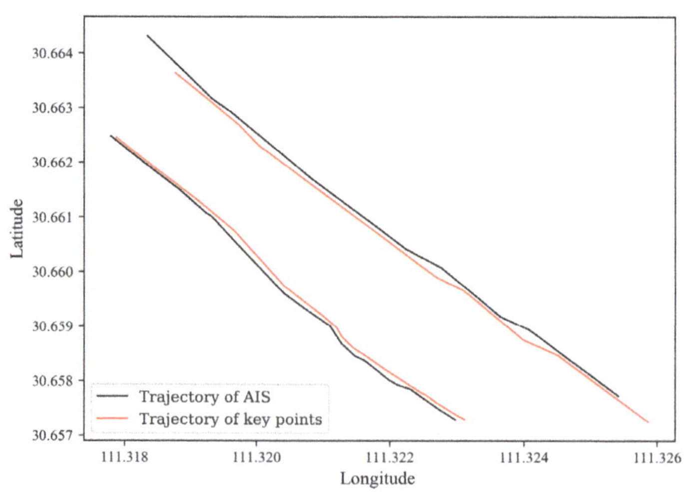

Figure 9. Test results of homography transformation model.

4.3. Ship Contour Extraction

Only images of bulk carriers and oil tankers were used, because the irregular shapes of passenger and container ships add unnecessary complexity to the development of the prototype model. We used surveillance cameras to obtain overhead views of vessels on the waterway to ensure that the deck surface matched the ship hull as much as possible. The object detection method was used to take automatic photos that were stored to create a semantic segmentation data set. The model training platform was a desktop computer with Windows 10 and an RTX3060Ti GPU using the Keras 2.2.5 framework. Target ship contours were labeled using Labelme software, and pixels were classified as either foreground or background. The ratio of the training set, validation set, and test set was 8:2:1. Online data enhancement was used to randomly amplify the image and label data in each batch in the training stage. The input data size for the model was 512 × 512.

The mean intersection over union (MIoU), which is a standard metric of the accuracy of a semantic segmentation algorithm, was used to assess the performance of the algorithm. The equation is as follows:

$$\text{MIoU} = \frac{1}{k+1}\sum_{i=0}^{k}\frac{p_{ii}}{\sum_{j=0}^{k} p_{ij} + \sum_{j=0}^{k} p_{ji} - p_{ii}} \tag{9}$$

where k is the number of pixel categories, p_{ij} is the number of pixels that originally belonged to category i but are predicted to be in category j, and MIoU is the average number of times the predicted value coincides with the actual value in each category. A greater value of MIoU indicates more accurate network prediction.

The FCN is commonly used for ship image semantic segmentation [37]. An FCN classifies images at a pixel level by selecting a sliding window for each pixel. Unet (unity networking), an improvement on the FCN [38], has been widely used in the field of transportation. Therefore, we compared these two models with DeepLabV3+. The IoU results for background and foreground after training for 50 epochs are shown in Table 4. It can be seen that DeepLabV3+ has clear advantages over FCN and Unet.

Table 4. Comparison test results of semantic segmentation.

	Background_IoU (%)	Boat_IoU (%)	Mean_IoU (%)
FCN	[illegible]	79.90	89.61
Unet	99.28	67.35	83.32
DeeplabV3+	**99.54**	**82.61**	**91.07**

The output images of DeepLabV3+ needed to be trimmed. Trimming was performed by calculating the number of pixels in the stern area along the transverse axis of the ship, and setting a threshold to eliminate scattered pixels to avoid individual misclassified pixels having undue impact on the corrected image. The segmentation results are shown in the first three columns of Figure 10. It can be seen from Figure 10 that the DeepLabV3+ model segmented the target area well without obstruction from the superstructure. The fourth column in Figure 10 shows the trimmed images that represent the actual ship contour region after coordinate mapping. The pixel size of the image corresponds to a real-world distance of 115 m. It is clear from these results that the size and heading of ships at different distances in the image were approximately estimated.

A comprehensive review of all the experimental results shows that the DeepLabV3+ semantic segmentation model can be successfully used on a high-performance server that receives automatic photos and uploading based on lightweight object detection and accomplishes contour extraction tasks on the cloud server, thus avoiding bandwidth usage for transmitting large amounts of remote video data. The object detection model will determine the rectangular region surrounding a vessel, increase the proportion of ship features in DeepLabV3+ input, and improve the pertinence of the semantic segmentation model.

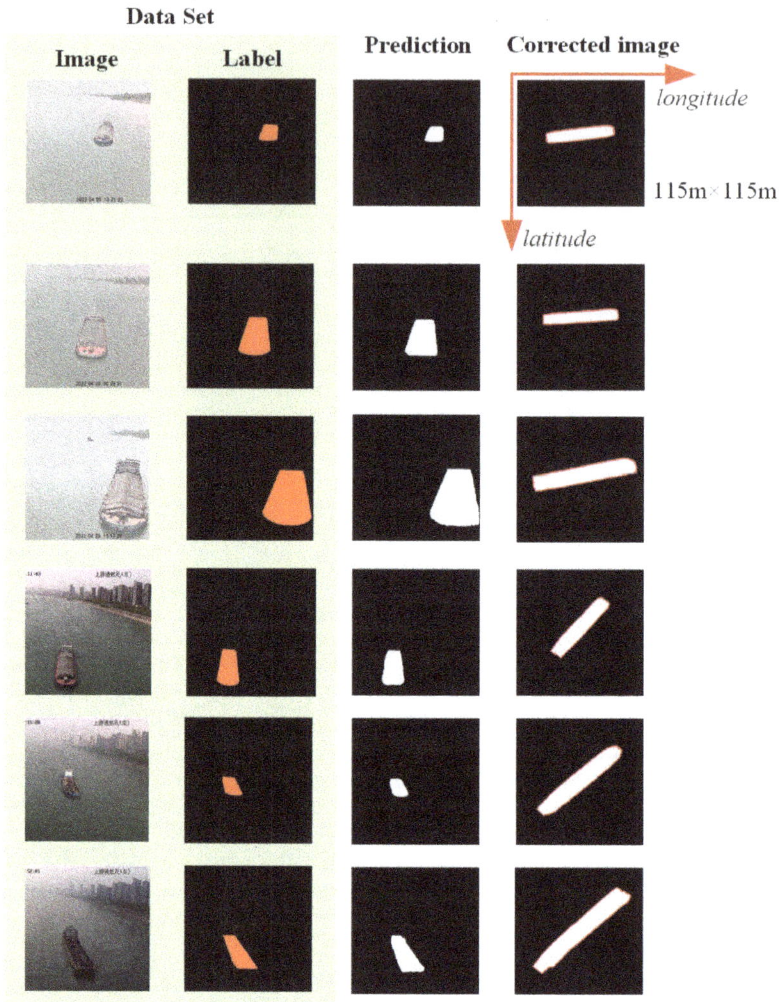

Figure 10. Ship contour extraction and size restoration results (The Non-English characters in the illustrations are meaningless).

We captured the video frame at the same time of receiving the AIS signal, and extracted the ship's aera with the semantic segmentation model. Based on the restored image, edge detection was combined with the Hough transform to further detect the minimum enclosing rectangle of the ship. We tested for several ships by estimating their dimensions, which we then compared with the actual dimensions; the results are shown in Table 5. We found that the average length relative error was 5.84%, and the average width relative error was 7.53%; these values are in a reasonable range. We also noted that the maximum length error and width error were 14.2% and 16.5%, respectively; this was because the color and texture features of the ship are not obvious enough to cause semantic segmentation errors, and were further amplified by image correction.

Table 5. Ship dimension data calculation results and errors.

MMSI	Length (m)			Width (m)		
	Actual	Calculated	Error (%)	Actual	Calculated	Error (%)
413773165	87	90.3	3.8	14	14.6	4.3
413774959	87	84.2	3.2	15	16.3	8.7
413779378	90	96.7	7.4	15	17.0	13.3
413781326	106	110.6	4.3	17	18.2	7.1
413783151	107	115.2	7.7	16	17.1	6.9
413803847	90	96.8	7.6	15	15.4	2.7
413811188	100	97.5	2.5	16	16.9	5.6
413819165	100	114.2	14.2	17	19.8	16.5
413831856	110	113.6	3.3	19	20.0	5.3
413801536	107	103.1	3.6	16	16.6	3.8
Mean	-	-	5.8	-	-	7.4

According to the experimental results of Park et al. [4] using satellite-observed ships, the RMS errors for the length and width were 12.1 m and 6.8 m, respectively. We further calculated that the RMS error according to Table 5 and obtained the corresponding results as 6.6 m and 1.4 m, respectively, which are obviously better. This final result is valuable for creating and improving AIS data, and also provides a foundation for calculating ship collision risk.

5. Conclusions

Ship dimension information is important at the micro level in ship collision risk calculations. To avoid having to work with missing or incorrect ship dimension data, deep learning algorithms were used to extract ship contours from inland waterway surveillance video and real-time AIS information. According to the experimental results, the proposed object detection model and semantic segmentation model were very accurate in our experimental trials, and thus successfully resolved the missing AIS data issue.

The easily used lightweight ship object detection model that we developed for edge computing processed at a high frame rate without GPU acceleration, and facilitated the automatic acquisition and uploading of ship images. Therefore, it provides crucial support for shipping control on inland waterways.

In this study, AIS real-time position information [illegible] point to create a coordinate mapping model that rapidly and accurately mapped sensor image coordinates onto water surface coordinates without the need to rely on various projection parameters. Therefore, this method merits future development to further promote the combination of video, AIS, and radar information.

The combination of the object detection model and the coordinate mapping model, together with the use of the semantic segmentation algorithm, allowed us to extract the ship contour from the image and predict its actual size. The experimental trial results suggest the method is effective, and demonstrates the capability to extract ship dimension information automatically. This capability will benefit inland waterway regulatory authorities by enabling them to improve management of ship navigation.

The method still has some limitations. On the one hand, it has requirements regarding camera heights and shooting angles for coordinate mapping, as described in Section 3.2; therefore, a camera is best installed in the midspan of an inland river bridge with a bridge floor elevation of more than 50 m. On the other hand, the contour extraction method is not suitable for passenger ships and container ships, because their true contours are poorly reflected in the image.

Errors in ship contour extraction mainly come from the coordinate mapping model and semantic segmentation model. In our future research, we will add more ship images to the semantic segmentation dataset to improve the generalizability of the dataset so that the model can more exactly segment images of different types of ships. Vessel GPS positioning generally exists within 1 m of error, which can affect the effectiveness of the coordinate mapping method. Thus, we will explore the error correction method based on constructing the trajectory equation to predict the GPS positioning coordinates according to the characteristics of GPS positioning coordinates obeying the Gaussian distribution at a certain moment. We will also endeavor to constantly use ships to simulate the control points during the experiment, in order to continuously update the optimal homography matrix.

Author Contributions: Methodology, validation, writing, original draft—Z.H. and Q.M.; data curation—L.L. and C.Y.; review and editing, funding acquisition—Q.H. All authors have read and agreed to the published version of the manuscript.

Funding: This research is supported by the National Natural Science Foundation of China (Grant No. 52372316).

Institutional Review Board Statement: Not applicable.

Informed Consent Statement: Informed consent was obtained from all subjects involved in the study.

Data Availability Statement: Not applicable.

Conflicts of Interest: The authors declare no conflict of interest.

References

1. Bakdi, A.; Glad, I.K.; Vanem, E.; Engelhardtsen, Y. Ais-based multiple vessel collision and grounding risk identification based on adaptive safety domain. *Materials* **2019**, *8*, 5. [CrossRef]
2. Liu, W.R.; Guo, Y.; Nie, J.; Hu, Q.; Xiong, Z.; Yu, H.; Guizani, M. Intelligent Edge-Enabled Efficient Multi-Source Data Fusion for Autonomous Surface Vehicles in Maritime Internet of Things. *IEEE Trans. Green Commun. Netw.* **2022**, *6*, 1574–1587. [CrossRef]
3. Huang, Z.S.; Hu, Q.Y.; Mei, Q.; Yang, C.; Wu, Z. Identity recognition on waterways: A novel ship information tracking method based on multimodal data. *J. Navig.* **2021**, *74*, 1336–1352. [CrossRef]
4. Park, J.J.; Park, K.A.; Foucher, P.Y.; Lee, M.; Oh, S. Estimation of ship size from satellite optical image using elliptic characteristics of ship periphery. *Int. J. Remote Sens.* **2020**, *41*, 5905–5927. [CrossRef]
5. Yan, Z.; Yan, X.; Xie, L. Inland Ship Edge Detection Algorithm based on Improved Canny Operator. *J. Converg. Inf. Technol.* **2012**, *7*, 567–575.
6. Gu, D.; Xu, X. Multi-feature extraction of ships from SAR images. In Proceedings of the 2013 6th International Congress on Image and Signal Processing (CISP), Hangzhou, China, 16–18 December 2013; pp. 454–458.
7. Zhu, J.; Qiu, X.; Pan, Z.; Zhang, Y.; Lei, B. Projection shape template-based ship target recognition in TerraSAR-X images. *IEEE Geosci. Remote Sens. Lett.* **2017**, *14*, 222–226. [CrossRef]
8. Nie, Y.Y.; Fan, S.C.; Shui, P.L. Fast ship contour extraction in SAR images. *J. Eng.* **2019**, *19*, 5885–5888. [CrossRef]
9. Simonyan, K.; Zisserman, A. Very Deep Convolutional Networks for Large-Scale Image Recognition. *arXiv* **2014**, arXiv:1409.1556.
10. Szegedy, C.; Liu, W.; Jia, Y.; Sermanet, P.; Rabinovich, A. Going deeper with convolutions. In Proceedings of the 2015 IEEE Conference on Computer Vision and Pattern Recognition (CVPR), Boston, MA, USA, 7–12 June 2015; p. 15523970.
11. Ioffe, S.; Szegedy, C. Batch normalization: Accelerating deep network training by reducing internal covariate shift. In Proceedings of the 32nd International conference on machine learning, Lile, France, 6–11 July 2015; pp. 448–456.
12. He, K.; Zhang, X.; Ren, S.; Sun, J. Deep residual learning for image recognition. In Proceedings of the IEEE Conference on Computer Vision and Pattern Recognition (CVPR), Las Vegas, NV, USA, 27–30 June 2016; p. 16541111.
13. Lin, H.; Shi, Z.; Zou, Z. Fully Convolutional Network with Task Partitioning for Inshore Ship Detection in Optical Remote Sensing Images. *IEEE Geosci. Remote Sens. Lett.* **2017**, *14*, 1665–1669. [CrossRef]
14. Bovcon, B.; Kristan, M. A water-obstacle separation and refinement network for unmanned surface vehicles. *arXiv* **2020**, arXiv:2001.01921.
15. Ust, L.; Kristan, M. Learning maritime obstacle detection from weak annotations by scaffolding. *arXiv* **2022**, arXiv:2108.00564.
16. Kelm, A.P.; Zolzer, U. Walk the Lines: Object Contour Tracing CNN for Contour Completion of Ships. In Proceedings of the 25th International Conference on Pattern Recognition (ICPR), Milan, Italy, 13–18 September 2020; pp. 3993–4000.
17. Howard, D.; Roberts, S.C.; Brankin, R. Evolution of Ship Detectors for Satellite SAR Imagery. In *Genetic Programming*; Springer: Berlin/Heidelberg, Germany, 1999; Volume 1598, pp. 135–148.
18. Kim, H.T.; Park, J.S.; Yu, Y.S. Ship detection using background estimation of video and AIS informations. *J. Korea Inst. Inf. Commun. Eng.* **2010**, *14*, 2636–2641.
19. Fefilatyev, S.; Goldgof, D.; Shreve, M.; Lembke, C. Detection and tracking of ships in open sea with rapidly moving buoy-mounted camera system. *Ocean Eng.* **2012**, *54*, 1–12. [CrossRef]
20. Yang, T.J.; Zhang, S.; Zhou, G.Q.; Jiang, C.X. Design of a real-time system of moving ship tracking on-board based on FPGA in remote sensing images. In Proceedings of the International Conference on Intelligent Earth Observing and Applications 2015, Guilin, China, 9 December 2015; p. 980804.

21. Ren, S.; He, K.; Girshick, R.; Sun, J. Faster R-CNN: Towards Real-Time Object Detection with Region Proposal Networks. *IEEE Trans. Pattern Anal. Mach. Intell.* **2017**, *39*, 1137–1149. [CrossRef] [PubMed]
22. He, K.; Gkioxari, G.; Dollár, P.; Girshick, R. Mask r-cnn. In Proceedings of the IEEE International Conference on Computer Vision (ICCV), Venice, Italy, 22–29 October 2017; pp. 2980–2988.
23. Redmon, J.; Divvala, S.; Girshick, R.; Farhadi, A. You only look once: Unified, real-time object detection. In Proceedings of the 2016 IEEE Conference on Computer Vision and Pattern Recognition (CVPR), Las Vegas, NV, USA, 27–30 June 2016; pp. 779–788.
24. Liu, W.; Anguelov, D.; Erhan, D.; Szegedy, C.; Reed, S.; Fu, C.Y.; Berg, A.C. SSD: Single Shot MultiBox Detector. In Proceedings of the European Conference on Computer Vision (ECCV), Amsterdam, the Netherlands, 10–16 October 2016; pp. 21–37.
25. Ma, J.; Shao, W.; Ye, H.; Wang, L.; Wang, H.; Zheng, Y.; Xue, X. Arbitrary-oriented scene text detection via rotation proposals. *IEEE Trans. Multimed.* **2018**, *20*, 3111–3122. [CrossRef]
26. He, L.; Yi, S.; Mu, X.; Zhang, L. Ship Detection Method Based on Gabor Filter and Fast RCNN Model in Satellite Images of Sea. In Proceedings of the 3rd International Conference on Computer Science and Application Engineering (CSAE), Sanya, China, 22–24 October 2019; pp. 1–7.
27. Zhang, S.; Wu, R.; Xu, K.; Wang, J.; Sun, W. R-cnn-based ship detection from high resolution remote sensing imagery. *Remote Sens.* **2019**, *11*, 631. [CrossRef]
28. Guo, H.; Yang, X.; Wang, N.; Song, B.; Gao, X. A rotational libra r-cnn method for ship detection. *IEEE Trans. Geosci. Remote Sens.* **2020**, *58*, 5772–5781. [CrossRef]
29. Einhorn, E.; Schroter, C. Attention-driven monocular scene reconstruction for obstacle detection, robot navigation and map building. *Robot. Auton. Syst.* **2011**, *59*, 296–309. [CrossRef]
30. Chen, S.; Mei, S.; Jia, G. KFPA Monocular Ranging Algorithm Design and Application in Mobile edge Computing. *J. Internet Technol.* **2021**, *22*, 1131–1142.
31. Raza, M.; Chen, Z.; Rehman, S.U. Framework for estimating distance and dimension attributes of pedestrians in real-time environments using monocular camera. *Neurocomputing* **2018**, *275*, 533–545. [CrossRef]
32. Huang, L.; Zhe, T.; Wu, J.; Wu, Q.; Pei, C.; Chen, D. Robust inter-vehicle distance estimation method based on monocular vision. *IEEE Access.* **2019**, *7*, 46059–46070. [CrossRef]
33. Zhe, T.; Huang, L.; Wu, Q.; Zhang, J.; Pei, C.; Li, L. Inter-vehicle distance estimation method based on monocular vision using 3d detection. *IEEE Trans. Veh. Technol.* **2020**, *69*, 4907–4919. [CrossRef]
34. Jie, H.; Li, S.; Gang, S.; Albanie, S. Squeeze-and-excitation networks. *IEEE Trans. Pattern Anal. Mach. Intell.* **2017**, *8*, 2011–2023.
35. Chen, L.C.; Papandreou, G.; Kokkinos, L. Deeplab: Semantic image segmentation with deep convolutional nets, atrous convolution, and fully connected CRFs. *IEEE Trans. Pattern Anal. Mach. Intell.* **2018**, *40*, 834–848. [CrossRef] [PubMed]
36. Chollet, F. Xception: Deep learning with depthwise separable convolutions. In Proceedings of the IEEE Conference on Computer Vision and Pattern Recognition (CVPR), Honolulu, HI, USA, 21–26 July 2017; pp. 1800–1807.
37. Long, J.; Shelhamer, E.; Darrell, T. Fully Convolutional Networks for Semantic Segmentation. *IEEE Trans. Pattern Anal. Mach. Intell.* **2015**, *39*, 640–651.
38. Ronneberger, O.; Fischer, P.; Brox, T. U-net: Convolutional networks for biomedical image segmentation. In Proceedings of the International Conference on Medical Image Computing and Computer-Assisted Intervention, Munich, Germany, 5–9 October 2015, pp. 234–241.

Journal of
Marine Science and Engineering

MDPI

Article

Probabilistic Modeling of Maritime Accident Scenarios Leveraging Bayesian Network Techniques

Shiguan Liao [1], Jinxian Weng [2,*], Zhaomin Zhang [1,*], Zhuang Li [3] and Fang Li [2]

1 School of Management, Shenzhen Polytechnic, Shenzhen 518055, China; sgliao113@gmail.com
2 College of Transport and Communications, Shanghai Maritime University, Shanghai 201306, China
3 Naval Architecture and Shipping College, Guangdong Ocean University, Zhanjiang 524088, China; zhuang.li@gdou.edu.cn
* Correspondence: jxweng@shmtu.edu.cn (J.W.); zhzhmin@szpt.edu.cn (Z.Z.)

Abstract: This paper introduces a scenario evolution model for maritime accidents, wherein Bayesian networks (BNs) were employed to predict the most probable causes of distinct types of maritime incidents. The BN nodes encompass factors such as accident type, life loss contingency, accident severity, quarter and time period of the accident, and type and gross tonnage of the involved ships. An analysis of 5660 global maritime accidents spanning the years 2005 to 2020 was conducted. Using Netica software, a tree augmented network (TAN) model was constructed, thus accounting for interdependencies among risk-influencing factors. To confirm these results, a validation process involving sensitivity analysis and historical accident records was performed. Following this, both forward causal inference and reverse diagnostic inference were carried out on each node variable to scrutinize the accident development trend and evolution process under preset conditions. The findings suggest that the model was competent in effectively predicting the likelihood of various accident scenarios under specific conditions, as well as extrapolating accident consequences. Forward causal reasoning unveiled that general cargo ships with a gross tonnage of 1–18,500 t were most prone to experiencing collision and stranding/grounding accidents in the first quarter. Reverse diagnostic reasoning indicated that, in the early morning hours, container ships, general cargo ships, and chemical ships with a tonnage of 1–18,500 t were less likely to involve life loss in the event of collision accidents.

Keywords: maritime traffic safety; maritime accident; Bayesian network (BN); accident scenario [illegible]

Citation: Liao, S.; Weng, J.; Zhang, Z.; Li, Z.; Li, F. Probabilistic Modeling of Maritime Accident Scenarios Leveraging Bayesian Network Techniques. *J. Mar. Sci. Eng.* **2023**, *11*, 1513. https://doi.org/10.3390/jmse11081513

Academic Editor: Mihalis Golias

Received: 18 June 2023
Revised: 14 July 2023
Accepted: 28 July 2023
Published: 29 July 2023

1. Introduction

Intricately variable and multifaceted climatic conditions, endemic to an expansive marine environment, have perpetually underscored maritime transport as a vocation of considerable risk. Incidents disrupting maritime transit, which encompass a spectrum of occurrences from vessel collisions to groundings, and from onboard fires to devastating explosions, bear the hallmark of low frequency yet are marked by their profoundly destructive aftermath [1]. Upon the unfortunate manifestation of a maritime accident, a tsunami of undesirable outcomes typically ensues, including, notably, substantial financial loss, a tragic toll of human casualties, or even the insidious onset of extensive environmental pollution [2]. As an imperative and fundamental step toward mitigating the inherent perils of maritime navigation, a comprehensive exploration into the multifactorial etiology of these traffic mishaps proves indispensable [3–5]. Moreover, it becomes increasingly pivotal to meticulously undertake a dynamic risk evaluation, focusing on the myriad facets of maritime operations [6–8]. Complementing this, the development and implementation of robust predictive models, which can potentially forecast the calamitous conjunction of human fatality and its probability in the event of an accident, could contribute significantly toward minimizing future maritime disasters [9–12].

Over the years, an impressive corpus of scholarly efforts has been devoted to enhancing our understanding of maritime traffic safety, including explorations into accident causation analysis [3–5,13], accident consequence assessment [14–16], and accident loss computation [17,18]. These endeavors have given rise to an array of innovative evaluation methodologies. In one notable study, Hu et al. [3] skillfully harnessed the capabilities of the Human Failure Analysis and Classification System (HFACS) in tandem with structural equation modeling (SEM) to disentangle an intricate web of causal factors underpinning marine traffic accidents (MTAs). Chou et al. [4], in a synergistic integration of technologies, amalgamated the Automatic Identification System, Geographic Information System, and an electronic chart (e-chart) to scrutinize the interplay between environmental factors, geographical locations, and the common causes of marine mishaps. By overlaying vessel traffic flows, accident sites, and environmental data on a shared e-chart, their research unfurled valuable insights into port authorities when streamlining ship traffic flow and curtailing the prevalence of marine accidents in the vicinity of ports. Meanwhile, Xue et al. [5] proffered a comprehensive analytical framework for investigating the peculiarities and causative factors of ship accidents, utilizing a decade's worth of historical data that were harvested from the capriciously fluctuating backwater expanse of the Three Gorges Reservoir region. Their extensive work yielded a thorough summary and visualization of vessel accident categories and severity, involved vessel types, spatial–temporal distribution characteristics, and vessel accident loss, along with the underlying causes and lessons gleaned from pertinent accidents that were achieved through a rigorous statistical and comparative analysis of historical data. Elsewhere, Fu et al. [13] engineered a bivariate probit model to delve into an array of 311 Arctic ship accidents spanning from 1998 to 2017. Their study brought to the fore influential factors such as gross tonnage, ship type, ship age, accident type, accident year, accident location, wind, and sea ice as the primary contributors to accident severity. Simultaneously, their research unveiled an intriguing negative correlation between serious accidents and those resulting in pollution. As research on maritime traffic accidents has illuminated a gamut of potential causative factors, the increased granularity of available accident data has spurred a growing number of scholars to concentrate on the ramifications of these mishaps, specifically on the evaluation of accident consequences and loss computation. Such undertakings have risen to prominence, particularly in the eyes of managers concerned with incidents that yield significant economic damage and human casualties. For instance, Chen et al. [15] presented an evidence-based Fuzzy Bayesian network methodology to erect probabilistic models of marine accidents, thereby enabling the appraisal of accidents that were likely to spawn severe consequences. In a similar vein, Ventikos and Giannopoulos [16] introduced a criterion to assess the risks and repercussions within the maritime transport sector from a societal perspective, thereby formulating a novel framework for the marine risk assessment, which facilitated a comparison of disparate accident scales and characteristics, while accurately mirroring the risk threshold society was prepared to tolerate. Chen et al. [17] pioneered an enhanced entropy weight-TOPSIS model to furnish a holistic analysis and appraisal of the marine total loss incidents, encompassing a global scope from 1998 to 2018. These studies, though highly impactful, predominantly undertake analyses either from the standpoint of accident causation or the evaluation of accident consequences. Rarely do these scholarly pursuits straddle both domains in a bidirectional inquiry.

In the realm of accident scenario analysis, methodological constructs like event tree analysis and accident tree analysis are frequently utilized in the assembly of traffic accident scenario evolution models [19–23]. However, the breadth of most accident cause analyses often overshadows their specificity, impeding their ability to yield targeted recommendations to forestall analogous events [24]. To bridge this gap, scholars could employ a Bayesian network-based maritime accident scenario modeling approach. Bayesian networks stand as a form of a probabilistic graphical model, which is deftly equipped to encapsulate and deliberate over uncertain knowledge and nebulous relationships among variables. This versatile modeling approach, designed to embrace the labyrinthine and dynamic character of maritime activities, excels at discerning the contributory factors that precipitate maritime accidents [3,5]. Employing a synergistic blend of historical data and expert acumen, this model could approximate both the likelihood of an accident's occurrence and the potential fallout arising from a range of accident scenarios [14,16]. Bayesian networks (BN) find broad application in confronting uncertain multi-factor causality inference, accident causation analysis, and scenario prediction, making them invaluable tools in road and waterway transportation sectors [25–32]. Various scholars have employed these tools in diverse studies: Zou and Yue [33] melded the probabilistic risk analysis with the BN theory to explore the origins of road traffic accidents; Yuan et al. [34] constructed a scenario-derived prediction model for the repercussions of fire accidents in oil and gas storage and transportation emergency processes, leveraging a defuzzification method and a dynamic BN model. Other researchers, such as Zhao et al. [35], have used the ISM-BN model to assess the impact of varying factors on maritime safety, successfully pinpointing the critical risk components for different accident types. Afenyo et al. [36] utilized a BN model to sketch an Arctic shipping accident scenario and illuminated the crucial causative elements of a potential accident scenario. Similarly, Jiang et al. [37] proposed a Bayesian network-based risk analysis strategy to evaluate maritime accidents along the 21st-century Maritime Silk Road (MSR), identifying the principal influencing factors that could bolster accident prevention measures and ensure maritime transportation safety and sustainability. In a more focused study, Si et al. [38] employed a BN structure learning algorithm that paired the kernel density estimation with a model weighted average strategy to dissect the causative elements of container ship collisions, basing their analysis on a limited set of container ship collision sample data. Other studies like Fan et al. [39] and Hänninen et al. [40] proposed similar Bayesian network-based risk analysis approaches to understand the contributing factors to maritime transport accidents, with the latter focusing more on maritime safety management and its relationship with maritime traffic safety. Despite these successes, these aforementioned studies suffer from a triad of limitations: (1) a paucity of sample data from maritime accidents, (2) a labor-intensive and time-consuming data collection process, and (3) the inherent difficulty of obtaining accident loss records. Summarily, while waterway transportation research has honed its focus on accident causality reasoning and accident causation analysis, there remains a conspicuous void in the research landscape pertaining to accident scenario modeling.

In light of this, this paper aimed to build a BN model for the evolution of maritime accident scenarios using global maritime accident data. These data derived from the Global Integrated Shipping Information System (GISIS) and established by the International Maritime Organization (IMO) have been widely used by scholars in maritime accident studies [41–47]. The novelty of this research lies in the use of a BN-based approach to model maritime traffic accident scenarios. This is a unique method of analyzing the causes of maritime traffic accidents through performing dynamic risk assessments on shipping activities and predicting the probability of accident occurrence and its consequences. This innovative approach enables the identification and simulation of influencing factors across a range of accident scenarios, providing an intricate understanding of the complexities associated with maritime traffic accidents.

This study provides comprehensive analysis and valuable insights into 5660 global maritime accidents from 2005 to 2020. The accident data were well sampled, non-manually collected, open, and, more importantly, provided a high number of data fields in relation to accident losses. This made it possible to compensate for data limitations that have existed in previous studies. This study had two main contributions. First, a tree augmented network (TAN) model was developed to construct BN and train the data, and a data-driven BN-based method was proposed that could effectively predict the probability and consequences of accidents. Second, the proposed model was able to predict the causal factors that were most likely to lead to specific accident consequences; this could help maritime stakeholders implement effective preventive measures to improve maritime transportation safety.

The rest of this article is structured as follows. Section 2 briefly introduces the structure and construction method of BN and further introduces the method TAN driven by the data. Section 3 builds the TAN model based on the data of 5660 maritime accidents and carries out sensitivity analysis and simulation verification on the built model. Section 4 uses the two-way reasoning ability of the TAN model to predict the accident chain and analyze the accident causes. Finally, the fifth part summarizes the full text.

2. BN Structure Learning—TAN

BN is a directed acyclic graph (DAG) that is composed of nodes and directed edges and is widely employed to illustrate the interdependence and strength of associations between variables. As shown in Figure 1, in DAG $S = \{X, E\}$, X denotes the set of nodes in the network, $X_i \in X$ denotes the random variable in the domain of the definition of this restriction, and E denotes a set of directed edges in this network. The network represents the interrelationship between variables through vectorial arcs, with the intensity of each association specified by a table of conditional probabilities.

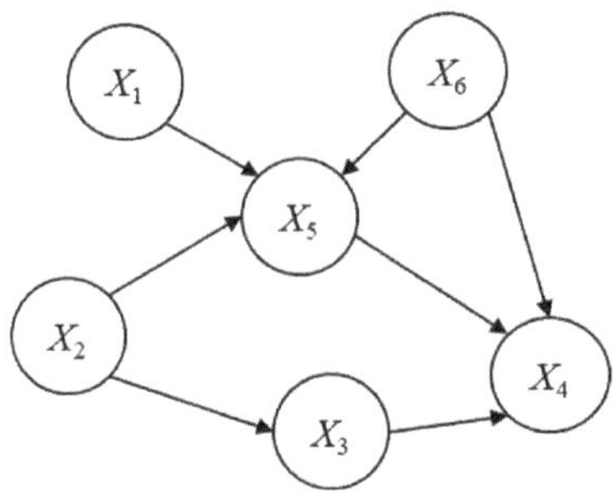

Figure 1. Graph of a valid BN with directed acyclic graph structure.

There are two primary approaches to the generation of BN structures: (1) the expert knowledge method and (2) the data-driven method. In the expert knowledge method, the BN structure was built by subjectively evaluating the causal relationships between variables. Conversely, the data-driven method was employed to uncover the interdependence between variables based on the learning algorithm of the BN model and data correlations. In this study, since sufficient sample data were collected, the data-driven method was used to construct the BN structure.

Data-driven Bayesian approaches could be classified into three main categories: (1) the naive Bayesian network (NBN), (2) the augmented naive Bayesian network (ABN), and (3) the tree augmented network (TAN). Among these, TAN learning effectively combined the simplicity and robustness of NBN computation with its ability to characterize interaction dependencies among variables, thus providing insights into the key factors leading to the outcomes of specific accidents [48]. Therefore, this paper employed the data-driven TAN approach to construct the BN structure.

BN encodes the joint probability distribution over a set of random variables U. We let $U = \{A_1, \cdots A_n, C\}$, where n denoted the number of influencing factors, where $A_1, \cdots A_n$ represent the influencing factors, and C is a class variable (accident type). It was established that the set of parent nodes of C in U was empty, meaning $\prod C = \varnothing$. Moreover, $\prod A_i$ had at most one other node in addition to C that could have an associated edge pointing to it. The joint probability density distribution adhered to the following equation:

$$P(A_1, \cdots A_n, C) = P(C) \cdot \prod_{i=1}^{n} P(A_i|C) \tag{1}$$

In the process of learning the TAN structure, Chow and Liu [49] proposed an approach to optimize and construct the BN structure using the conditional mutual information of each attribute pair. This function was defined as:

$$I_P(A_i, A_j \mid C) = \sum\nolimits_{a_{ii}, a_{ji}, c_i} P(a_{ii}, a_{ji}, c_i) log \frac{P(a_{ii}, a_{ji} \mid c_i)}{P(a_{ii} \mid c_i) P(a_{ji} \mid c_i)} \tag{2}$$

where I_P denotes the conditional mutual information; a_{ii} is the i-th state of the influencing factor A_i; and a_{ji} is the i-th state of the influencing factor A_j.

3. Global Maritime Accident TAN Model

3.1. Data Collection

This paper utilized the Marine Casualties and Incidents (MCI) database in GISIS, which is managed by IMO [50]. GISIS is a comprehensive, global maritime information system. In accordance with IMO regulations, every country with sovereignty over its territorial sea is required to report maritime accidents that occur within its waters to the IMO. The MCI database contains two types of information in relation to global maritime accidents: first, the factual data gathered from various sources, and second, detailed data obtained from casualty investigation reports submitted to the IMO.

The MCI database houses global maritime accident data dating back to 1973. Between 1973 and 2000, the annual number of recorded maritime accidents was quite limited. From 2001 onward, the number of accidents documented in the MCI database has been more consistent. However, the accident timestamps during 2001–2004 are only accurate to the day, which is not sufficient for studying the specific time periods in which these accidents occurred. Consequently, low-quality data from the early years have been excluded, and a total of 5660 maritime accidents recorded from 2005 to 2020 were utilized to construct the BN model.

3.2. Node Variable Definitions

Based on the literature's studies on maritime accident factor analysis [1,43,51,52], there are 16 primary factors that contribute to maritime accidents, including the ship type, hull type, ship's age, length, gross tonnage, operation, voyage segment, ship's speed, condition, equipment or device condition, ship's design, interaction information, weather conditions, ocean conditions, time period, and channel traffic [illegible] the information available in the MCI database, seven node variables for the BN model were selected; these included the accident quarter, accident period, accident type, ship type involved, total tonnage of the ship involved, life loss contingency, and accident severity.

Given the requirement of discrete variables for BN nodes, it was necessary to discretize continuous variables in the accident statistics. The division of accident occurrence quarters into the first quarter (January, February, and March), second quarter (April, May, and June), third quarter (July, August, and September), and fourth quarter (October, November, and December) were conducted. The categorization of accident periods was made during dawn (0:00–5:59), early morning (6:00–8:59), morning (9:00–11:59), noon (12:00–13:59), afternoon (14:00–16:59), early evening (17:00–19:59), and evening (20:00–23:59). To discretize the gross tonnage of the ships involved, the collected data and the centroid clustering (CC) algorithm were utilized for their classification. The CC algorithm, which uses the minimization error sum of squares as the objective function, was employed and terminated when the number of iterations reached a preset maximum of 5000 iterations. The optimal classification results yielded four groups based on the gross tonnage of the ships involved: (1–18,500 t), (18,501–57,500 t), (57,501–120,000 t), and (120,001–403,342 t). Among these, 403,342 t represented the maximum total tonnage of the ships involved in the collected data.

Furthermore, in this paper, we classified non-routine accidents, such as missing ships, life-saving equipment accidents, and numerous accident types with irregular or rare records, accounting for no more than 5% as "others" [53]. Multipurpose ships, tugboats, supply, and offshore vessels, unspecified ship types, and other ship types represented no more than 10% were categorized as "others" [53]. Table 1 presents the names, classifications, frequency of occurrence, and percentages of each discrete variable category. Specifically, the "quarter of accident" is a variable divided into four categories, corresponding to the four quarters of the year. Category "a" represents accidents that occurred in the first quarter (January to March), with a frequency of 1539, accounting for 27.19% of the total occurrences. Similarly, "b" is for the second quarter (April to June), "c" for the third quarter (July to September), and "d" for the fourth quarter (October to December), each with their respective frequencies and percentages. The "ship type" is a variable that has seven categories. For instance, "a" represents general cargo ships, which were involved in accidents 989 times, making up 17.47% of the total occurrences. "b" stands for bulk carriers, "c" for container ships, "d" for chemical tankers/oil tankers, "e" for passenger ships, "f" for fishing ships, and "g" for others. Each category has its corresponding frequencies and percentages of occurrence. The "accident type" is a variable that categorizes the types of accidents that occur. For example, category "a" denotes collisions, which occurred 1016 times, representing 17.95% of the total accidents. Similarly, "b" stands for stranding/grounding, "c" for fire/explosions, "d" for capsize, "e" for machinery damage, "f" for contact, and "g" for others, each with their respective frequencies and percentages. Each of the remaining variables in Table 1 followed a similar pattern, wherein specific categories were defined for each variable, along with their frequency of occurrence and corresponding percentages.

Table 1. Variables for building BN.

Variable Name	Classification	Frequency	Percentage/%	Variable Name	Classification	Frequency	Percentage/%
Quarter of accident	a (the first quarter)	1539	27.19	Ship type	a (general cargo ship)	989	17.47
	b (the second quarter)	1353	23.90		b (bulk carrier)	255	4.50
	c (the third quarter)	1406	24.84		c (container ship)	370	6.54
	d (the fourth quarter)	1362	24.06		d (chemical tanker/oil tanker)	537	9.49
Period of accident	a (dawn 0–5 a.m.)	1954	34.52		e (passenger ship)	453	8.00
	b (early morning 5–8 a.m.)	562	9.93		f (fishing ship)	634	11.20
	c (morning 8–11 p.m.)	693	12.24		g (others)	2422	42.79
	d (noon 11–13 p.m.)	427	7.54	Gross tonnage	a (gross tonnage [1,18,500])	4011	70.87
	e (afternoon 13–16 p.m.)	647	11.43		b (gross tonnage [18,501,57,500])	1219	21.54
	f (early evening 16–19 p.m.)	540	9.01		c (gross tonnage [57,501,120,000])	340	6.00
	g (evening 19–24 p.m.)	837	14.79		d (gross tonnage [120,001,403,342])	90	1.59
Accident type	a (collision)	1016	17.95	Life loss contingency	a (life loss)	1651	29.17
	b (stranding/grounding)	823	14.54		b (no life loss)	4009	70.83
	c (fire/explosion)	754	13.32	Severity of accident	a (particularly serious accidents)	2837	50.12
	d (capsize)	365	6.45		b (serious accidents)	2034	35.94
	e (machinery damage)	287	5.07		c (general accident)	622	10.99
	f (contact)	281	4.96		d (unspecified accident)	167	2.95
	g (others)	2134	37.70				

3.3. TAN Modeling

Based on the data processing results, the relationship between the six influencing factors and accident consequences was examined. Netica software with a "learning network" function was employed to develop a TAN model, which was grounded on Equation (2), ensuring that all connections between nodes were meaningful. The BN qualitative structure was trained by data, followed by a rigorous review conducted by domain experts to confirm the significance of the links between these nodes. In this study, no changes were made during the finetuning process, as all the interrelationships suggested by the data were in alignment with reality. The initial structure of TAN, which is depicted in Figure 2, was based on the data-driven TAN training results that showcased the realistic correlations between variables. The numbers depicted in Figure 2 represent the initial results of the TAN model. For instance, if the type of accident is divided into seven categories, the initial proportion of each category after initialization is approximately 14.3%. Therefore, the sum of the proportions for all categories would equate to 100%. This explanation is applicable to all other variables depicted in Figure 2 as well.

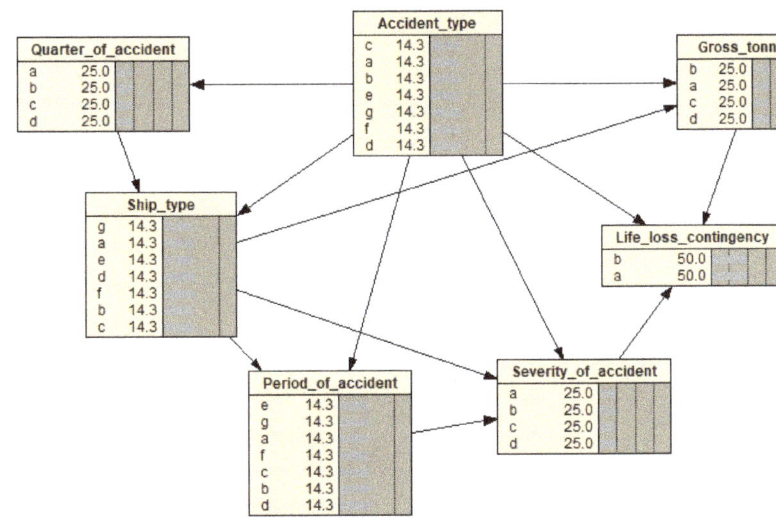

Figure 2. Initial structure of TAN model.

Utilizing the TAN model, Netica software employed basis functions to create a structure learning module and a parameter learning module, which automatically learned the conditional probability table (CPT) parameters from the sample dataset. The construction of TAN and the obtainment of CPT facilitated the calculation of the posterior probability of each variable. The statistical results of these probabilistic variables were instrumental in the analysis of maritime safety considerations and the facilitation of accident prevention. Figure 3 presents the TAN results for the random variables of interest.

3.4. Sensitivity Analysis and Model Validation

3.4.1. Sensitivity Analysis

In the Netica software, the accident type was selected as the target node, and sensitivity analysis on this node was conducted to identify the factors with the greatest influence on the target node within the TAN model.

The mutual information value represents the sensitivity level between two random variables; a higher value indicates the greater sensitivity of the influencing factor to the target node and, conversely, its lower sensitivity. The sensitivity analysis function in Netica software was used to calculate the mutual information value, percentage, and variance for each influencing factor and accident type, as displayed in Table 2. According to Table 2, the accident consequence and accident severity were the factors most sensitive to the accident type performance, with mutual information values of 0.14246 and 0.14033, respectively; these were notably higher than those of the other four factors. These results revealed how accident consequence and accident severity were the two most intuitive factors for determining the type of accident, followed by ship type, gross tonnage of the ship, time period, and quarter.

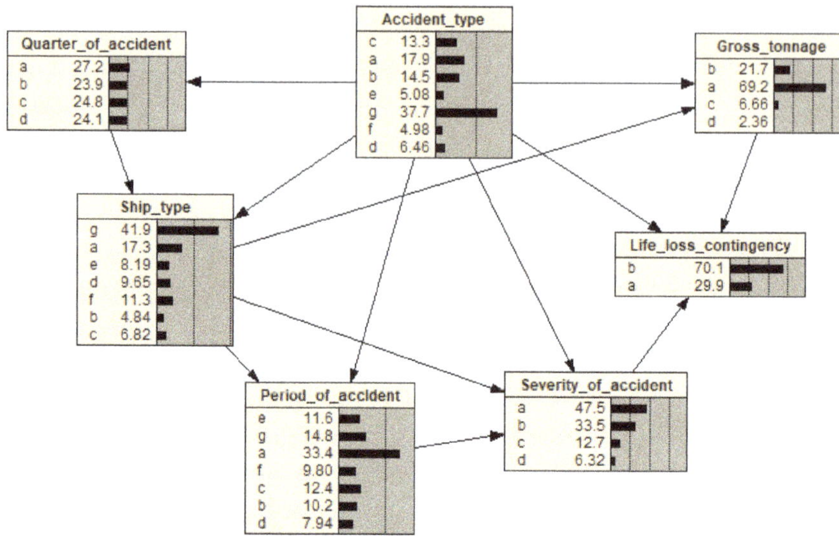

Figure 3. TAN of vessel navigation risk.

Table 2. Mutual information shared with "accident type".

Nodes	Mutual Information Value	Percentage/%	Variance
Life loss contingency	0.14246	5.800	0.0176774
Accident severity	0.14033	5.710	0.0088289
Ship type	0.04235	1.720	0.0013155
Vessel gross tonnage	0.02096	0.853	0.0004918
Time period	0.02006	0.817	0.0012170
Quarter	0.00421	0.171	0.0000869

Concerning the pivotal factors impacting a variety of accident types, the subsequent step entailed a discernment of how these factors, or the states of these factors, influence the intended accident category. This was conducted by calculating the joint probability of each factor and the "accident category", as depicted in Table 3.

According to Table 3, the state of each factor that exerted the most significant impact on an accident category is disclosed (in bold value). For instance, in the "life loss contingency" category, when in the state "life loss", the highest likelihood was for accident type "collision" (22.2%), whereas in the state "no life loss", there was the lowest probability to be "collision" (7.9%). In the "severity of accident" category, state "serious accidents" demonstrated the highest probability for accident type "stranding/grounding" (23.5%), while state "particularly serious accidents" exhibited the lowest probability for accident type "stranding/grounding" (7.09%). Looking at the "ship type", type "bulk carrier" showed the highest likelihood for accident type "collision" (25.9%), whereas type "fishing ship" indicated the lowest probability for accident type "collision" (8.53%), but had the highest probability to be in "fire/explosion" (21.9%) and "capsize" (12.4%). Although "gross tonnage" and "quarter of accident" showed little difference in the probability of influencing "accident type", the probability of "collision" was the highest. In addition, both "collision" and "capsize" showed the highest probability at nighttime.

Table 3. The joint probability of the TAN model.

Life loss contingency							
	a	b	c	d	e	f	g
a	7.90	3.34	12.10	**10.30**	2.31	1.94	**62.10**
b	**22.20**	**19.30**	**13.80**	4.83	**6.27**	**6.27**	27.20
Severity of accident							
	a	b	c	d	e	f	g
a	14.50	7.09	13.10	**9.65**	2.18	1.94	**51.60**
b	**22.00**	**23.50**	**14.90**	2.77	7.07	6.73	23.00
c	20.50	17.60	11.40	3.46	**9.18**	**10.30**	27.60
d	17.40	17.10	10.60	8.03	8.19	7.87	30.90
Ship type							
	a	b	c	d	e	f	g
a	18.80	19.90	7.76	7.68	5.96	5.39	34.50
b	**25.90**	**21.90**	5.63	2.38	**6.01**	4.98	33.20
c	21.70	10.80	13.00	2.17	4.73	4.73	**42.90**
d	22.60	13.10	21.00	2.04	5.68	4.51	31.00
e	10.70	14.80	15.40	6.02	5.14	**9.23**	38.70
f	8.53	10.30	**21.90**	**12.40**	4.70	2.57	39.70
g	18.90	13.50	12.10	6.63	4.62	4.77	39.50
Gross tonnage							
	a	b	c	d	e	f	g
a	16.80	**15.20**	13.60	**8.52**	5.43	4.60	35.90
b	**21.00**	14.30	12.10	1.44	3.34	5.48	**42.30**
c	20.10	11.00	13.10	1.98	6.06	5.56	42.20
d	18.20	7.96	**18.00**	4.87	**8.20**	**9.55**	33.20
Period of accident							
	a	b	c	d	e	f	g
a	19.20	16.00	14.30	**7.39**	5.19	3.16	34.70
b	21.80	**17.30**	11.40	4.70	3.88	5.83	35.10
c	12.60	9.42	13.90	5.22	**5.90**	**6.63**	**46.40**
d	17.00	11.40	12.80	7.25	4.38	5.72	41.50
e	[illegible]	11.30	**16.60**	7.15	4.95	6.12	39.30
f	12.90	16.50	10.50	7.29	5.00	[illegible]	[illegible]
g	**23.50**	16.50	11.50	5.07	5.47	5.23	32.70
Quarter of accident							
	a	b	c	d	e	f	g
a	17.40	**16.90**	13.30	5.99	4.56	4.82	37.00
b	**18.80**	12.50	**14.90**	5.86	4.60	5.48	37.80
c	17.10	14.60	13.80	6.62	**5.98**	3.87	**38.10**
d	18.60	13.80	11.20	**7.42**	5.22	**5.80**	37.90

This analysis underscores the influence that a particular state had on a single factor in an accident category. Additionally, it demonstrates how different states of a single factor contributed to the probability of a specific accident category. Generally, more attention should be paid to those conditions that display high probabilities of accidents due to the state of the single factor under an accident type.

3.4.2. Model Validation

To validate the effectiveness of the TAN model, three offshore accident cases from 2021 were randomly selected, each with varying accident consequences and severities, and were labeled as events 1, 2, and 3. The case data were input into the model for scenario analysis, and Table 4 presents the relevant data information for these accident cases.

Table 4. State values of real event factor variables.

Variables	Event Number		
	1	2	3
Quarterly	c	a	b
Time period	e	b	g
Ship type	g	a	g
Life loss contingency	a	b	b
Accident severity	a	b	c
Vessel gross tonnage	b	a	b
Accident type	g	b	a
Accident probability	75.1%	38.0%	44.4%

Based on the data from three randomly selected events, the probability of the known nodes, such as the quarter, time period, vessel type, accident consequence, accident severity, and gross tonnage of the vessel, was set to 100%. The types and probabilities of the predicted accidents were then observed. As illustrated in Table 4, the probability of other accident types occurring in event 1 was 75.1%; the probability of stranding/grounding in event 2 was 38.0%; and the probability of collision in event 3 was 44.4%. When compared to the original data's accident types, the predicted accident types for the three events matched, indicating that the model's predictions were accurate to some extent. Since the occurrence probability of other accident types in the original data was significantly higher than that of collision and stranding/grounding, the data-driven TAN model's simulation results demonstrated better performance in predicting the occurrence probability of other accident types (e.g., Figure 4a) and for average results in predicting collision and stranding/grounding accidents (e.g., Figure 4b,c).

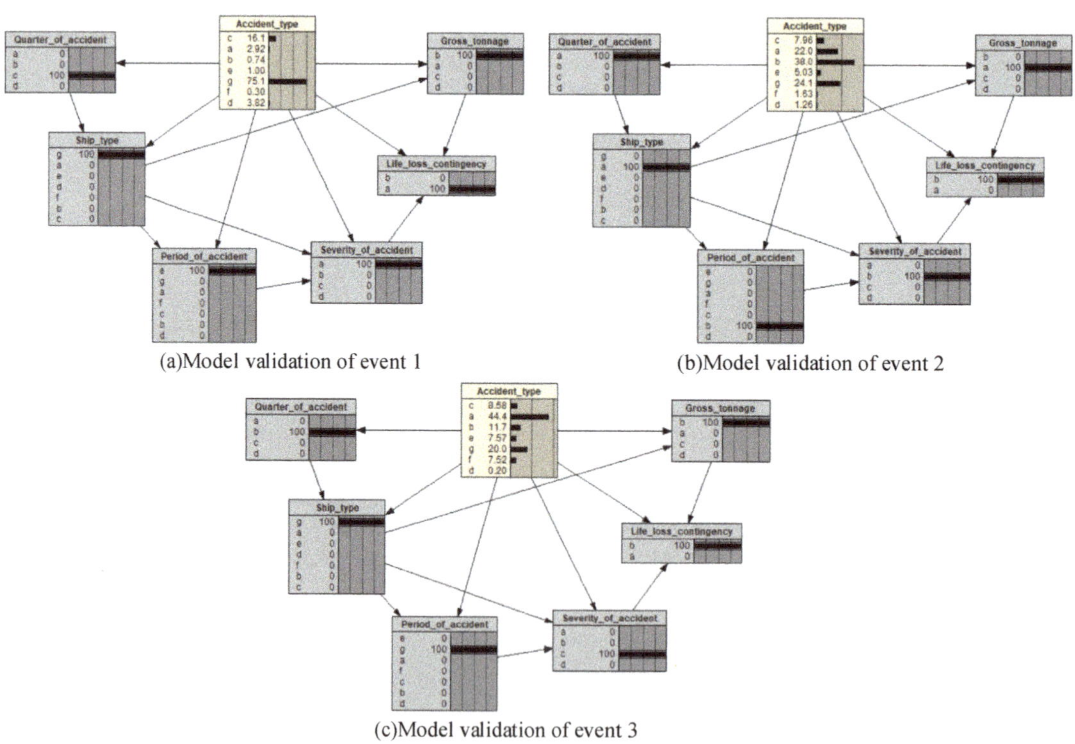

Figure 4. Event model validation.

4. Results and Discussion

The TAN model has the ability to reason bidirectionally and can help explain the most likely scenarios that are associated with a specific accident type. The data-driven TAN-based model examined the correlations between various influencing factors of maritime accidents and accident types, as well as accident consequences. This analysis enabled the prediction of the likelihood of various accident scenarios and the extrapolation of accident consequences under specific conditions.

4.1. Accident Chain Forecast

After the manipulation of the TAN model using Netica software, the relationships between the influencing factors and the accident type, including life loss contingency and the probability of each node, were obtained. By adjusting the placement bar of a single node or multiple nodes, the target node's probability trend was observed; therefore, a judgment could be formed of the potential trends and consequences of the accident.

The parameters of the conditions for maritime accidents were first simulated by changing a single node and observing the changes in the target node. When changing the ship type, more significant changes occurred in the probability of each accident type. For example, when the ship type was set to a chemical ship, the probability of a fire/explosion accident type increased significantly. When the ship type was a bulk carrier, the probability of the collision accident type increased notably. This study showed that different ship types could lead to significant differences in the occurrence of accident types. Additionally, the ship's gross tonnage and the accident's quarter and time also impacted the accident type.

Since the accident type was influenced by the joint decision of several nodes, the influence of a single node on the accident type was more one-sided. Therefore, the accident quarter was set to the first quarter (with the variable node's confidence bar set to 100%), the ship type was set to a general cargo ship, and the gross tonnage to 1–18,500 t, as shown in Figure 5. The change in the accident type and accident severity node probability from the early morning to evening is shown in Figure 6. As seen in Figure 6, among the types of maritime accidents throughout the day, the probability of a fire/explosion on ships was low, except for the afternoon time period, which was 18.2%; the probabilities of capsize, machinery damage, and contact were also low below 10%. Among other accidents, the probabilities of ship collision and stranding/grounding accidents were significantly higher at around 20%. Additionally, it was observed that the occurrence probability of stranding/grounding accidents was significantly higher during the dawn and evening than in other periods.

Unlike previous studies, this paper focused specifically on the question of whether or not the consequences of an accident could involve a loss of human life when an accident occurred under this scenario. As shown in Figure 7, the change in the probability of the "life loss contingency" node from the early morning to the evening showed that the probability of an accident consequence that did not involve loss of life was much higher than the probability of an accident consequence that involved loss of life throughout the day in this scenario. The results of the study indicate that the probability of potential loss of life is low for all accident types in this scenario. Therefore, accidents involving collisions and fires/explosions do not necessarily result in loss of life outcomes either and may need to be combined with additional accident causation in order to obtain more reliable conclusions.

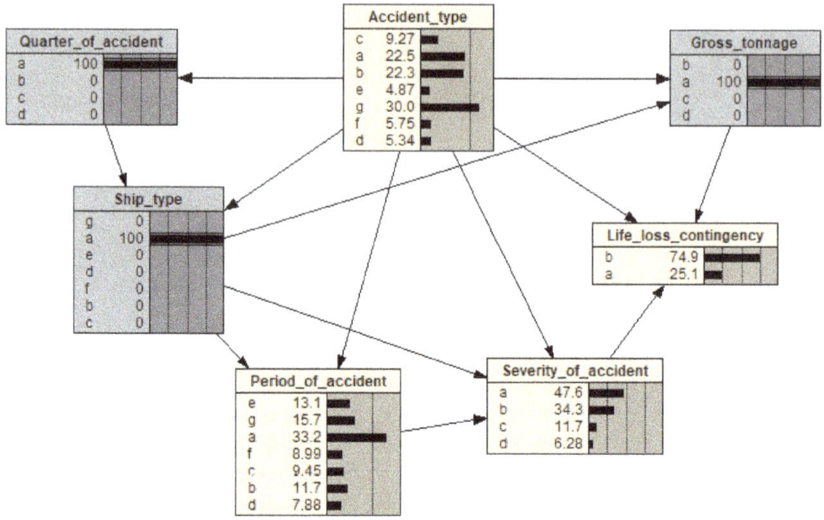

Figure 5. Analysis of the accident chain with the characteristics "first quarter", "general cargo", and "gross ship tonnage set to 1–18,500 t".

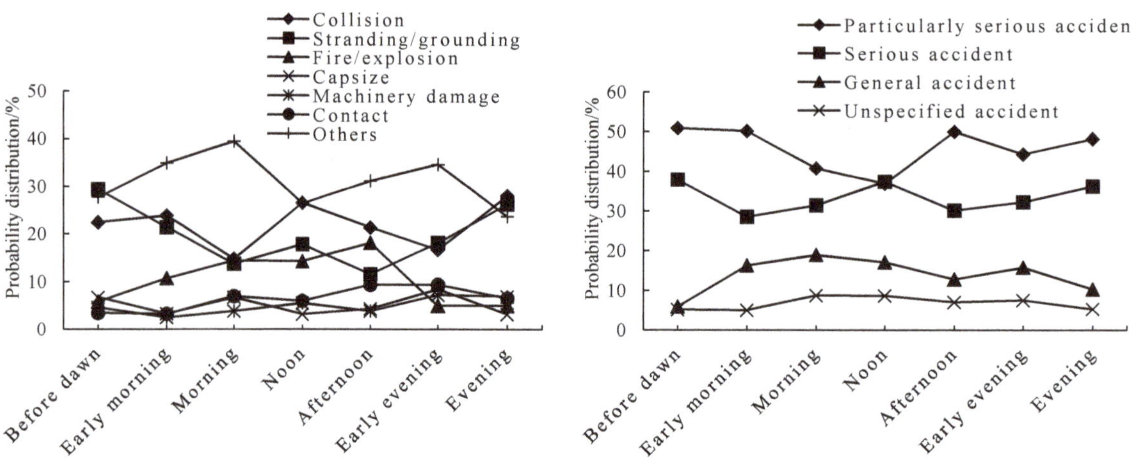

Figure 6. The posterior probability of "accident type" and "severity of accident" in specific accident scenarios.

In summary, the highest probabilities of collision and stranding/grounding occurred at dawn, with the accident severity for this appearing particularly serious. A collision was most likely to occur at noon, with high accident severity. Particularly severe collision and stranding/grounding accidents were more likely to occur in the evening. It is worth mentioning that although the probabilities of collision and stranding/grounding of ships were higher in this scenario, the probability of life loss was relatively low, and the accident consequences were less affected by the time of the accidents.

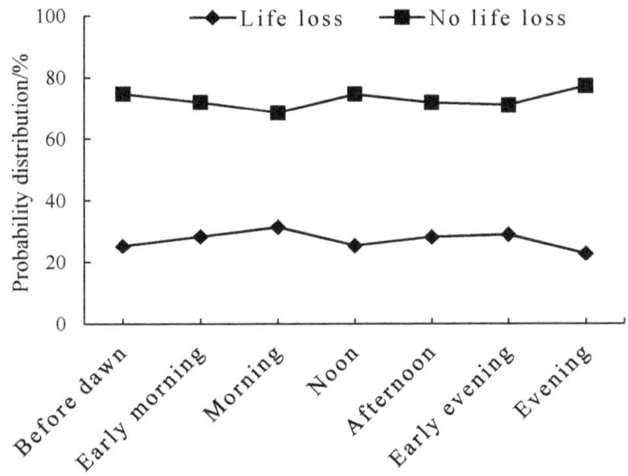

Figure 7. The posterior probability of "life loss contingency" in specific accident scenarios.

4.2. Accident Cause Analysis

In this study, the bidirectional reasoning of the TAN model was employed, allowing for both causal reasoning and diagnostic reasoning. The application of causal reasoning aided in the prediction of accident chains, while diagnostic reasoning assisted in the analysis of the accident causes. By determining the type, consequence, and severity of an accident, a more intuitive comprehension of the causes and mechanisms behind maritime accidents could be achieved.

As demonstrated in Figure 8, certain patterns were identified from the research data. For instance, when the accident type was classified as a collision, with severity as a general accident and no involvement of life loss, there was a significantly higher probability of container ships, general cargo ships, and chemical ships that were involved compared to other ship types. This indicates a necessity for focusing on the safety measures of these types of vessels, given their higher likelihood of being involved in collision accidents. The results also revealed a correlation between ship tonnage, time of the accident, and frequency of collision accidents. Ships with a tonnage between 1 and 18,800 t [illegible] in collisions during dawn hours. A plausible explanation for this might be the combined influence of lower visibility conditions, potential crew fatigue, and less active navigation during these hours.

Furthermore, the data suggest that accidents involving container ships, general cargo ships, and chemical ships of such tonnage typically have a lower probability of causing life loss. This might be attributed to the relative ease with which personnel can escape from smaller ships in distress or potentially the higher success rate of rescue operations due to the manageability of these smaller vessels.

These findings provide essential insights into maritime accident patterns. By identifying specific circumstances and ship types that are associated with a higher risk of accidents, it could be possible to develop more targeted safety protocols and preventive measures. It also highlights the usefulness of predictive models, such as the TAN model, for risk management in the maritime industry.

In conclusion, these findings emphasize the intricate nature of maritime accidents and the numerous variables involved. Through the bidirectional reasoning of the TAN model, a more thorough understanding of these accidents could be obtained, potentially leading to the development of more effective accident prevention strategies and, ultimately, the enhancement of maritime safety.

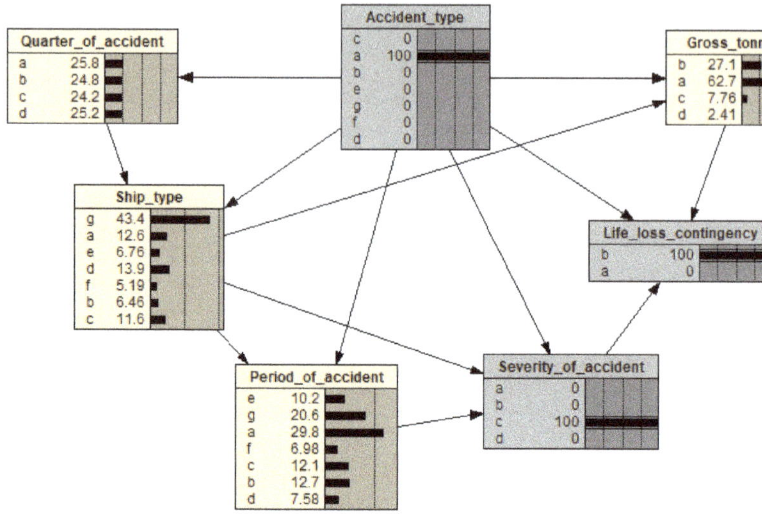

Figure 8. Analysis of accident scenarios with the characteristics "collision", "general accident", and "no life loss".

5. Conclusions

Ship safety has always been a major concern in the maritime transportation industry. In this paper, a TAN model for a maritime traffic accident risk assessment was constructed to analyze the relationship between the consequences of maritime accidents and various influencing factors and to use model simulation to analyze how different risk factors can affect different types of maritime accidents.

The TAN model was constructed based on data from a total of 5660 maritime accidents from 2005 to 2020. In addition to other accident types, the accident type with the highest probability of occurrence among maritime traffic accidents included collision, followed by stranding/grounding, and then fire/explosion.

The sensitivity analysis and simulation validation of the constructed model showed that accident consequences and accident severity are the two most intuitive factors when determining the type of accident occurrence, followed by the ship type, gross tonnage of the ship, time period, and season. The constructed model effectively predicted the likelihood of various accident scenarios and accident consequence projections under specific conditions.

According to the causal reasoning analysis of the TAN model and under the conditions of "first quarter", "general cargo ship", and "ship's gross tonnage of 1–18,500 t," the probability of ship collision and stranding/grounding accidents was higher, while the probability of life loss was relatively low, and the consequences of this accident were less affected by the time of the accident. According to the analysis of the model's diagnostic reasoning, in the general collision accident chain without loss of life, container ships, general cargo ships, and chemical ships were the main types of ships involved in such accidents. Ships with a tonnage of 1–18,500 t were more likely to have such accidents during the dawn; however, their probability of causing loss of life was lower. These findings carry significant implications for enhancing safety measures in the maritime transportation industry. By understanding the frequency, severity, and common conditions of various types of accidents, stakeholders could develop more targeted and effective accident prevention strategies.

Despite utilizing a substantial volume of publicly available accident data to achieve reliable predictive outcomes, we acknowledge the limitations of our study. It is plausible that the introduction of more variables could alter these results. Future work should focus on expanding this model to include additional variables such as environmental factors (weather conditions, sea state), ship design and maintenance factors, human factors (crew experience, fatigue), and others that could impact the risk and consequences of maritime accidents. Additionally, more in-depth research should be carried out to investigate the different patterns of accidents associated with different types of ships at various times of day to refine preventative measures accordingly.

Author Contributions: Conceptualization, methodology, software, writing—original draft, S.L.; supervision, J.W.; writing—review and editing, Z.Z.; data curation, F.L. and Z.L.; visualization, Z.L. All authors have read and agreed to the published version of the manuscript.

Funding: This research was supported by the National Natural Science Foundation of China under grant 52072237, the Guangdong Basic and Applied Basic Research Foundation of China under grant 2022A1515110776, and the Project of Educational Commission of Guangdong Province of China under grant 2021WTSCX222.

Institutional Review Board Statement: Not applicable.

Informed Consent Statement: Not applicable.

Data Availability Statement: Not applicable.

Conflicts of Interest: The authors declare no conflict of interest.

References

1. Fan, S.; Blanco-Davis, E.; Yang, Z.; Zhang, J.; Yan, X. Incorporation of human factors into maritime accident analysis using a data-driven Bayesian network. *Reliab. Eng. Syst. Saf.* **2020**, *203*, 107070. [CrossRef]
2. Zhang, G.; Thai, V.V. Expert elicitation and Bayesian Network modeling for shipping accidents: A literature review. *Saf. Sci.* **2016**, *87*, 53–62. [CrossRef]
3. Hu, S.; Li, Z.; Xi, Y.; Gu, X.; Zhang, X. Path Analysis of Causal Factors Influencing Marine Traffic Accident via Structural Equation Numerical Modeling. *J. Mar. Sci. Eng.* **2019**, *7*, 96. [CrossRef]
4. Chou, C.-C.; Wang, C.-N.; Hsu, H.-P.; Ding, J.-F.; Tseng, W.-J.; Yeh, C.-Y. Integrating AIS, GIS and E-Chart to Analyze the Shipping Traffic and Marine Accidents at the Kaohsiung Port. *J. Mar. Sci. Eng.* **2022**, *10*, 1543. [CrossRef]
5. Xue, J.; Papadimitriou, E.; Reniers, G.; Wu, C.; Jiang, D.; van Gelder, P.H.A.J.M. A comprehensive statistical investigation framework for characteristics and causes analysis of ship accidents: A case study in the fluctuating backwater area of Three Gorges Reservoir region. *Ocean Eng.* **2021**, *229*, 108981. [CrossRef]
6. Göksu, S.; Arslan, Ö. Quantitative Analysis of Dynamic Risk Factors for Shipping Operations. *J. ETA Marit. Sci.* **2020**, *8*, 86–97. [CrossRef]
7. Li, Z.; Y[illegible], C.; Zhu, Y.; Cao, G.; Hu, S. A decision support model for ship navigation in Arctic waters based on dynamic risk assessment. *Ocean Eng.* **2022**, *244*, 110427. [CrossRef]
8. Guo, Y.; Jin, Y.; Hu, S.; Yang, Z.; Xi, Y.; Han, B. Risk evolution analysis of ship pilotage operation by an integrated model of FRAM and DBN. *Reliab. Eng. Syst. Saf.* **2023**, *229*, 108850. [CrossRef]
9. Zhang, L.; Wang, H.; Meng, Q.; Xie, H. Ship accident consequences and contributing factors analyses using ship accident investigation reports. *Proc. Inst. Mech. Eng. Part O J. Risk Reliab.* **2019**, *233*, 35–47. [CrossRef]
10. Otay, E.N.; Özkan, S. Stochastic Prediction of Maritime Accidents in the strait of Istanbul. In Proceedings of the 3rd International Conference on Oil Spills in the Mediterranean and Black Sea regions, Istanbul, Turkey, 1 September 2003; pp. 92–104.
11. Koromila, I.; Nivolianitou, Z.; Giannakopoulos, T. Bayesian network to predict environmental risk of a possible ship accident. In Proceedings of the 7th International Conference on PErvasive Technologies Related to Assistive Environments, Rhodes Greece, 27–30 May 2014; pp. 1–5.
12. Zhang, C.; Zou, X.; Lin, C. Fusing XGBoost and SHAP models for maritime accident prediction and causality inter-pretability analysis. *J. Mar. Sci. Eng.* **2022**, *10*, 1154. [CrossRef]
13. Fu, S.; Liu, Y.; Xi, Y.; Zhang, M.; Hu, S. Analysis of influencing factors on severity of ship accidents in ice-covered Arctic waters. *China Saf. Sci. J.* **2019**, *29*, 84. (In Chinese)
14. Baksh, A.-A.; Abbassi, R.; Garaniya, V.; Khan, F. Marine transportation risk assessment using Bayesian Network: Application to Arctic waters. *Ocean Eng.* **2018**, *159*, 422–436. [CrossRef]
15. Chen, P.; Zhang, Z.; Huang, Y.; Dai, L.; Hu, H. Risk assessment of marine accidents with Fuzzy Bayesian Networks and causal analysis. *Ocean Coast. Manag.* **2022**, *228*, 106323. [CrossRef]
16. Ventikos, N.P.; Giannopoulos, I.F. Assessing the consequences from marine accidents: Introduction to a risk acceptance criterion for Greece. *Hum. Ecol. Risk Assess. Int. J.* **2013**, *19*, 699–722. [CrossRef]
17. Chen, J.; Bian, W.; Wan, Z.; Yang, Z.; Zheng, H.; Wang, P. Identifying factors influencing total-loss marine accidents in the world: Analysis and evaluation based on ship types and sea regions. *Ocean Eng.* **2019**, *191*, 106495. [CrossRef]

18. Chen, J.; Zhang, F.; Yang, C.; Zhang, C.; Luo, L. Factor and trend analysis of total-loss marine casualty using a fuzzy matter element method. *Int. J. Disaster Risk Reduct.* **2017**, *24*, 383–390. [CrossRef]
19. Nivolianitou, Z.; Leopoulos, V.; Konstantinidou, M. Comparison of techniques for accident scenario analysis in hazardous systems. *J. Loss Prev. Process. Ind.* **2004**, *17*, 467–475. [CrossRef]
20. Wang, W.; Jiang, X.; Xia, S.; Cao, Q. Incident tree model and incident tree analysis method for quantified risk assessment: An in-depth accident study in traffic operation. *Saf. Sci.* **2010**, *48*, 1248–1262. [CrossRef]
21. Liu, P.; Yang, L.; Gao, Z.; Li, S.; Gao, Y. Fault tree analysis combined with quantitative analysis for high-speed railway accidents. *Saf. Sci.* **2015**, *79*, 344–357. [CrossRef]
22. Ung, S.-T. Evaluation of human error contribution to oil tanker collision using fault tree analysis and modified fuzzy Bayesian Network based CREAM. *Ocean Eng.* **2019**, *179*, 159–172. [CrossRef]
23. Ahn, Y.-J.; Yu, Y.-U.; Kim, J.-K. Accident Cause Factor of Fires and Explosions in Tankers Using Fault Tree Analysis. *J. Mar. Sci. Eng.* **2021**, *9*, 844. [CrossRef]
24. Kim, K.-I.; Jeong, J.S.; Lee, B.-G. Study on the Analysis of Near-Miss Ship Collisions Using Logistic Regression. *J. Adv. Comput. Intell. Intell. Inform.* **2017**, *21*, 467–473. [CrossRef]
25. Yu, H.; Khan, F.; Veitch, B. A Flexible Hierarchical Bayesian Modeling Technique for Risk Analysis of Major Accidents. *Risk Anal.* **2017**, *37*, 1668–1682. [CrossRef] [PubMed]
26. Hänninen, M. Bayesian networks for maritime traffic accident prevention: Benefits and challenges. *Accid. Anal. Prev.* **2014**, *73*, 305–312. [CrossRef] [PubMed]
27. Abistado, K.G.; Arellano, C.N.; Maravillas, E.A. Weather forecasting using artificial neural network and Bayesian network. *J. Adv. Comput. Intell. Intell. Inform.* **2014**, *18*, 812–817. [CrossRef]
28. Wang, J.; Zhang, M.; Huang, X.; Chen, J. Scenario analysis of road transportation accidents of inflammable and explosive hazardous chemicals. *China Saf. Sci. J.* **2019**, *29*, 171. (In Chinese)
29. Qiao, W.; Liu, Y.; Ma, X.; Liu, Y. Human factors analysis for maritime accidents based on a dynamic fuzzy Bayesian network. *Risk Anal.* **2020**, *40*, 957–980. [CrossRef]
30. Zhang, G.; Thai, V.V.; Yuen, K.F.; Loh, H.S.; Zhou, Q. Addressing the epistemic uncertainty in maritime accidents modelling using Bayesian network with interval probabilities. *Saf. Sci.* **2018**, *102*, 211–225. [CrossRef]
31. Li, K.X.; Yin, J.; Bang, H.S.; Yang, Z.; Wang, J. Bayesian network with quantitative input for maritime risk analysis. *Transp. A Transp. Sci.* **2012**, *10*, 89–118. [CrossRef]
32. Wang, L.; Yang, Z. Bayesian network modelling and analysis of accident severity in waterborne transportation: A case study in China. *Reliab. Eng. Syst. Saf.* **2018**, *180*, 277–289. [CrossRef]
33. Zou, X.; Yue, W.L. A Bayesian Network Approach to Causation Analysis of Road Accidents Using Netica. *J. Adv. Transp.* **2017**, *2017*, 2525481. [CrossRef]
34. Yuan, C.; Hu, Y.; Zhang, Y.; Zuo, T.; Wang, J.; Fan, S. Evaluation on consequences prediction of fire accident in emergency processes for oil-gas storage and transportation by scenario deduction. *J. Loss Prev. Process. Ind.* **2021**, *72*, 104570. [CrossRef]
35. Zhao, J.; Xie, L.; Yang, Y.; Hu, X.; Ou, C.; Zeng, R. An ISM-BN based research on navigation risk factors of inland waterway vessels. *China Saf. Sci. J.* **2022**, *32*, 37. (In Chinese)
36. Afenyo, M.; Khan, F.; Veitch, B.; Yang, M. Arctic shipping accident scenario analysis using Bayesian Network approach. *Ocean Eng.* **2017**, *133*, 224–230. [CrossRef]
37. Jiang, M.; Lu, J.; Yang, Z.; Li, J. Risk analysis of maritime accidents along the main route of the Maritime Silk Road: A Bayesian network approach. *Marit. Policy Manag.* **2020**, *47*, 815–832. [CrossRef]
38. Si, D.; Zhang, J.; Lang, K. Causation analysis of container ship collision accidents based on improved BN. *China Saf. Sci. J.* **2019**, *29*, 31. (In Chinese)
39. Fan, S.; Yang, Z.; Blanco-Davis, E.; Zhang, J.; Yan, X. Analysis of maritime transport accidents using Bayesian networks. *Proc. Inst. Mech. Eng. Part O J. Risk Reliab.* **2020**, *234*, 439–454. [CrossRef]
40. Hänninen, M.; Banda, O.A.V.; Kujala, P. Bayesian network model of maritime safety management. *Expert Syst. Appl.* **2014**, *41*, 7837–7846. [CrossRef]
41. Zhao, X.; Yuan, H.; Yu, Q. Autonomous vessels in the Yangtze river: A study on the maritime accidents using data-driven bayesian networks. *Sustainability* **2021**, *13*, 9985. [CrossRef]
42. Acharya, T.D.; Yoo, K.W.; Lee, D.H. GIS-based Spatio-temporal Analysis of Marine Accidents Database in the Coastal Zone of Korea. *J. Coast. Res.* **2017**, *79*, 114–118. [CrossRef]
43. Uğurlu, Ö.; Köse, E.; Yıldırım, U.; Yüksekyıldız, E. Marine accident analysis for collision and grounding in oil tanker using FTA method. *Marit. Policy Manag.* **2015**, *42*, 163–185. [CrossRef]
44. Antão, P.; Teixeira, A.; Soares, C.G. Statistical characterization of risk influencing factors in ship collision accidents. In *Developments in Maritime Technology and Engineering*; CRC Press: Boca Raton, FL, USA, 2021; pp. 221–229.
45. Magda, B. Fires as a cause of ship accidents—A statistical approach. *Saf. Fire Technol.* **2015**, *37*, 171–180.
46. Li, H.; Ren, X.; Yang, Z. Data-driven Bayesian network for risk analysis of global maritime accidents. *Reliab. Eng. Syst. Saf.* **2023**, *230*, 108938. [CrossRef]
47. Huang, D.-Z.; Hu, H.; Li, Y.-Z. Spatial Analysis of Maritime Accidents Using the Geographic Information System. *Transp. Res. Rec. J. Transp. Res. Board* **2013**, *2326*, 39–44. [CrossRef]

48. Friedman, N.; Geiger, D.; Goldszmidt, M. Bayesian network classifiers. *Mach. Learn.* **1997**, *29*, 131–163. [CrossRef]
49. Chow, C.; Liu, C. Approximating discrete probability distributions with dependence trees. *IEEE Trans. Inf. Theory* **1968**, *14*, 462–467. [CrossRef]
50. International Maritime Organization-Global Integrated Shipping Information System. Marine Casualties and Incidents. Available online: https://gisis.imo.org/Public/MCI/Default.aspx (accessed on 21 June 2022).
51. Zhu, L.; Lu, L.; Zhang, W.; Zhao, Y.; Song, M. Analysis of Accident Severity for Curved Roadways Based on Bayesian Networks. *Sustainability* **2019**, *11*, 2223. [CrossRef]
52. Liu, L.; Ye, X.; Wang, T.; Yan, X.; Chen, J.; Ran, B. Key Factors Analysis of Severity of Automobile to Two-Wheeler Traffic Accidents Based on Bayesian Network. *Int. J. Environ. Res. Public Health* **2022**, *19*, 6013. [CrossRef]
53. Zhang, Y.; Sun, X.; Chen, J.; Cheng, C. Spatial patterns and characteristics of global maritime accidents. *Reliab. Eng. Syst. Saf.* **2020**, *206*, 107310. [CrossRef]

Journal of
Marine Science and Engineering

MDPI

Article

A COLREGs-Compliant Ship Collision Avoidance Decision-Making Support Scheme Based on Improved APF and NMPC

Haibin Li, Xin Wang *, Tianhao Wu and Shengke Ni

College of Navigation, Dalian Maritime University, Dalian 116026, China; 18742510935@163.com (H.L.); Luowang0602@163.com (T.W.); shengkeni@dlmu.edu.cn (S.N.)
* Correspondence: xin.wang@dlmu.edu.cn

Abstract: In this paper, combined with the improved artificial potential field (IAPF) method and the nonlinear model predictive control (NMPC) algorithm, a collision avoidance decision-making support scheme considering ship maneuverability and the International Regulations for Preventing Collisions at Sea (COLREGs) is proposed. First, to comply with the requirements of COLREGs, an improved repulsive potential field is presented for different encounter scenarios when the ship detects the risk of collision, and the coordinated ship domain is applied to provide safety criteria for collision avoidance. Then, by transforming the MMG model to a discrete-time nonlinear system, the NMPC is utilized to predict the future state of the ship according to the current state, and the IAPF method is incorporated to calculate the potential field in each future state as the objective function. Following this approach, the action taken to avoid collision is more effective, the ship motion in avoiding collision is more accurate, and the collision avoidance decision making is more reasonable. Finally, two simulation examples of multi-ship encounter scenarios are applied to illustrate the merits and effectiveness of the proposed collision avoidance decision-making support scheme.

Keywords: collision avoidance; improved artificial potential field; nonlinear model predictive control; ship maneuverability; COLREGs

Citation: Li, H.; Wang, X.; Wu, T.; Ni, S. A COLREGs-Compliant Ship [illegible] Decision-Making Support Scheme Based on Improved APF and NMPC. *J. Mar. Sci. Eng.* **2023**, *11*, 1408. https://doi.org/10.3390/jmse11071408

Academic Editor: Sergei Chernyi

Received: 10 June 2023
Revised: 4 July 2023
Accepted: 11 July 2023
Published: 13 July 2023

1. [illegible]

In the past decades, due to the high incidence and serious consequences of ship collision, preventing collision accidents has always been in the spotlight among practitioners and researchers. Through the analysis and investigation of a large number of accident reports, researchers have come to a common conclusion that human factors are the main cause of ship collision accidents [1]. To mitigate or even eliminate the impact of human factors, the research on ship anti-collision mainly focuses on assisting human and autonomous collision avoidance [2].

Ship path planning methods are always presented to realize autonomous collision avoidance and navigation. The application of A-star [3], rapid-exploring random tree [4] and other algorithms [5] in ship path planning has been developed for many years. Comparatively, these studies take into account static obstacles and ignore dynamic obstacles. The artificial potential field (APF) method, which was first formally applied [6] to real-time collision avoidance of robots in known static environments, has gradually attracted the attention of scholars and is widely used in path planning, such as unmanned aerial vehicle (UAV) [7], autonomous vehicle [8], etc. Due to its ability to deal with static and dynamic obstacles, the APF method is also applied in ship collision avoidance [9,10]. In the process of collision avoidance, the requirements of some key rules in COLREGs are usually taken into account. A COLREGs-constrained multi-ship real-time autonomous collision avoidance decision-making algorithm based on improved APF was proved to have the advantages of fast calculation speed and strong robustness [11]. The repulsive potential field was modified depending on the relevant descriptions of COLREGs for different encounter scenarios, and a guidance strategy was proposed for an underactuated unmanned surface vehicle (USV) based on improved APF [12]. However, the traditional APF method is prone to falling into local optima in path planning. Some researchers gradually focus on combining APF and other algorithms to overcome the drawbacks of traditional APF. A collision avoidance scheme based on APF and deep Q-learning network (DQN) was presented, incorporating the resultant force of APF and the requirements of COLREGs into the reward function of DQN [13]. To improve the feasibility and reasonability of an anti-collision scheme, a path planning algorithm considering the requirements of COLREGs and combining modified velocity obstacle (VO) and APF was proposed [14]. However, compared with the traditional APF, the repulsive potential field for three encounter scenarios should be developed and improved according to the rules of COLREGs.

At present, a lot of research works on path planning methods ignore whether the obtained path is prone to being tracked by the ship. In other words, path planning should also fully consider the ship motion control algorithm and tracking effect. Due to its outstanding performance in predicting the future state and dealing with multi-constraint problems, model predictive control (MPC) was widely utilized in ship motion control [15,16]. MPC is a control method based on the object, and its control accuracy is directly related to the model's accuracy. For a nonlinear system, linearizing the nonlinear system and then utilizing MPC control can simplify the control process and improve the calculation speed, but it may also lead to low control accuracy, poor controller robustness and other consequences [17]. For ship trajectory tracking and obstacle avoidance in uncertain external environmental disturbance, a trajectory tracking control method based on nonlinear model predictive control (NMPC) was proposed to ensure the high control accuracy and strong robustness of the controller [18]. A trajectory tracking with obstacle avoidance algorithm, incorporating an event-triggered mechanism into the NMPC design, was presented, which can ensure good obstacle avoidance effect and reduce the computational burden [19]. Obviously, to improve control accuracy and controller robustness, the NMPC scheme is more suitable for ship motion control than the MPC scheme.

Based on the excellent tracking performance of the MPC scheme, the combination of MPC and APF has gradually attracted scholars' attention in collision avoidance. A repulsive potential field was included in an NMPC method to avoid collision with obstacles and control multiple USV in arbitrary formations [20]. However, this scheme ignores the COLREGs and only considers static obstacle avoidance. Considering the constraints of ship maneuverability and the requirements of COLREGs, a novel motion planning method based on MPC and APF was proposed for multi-object collision avoidance [21], which can solve the problem of easily falling into local optima in traditional APF. It is worth noting that the ship kinematic model used was obtained by ignoring the influence of sway velocity components and simplifying the hydrodynamic parameters and derivatives of the model [22]. This model is mainly applied to describe the ship maneuvering characteristics of a small rudder angle, while the accuracy of describing the ship maneuvering characteristics of a large rudder angle is insufficient [23].

In particular, the maneuvering modeling group (MMG) model is one of the famous and high-precision mathematical models for ship maneuvering presented by the Japanese Towing Tank Conference (JTTC) [24]. To describe the ship maneuvering motion characteristics for a ship meeting at a close range, a collision avoidance dynamic support model was proposed [25] by combining a three-degrees-of-freedom (DOF) MMG model, a control algorithm and a collision avoidance parameter mathematical model, which demonstrates the importance and necessity of considering ship maneuvering motion characteristics in collision avoidance. In addition, ship maneuvering motion characteristics should be considered when providing the safety criteria for collision avoidance. The coordinated ship domain, which was proposed by taking into account ship maneuverability and mutual interaction of meeting ships [26], can provide more reasonable safety criteria than other ship domains.

Motivated by the above observation, this paper proposes a collision avoidance decision-making support scheme by combining the IAPF method and the NMPC algorithm, which considers the COLREGs and ship maneuverability. The main contributions are summarized as follows.

(1) According to some critical rules of COLREGs, the repulsive potential field for three encounter scenarios, i.e., head-on, crossing and overtaking situations, is developed and improved in this paper.
(2) A standard 3 DOF MMG model is applied to denote ship maneuvering motion characteristics in the process of collision avoidance, and then, a collision avoidance decision-making support scheme is proposed by incorporating the IAPF method into the NMPC design.
(3) The coordinated ship domain, which considers ship maneuverability and mutual interaction of meeting ships, is applied to determine the safety criteria in the process of collision avoidance.

The rest of the article is organized as follows. Section 2 describes the IAPF method considering the requirements of COLREGs and introduces the coordinated ship domain. Section 3 introduces the MMG model and gives the NMPC design procedure. Section 4 shows the effectiveness of the presented collision avoidance decision-making support scheme. Section 5 presents the conclusions.

2. Improved Artificial Potential Field Method

As a path planning method with simple principles and practical solid application, the APF is widely used in collision avoidance. The core idea of the APF method is to regard the motion of a ship in an actual environment as the motion in a virtual potential energy field, where the ship is affected by all kinds of forces. The attractive force of the goal point drives it toward the goal point. On the contrary, the obstacles in the environment produce a repulsive force to prevent the ship from colliding with obstacles. Therefore, the resultant force will bring the ship closer to the target point and away from the obstacles. In addition, when the APF is combined with MPC, instead of calculating the attractive and repulsive forces, the total potential field is calculated [21], and the path with the gradient of steepest descent in the total potential field is the optimal path.

2.1. Attractive Potential Field

The attractive potential $U_{att}(p)$ is defined as a function of the relative distance between the own ship (OS) and the goal point.

$$U_{att}(p) = \frac{1}{2}k_{att}\rho(p_{os}, p_g)^2 \tag{1}$$

where k_{att} is the attractive potential field coefficient; p_{os}, p_g are the position of the OS and the goal point, respectively. $\rho(p_{os}, p_g)$ is the distance between the own ship and the goal point.

2.2. COLREGs-Compliant Repulsive Potential Field

Collision risk assessment is a vitally important part of the process of collision avoidance. The closest point of approach (CPA) method is selected to assess collision risk and decide whether to take avoiding action [27]. Meanwhile, in the process of actual ship collision avoidance, the requirements of COLREGs cannot be ignored. Based on the above-mentioned statement [12], this paper proposes an improved repulsive potential field, which considers risk and complies with COLREGs. There are different ways to improve the different encounter situations. Primarily, according to Rule 14 of COLREGs [28], which describes the action of two power-driven vessels in a head-on situation, it can be found that the own ship shall alter course to the starboard and pass on the port side of the target ship (TS). Therefore, the repulsive potential field is constructed according to the change in the distance between the OS and the TS, and the distance from the OS to the longitudinal centerline of the TS. Then, some settings are added, so that the own ship's actions can avoid collision and comply with the requirements of COLREGs. The repulsive potential field generated by the TS is as follows:

$$\begin{cases} U_{rep} = \frac{1}{2}k_{rep}\frac{(d_s(p,p_L)+d_1)^2}{\rho(p_{os},p_t)^2} \\ \quad if \rho(p_{os},p_t) \le l_t, d_s(p,p_L) > -d_1 \; and \; \text{TCPA} \ge 0 \\ 0 \quad \text{others} \end{cases} \tag{2}$$

where k_{rep} is the repulsive potential field coefficient; $\rho(p_{os}, p_t)$ is the distance between the OS and the TS; l_t is the influence radius of the TS; d_1 is a preset reference distance in a head-on situation; $d_s(p, p_L)$ denotes the distance from the OS to the longitudinal centerline of the TS. If the OS is on the port side of the TS, $d_s(p, p_L)$ is a negative value; otherwise, it is a positive value.

Similarly, according to Rule 15 of COLREGs, which describes the action of two power-driven vessels in a crossing situation, it can be found that the give-way ship shall avoid crossing ahead of the other vessel. Assuming the OS is a give-way ship, the repulsive potential field is constructed according to the change in the distance from the OS to the TS and the distance from the OS to the transverse centerline of the TS. The repulsive potential field of the TS when the OS needs to avoid the TS in a crossing situation is as follows:

$$\begin{cases} U_{rep} = \frac{1}{2}k_{rep}\frac{(d_s(p,p_T)+d_2)^2}{\rho(p_{os},p_t)^2} \\ \quad if \rho(p_{os},p_t) \le l_0, d_s(p,p_T) > -d_2 \; and \; \text{TCPA} \ge 0 \\ 0 \quad \text{others} \end{cases} \tag{3}$$

where d_2 is a preset reference distance in a crossing situation; $d_s(p, p_T)$ denotes the distance from the OS to the transverse centerline of the TS. If the OS is on the stern side of the TS, $d_s(p, p_T)$ is a negative value; otherwise, it is a positive value.

For an overtaking situation, according to Rule 13 of COLREGs, it can be found that the give-way ship can alter the course to the starboard or the port depending on the navigation conditions. When the OS is a give-way ship, the repulsive potential field of the TS is constructed according to the change in the distance between the OS and the TS, and the distance from the OS to the transverse centerline of the TS as follows:

$$\begin{cases} U_{rep} = \frac{1}{2}k_{rep}\frac{(d_3-d_v(p,p_L))^2}{\rho(p_{os},p_t)^2} \\ \quad if \rho(p_{os},p_t) \le l_0, d_v(p,p_L) < d_3 \; and \; \text{TCPA} \ge 0 \\ 0 \quad \text{others} \end{cases} \tag{4}$$

where d_3 is a preset reference distance in an overtaking situation; $d_v(p, p_L)$ denotes the vertical distance between the OS and the longitudinal centerline of the TS, which is a positive value.

This paper aims to find a path with the gradient of steepest descent in the total potential field. Here, the total potential field at the ith moment is used for analysis, and the ship is assumed to be within the influence range of the N_t target ships. The total potential field $P_{f,i}$ at the ith moment is expressed as follows:

$$P_{f,i} = P_{g,i} + \sum_{n=1}^{N_t} P_{t,i,n} \tag{5}$$

where $P_{g,i}$ is the potential field of the goal at the ith moment; $P_{t,i,n}$ is the potential field of the target ship n at the ith moment.

2.3. Coordinated Ship Domain

Over the past decades, the research on the ship domain of marine traffic engineering has received much attention because the ship domain plays a vital role in the navigational safety of ships. The concept of the ship domain was first proposed as an "effective domain" [29], which means the domain around a ship under way, which most navigators of the following ships would avoid invading. Then, the ship domain was further regarded as the effective area around a ship, which a navigator would like to keep free for other ships and stationary obstacles [30]. Through years of progress, a large number of ship domains have been presented for various purposes, which were classified into three classes in a geometrical manner, i.e., circle, ellipse and polygon.

The shape and size of the ship domain model depend mainly on the characteristics of the ship, such as length and speed. However, the difference in the size of the ship domain between the two vessels can lead to a different identification of the danger level. Therefore, the coordinated ship domain is applied [26], which considers the cooperation between ships and the influence of the ship's advance on the setting of a safe distance. In addition, the distance from the middle of the ship to the bow of the ship is also a major factor and should be considered when ships meet at close quarters. Hence, the radius of the ship domain R_c is expressed as

$$R_c = D_{center} + TA_{\max} \tag{6}$$

where D_{center} denotes the distance between the center of the ship domain and the bow of the ship, which is equal to $0.5L_{\max}$; $TA_{\max}$ denotes the maximum ship length of the universal advance and the tactical diameter based on turning circle maneuver data, which is equal to $4.0L_{\max}$. Therefore, the size of the ship domain can be expressed as

$$R_c = 0.5L_{\max} + 4L_{\max} = 4.5L_{\max} \tag{7}$$

where $L_{\max}$ is the maximum length of the ships, which are involved in an encounter situation, which is expressed as follows

$$L_{\max} = \max(L_1, L_2, \ldots, L_n), (n = 1, 2, \ldots, N) \tag{8}$$

When multiple ships are meeting on the open sea, the coordinated ship domain is applied to the own ship and the target ships simultaneously to ensure passage at a safe distance, i.e., multiple meeting ships use the same ship domain.

3. Collision Avoidance Decision-Making Support Scheme Based on IAPF and NMPC

3.1. Maneuvering Modeling Group (MMG) Model

Figure 1 shows the body-fixed coordinate system o-xyz within the space-fixed coordinate system o_0-$x_0y_0z_0$, and the origin o is located in the mid-ship of the ship.

The variations in the ship heading angle ψ and ship position (x, y) in the space-fixed coordinate system are expressed as

$$\begin{cases} \dot{x} = u\cos\psi - v\sin\psi \\ \dot{y} = u\sin\psi + v\cos\psi \\ \dot{\psi} = r \end{cases} \tag{9}$$

$$v = v_m + x_G r \tag{10}$$

where u, v and r denote the surge velocity, sway velocity at the gravity and yaw rate, respectively; v_m and x_G are the sway velocity at mid-ship and the distance from the center of gravity to the mid-ship.

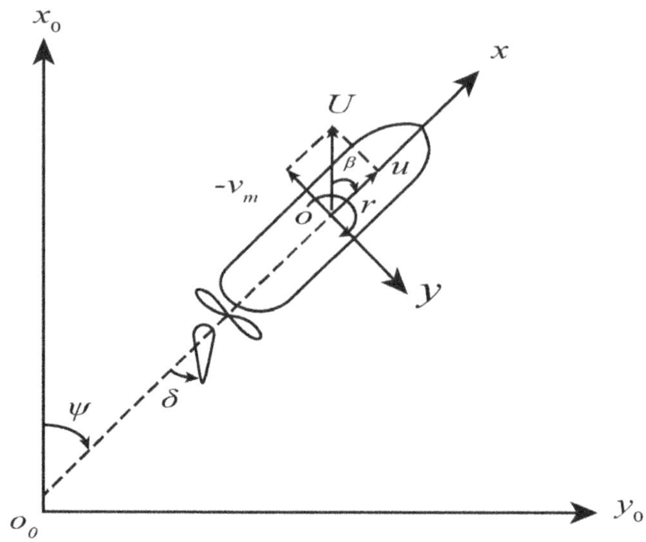

Figure 1. Coordinate systems.

Concretely, the standard 3 DOF MMG motion equations are defined as follows:

$$\begin{cases} (m+m_x)\dot{u}-(m+m_y)v_m r - x_G m r^2 = X_H + X_R + X_P \\ (m+m_y)\dot{v}_m + (m+m_x)ur + x_G m\dot{r} = Y_H + Y_R \\ (I_{zG}+x_G^2 m + J_z)\dot{r} + x_G m(\dot{v}_m + ur) = N_H + N_R \end{cases} \tag{11}$$

where m, m_x, m_y, J_z and I_{zG} are the ship's mass, the ship's added mass in the o–x axis direction, the added mass in the o–y axis direction, the added moment of inertia around the mid-ship and the moment of inertia of the ship around the center of gravity, respectively. X, Y and N are the surge force, lateral force and yaw moment around the mid-ship; the subscripts H, R and P denote the ship hull, rudder and propeller, respectively. The force and the moment acting on the hull X_H, Y_H and N_H are defined as follows:

$$\begin{cases} X_H = (1/2)\rho L_{pp} d U^2 X'_H(v'_m, r') \\ Y_H = (1/2)\rho L_{pp} d U^2 Y'_H(v'_m, r') \\ N_H = (1/2)\rho L_{pp}^2 d U^2 N'_H(v'_m, r') \\ U = \sqrt{u^2 + v_m^2} \end{cases} \tag{12}$$

where ρ, d, L_{pp} and U denote the water density, ship draft, ship length between the perpendiculars and the resultant velocity, respectively. v'_m is nondimensionalized by v_m/U, and r' is nondimensionalized by rL_{pp}/U. The force acting on the propeller X_P is defined as

$$\begin{cases} X_p = (1-t_P)\rho n_p^2 D_p^4 (k_{t2}J_p^2 + k_{t1}J_p + k_{t0}) \\ J_P = u\left[1 - w_{p0}\exp(C_0(\beta - x'_p r')^2)\right]/n_p D_p \\ \beta = \arctan(-v_m/u) \end{cases} \tag{13}$$

where δ, t_p, n_p, D_p and J_p denote the rudder angle, thrust deduction factor, propeller revolution, propeller diameter and propeller advanced ratio, respectively. k_{t0}, k_{t1}, k_{t2} are the propeller thrust open water characteristic coefficients; w_{p0}, C_0, β and x_p denote the wake coefficient of the propeller in a straight moving direction, the experimental constant, the drift angle and the longitudinal coordinate of the propeller position, respectively. The force and the moment acting on the rudder X_R, Y_R and N_R are defined as follows:

$$\begin{cases} X_R = -0.5(1-t_R)\rho A_R(u_R^2+v_R^2)f_\alpha \sin\alpha_R \sin\delta \\ Y_R = -0.5(1-a_H)\rho A_R(u_R^2+v_R^2)f_\alpha \sin\alpha_R \cos\delta \\ N_R = -0.5(x_R+a_H x_H)\rho A_R(u_R^2+v_R^2)f_\alpha \sin\alpha_R \cos\delta \\ \alpha_R = \delta - \tan^{-1}(v_R/u_R) \approx \delta - v_R/u_R \end{cases} \quad (14)$$

where t_R, a_H, x_H and x_R denote the steering resistance deduction factor, the increase factor of the rudder force, the distance from the additional lateral force component to the mid-ship and the distance from the rudder to the mid-ship, respectively. A_R, α_R, f_α, u_R and v_R represent the profile area, the effective inflow angle, the lift gradient coefficient, the longitudinal and lateral inflow velocity of the rudder, respectively.

Meanwhile, consider the actual performance of the rudder as follows

$$\dot{\delta} = -(1/T_E)\delta + (1/T_E)\delta_E \quad (15)$$

where T_E is the steering factor; δ_E denotes the order angle of the steering gear and also represents the actual input of the system.

3.2. Nonlinear Model Predictive Control Design

It can be seen from Equation (11) that the maneuvering equations are a system of first-order ordinary differential equations. The common numerical methods for solving such problems include Euler, Runge–Kutta, linear multi-step, etc. The Runge–Kutta method is commonly used in an actual calculation because it can improve the order of the algorithm to meet the different accuracy requirements [31]. Before solving the equation, combine the aforementioned maneuvering equations into a new system of ordinary differential equations, and transform it into the following form:

$$\begin{cases} \dot{u} = f_1(t,u,v_m,r,\delta,\psi) \\ \dot{v}_m = f_2(t,u,v_m,r,\delta,\psi) \\ \dot{r} = f_3(t,u,v_m,r,\delta,\psi) \\ \dot{\delta} = f_4(t,u,v_m,r,\delta,\psi) \\ \dot{\psi} = f_5(t,u,v_m,r,\delta,\psi) \\ \dot{x} = f_6(t,u,v_m,r,\delta,\psi) \\ \dot{y} = f_7(t,u,v_m,r,\delta,\psi) \end{cases} \quad (16)$$

If the value of each variable is known at time t_i, then at time $t_{i+1} = t_i + \tau$; the standard four-order Runge–Kutta formula for calculating each variable is

$$\begin{cases} u_{i+1} = u_i + \tau(K_{11}+2K_{12}+2K_{13}+K_{14})/6 \\ v_{m\ i+1} = v_{m\ i} + \tau(K_{21}+2K_{22}+2K_{23}+K_{24})/6 \\ r_{i+1} = r_i + \tau(K_{31}+2K_{32}+2K_{33}+K_{34})/6 \\ \delta_{i+1} = \delta_i + \tau(K_{41}+2K_{42}+2K_{43}+K_{44})/6 \\ \psi_{i+1} = \psi_i + \tau(K_{51}+2K_{52}+2K_{53}+K_{54})/6 \\ x_{i+1} = x_i + \tau(K_{61}+2K_{62}+2K_{63}+K_{64})/6 \\ y_{i+1} = y_i + \tau(K_{71}+2K_{72}+2K_{73}+K_{74})/6 \end{cases} \quad (17)$$

where τ is the time of each calculation step. The value of K_{ji} in the formula is calculated as follows:

$$\begin{cases} K_{ji} = f_j(t_i,u_i,v_{m\ i},r_i,\delta_i,\psi_i) \\ K_{j,i+1} = f_j(t_i+\frac{\tau}{2},u_i+\frac{\tau}{2}K_{1i},v_{m\ i}+\frac{\tau}{2}K_{2i},r_i+\frac{\tau}{2}K_{3i},\delta_i+\frac{\tau}{2}K_{4i},\psi_i+\frac{\tau}{2}K_{5i}) \\ K_{j,i+2} = f_j(t_i+\frac{\tau}{2},u_i+\frac{\tau}{2}K_{1,i+1},v_{m\ i}+\frac{\tau}{2}K_{2,i+1},r_i+\frac{\tau}{2}K_{3,i+1},\delta_i+\frac{\tau}{2}K_{4,i+1},\psi_i+\frac{\tau}{2}K_{5,i+1}) \\ K_{j,i+3} = f_j(t_i+\tau,u_i+\tau K_{1,i+2},v_{m\ i}+\tau K_{2,i+2},r_i+\tau K_{3,i+2},\delta_i+\frac{\tau}{2}K_{4,i+2},\psi_i+\tau K_{5,i+2}) \end{cases} \quad (18)$$

According to Equations (15)–(18), the general form of the ship's discrete-time nonlinear system can be expressed as follows:

$$\xi(i+1) = f(\xi(i)) + g_1\delta_E(i) \tag{19}$$

where $\xi(i) = [u(i), v_m(i), r(i), \delta(i), \psi(i), x(i), y(i)]^T$ represents the vector of the ship's state at the ith moment; $g_1 = 1/T_E$ stands for control gain, which is a constant. Define the predictive horizon N_p, the control horizon N_c, and $N_c \leq N_p$. According to the current state, the predicted state in the future N_p horizon can be obtained as follows:

$$\begin{cases} \xi(i+1|i) = f(\xi(i)) + g_1\delta_E(i) \\ \xi(i+2|i) = f(\xi(i+1|i)) + g_1\delta_E(i+1) \\ \quad\vdots \\ \xi(i+n|i) = f(\xi(i+n-1|k)) + g_1\delta_E(i+n-1) \\ \quad\vdots \\ \xi(i+N_c|i) = f(\xi(i+N_c-1|i)) + g_1\delta_E(i+N_c-1) \\ \quad\vdots \\ \xi(i+N_p|i) = f(\xi(i+N_p-1|i)) + g_1\delta_E(i+N_c-1) \end{cases} \tag{20}$$

where $\xi(i + n \mid i)$ is the vector of the predicted state at $i + n$ using the state $\xi(i)$ at the i moment. When $n \geq N_c$, $\delta_E(i + n) = \delta_E(i + N_c - 1)$. Thus, the optimal control sequence can be expressed as $U_c = [\delta_E(i), \delta_E(i + 1), \ldots, \delta_E(i + n - 1), \ldots, \delta_E(i + N_c - 1)]^T$, and only the first control action in the control sequence is finally applied to the plant in MPC applications [32].

In order to consider the influence of ship maneuverability in the process of collision avoidance, this paper combines the NMPC and APF methods to transform the ship motion planning problem into an optimization problem with constraints. Only the constraint of control input is considered here, i.e., the constraints of the rudder angle and rudder deflection speed are as follows:

$$\begin{cases} -35^\circ \leq \delta_i, \delta_{E\ i} \leq 35^\circ \\ -3^\circ/s \leq \dot{\delta} \leq 3^\circ/s \end{cases} \tag{21}$$

The process of solving optimization problems with the NMPC method is as follows. First, the objective function within the prediction horizon is obtained by predicting the future state of the ship from the current ship state. Then, the objective function is optimized to obtain the optimal input sequence at the current time. Moreover, select the first element of the optimal input sequence as system input at the current time to calculate the ship's state at the next time. Finally, repeat the above steps. In this process, combined with the improved APF, the objective function in the NMPC is defined as follows:

$$J = \sum_{n=1}^{N_p} P_{f,i}^n \tag{22}$$

where $P_{f,i}^n$ is the predictive total potential field at the ship's location at the moment i within the predictive horizon n, which is detailed in Equation (5).

4. Simulation and Analysis

In this section, two simulations are conducted to show the effectiveness of the presented collision avoidance decision-making support scheme. Simulation 1 is a multi-ship meeting scenario where only the own ship takes action to avoid other ships. Before this, different collision avoidance action distances are set for different encounter situations between two ships, and collision avoidance actions are taken only when two ships are close to the set distance. Simulation 2 is also a multi-ship encounter scenario where the own ship considers the risk of collision with all target ships and takes appropriate actions to avoid all target ships at once. Then, the proposed scheme and an existing method [12] are simulated, respectively, under the same scenario, and the advantages of the proposed scheme are analyzed based on the simulation results. The existing method, which is a COLREGs-compliant guidance strategy based on improved APF, can choose a suitable heading according to the change in the total potential field. Therefore, the heading and the trajectory obtained by the two methods are compared in the simulation.

The benchmark ship KVLCC2 tanker is selected as a sample ship from multiple meeting ships. The data of the sample ship [24] are given in Tables 1 and 2, respectively.

Table 1. Principal particulars of KVLCC2 tanker.

Symbol	Quantity	Symbol	Quantity
L_{pp} (m)	320.0	C_b	0.81
B (m)	58.0	D_P (m)	9.86
d (m)	20.8	H_R (m)	15.8
Displacement (m^3)	312,600	A_R (m)	112.5
x_G (m)	11.2		

Table 2. Hydrodynamics force coefficients of sample ship.

Symbol	Quantity	Symbol	Quantity	Symbol	Quantity
R'_0	0.022	N'_R	−0.049	w_{p0}	0.35
X'_{vv}	−0.040	N'_{vvv}	−0.030	C_0	−2.1
X'_{vr}	0.002	N'_{vvr}	−0.294	X'_p	−0.48
X'_{rr}	0.011	N'_{vrr}	0.055	t_R	0.387
X'_{vvvv}	0.771	N'_{rrr}	−0.013	a_H	0.312
Y'_v	−0.315	M'_x	0.022	X'_H	−0.464
Y'_R	0.083	M'_y	0.223	γ_R ($\beta_R < 0$)	0.395
Y'_{vvv}	−1.607	J'_z	0.011	γ_R ($\beta_R > 0$)	0.640
Y'_{vvr}	0.379	t_p	0.220	L'_R	−0.710
Y'_{vrr}	−0.391	k_{t0}	0.2931	ε	1.09
Y'_{rrr}	0.008	k_{t1}	−0.2753	κ	0.50
N'_v	−0.137	k_{t2}	−0.1385	f_α	2.747

Simulation 1. *In this simulation, a two-dimensional Cartesian coordinate system in nautical miles (nm) is utilized to express the position of the goal point and the ship. In the simulation studies, all ships sailed in open water with good visibility, and the initial states of the OS and the TS are shown in Table 3. The initial sway velocity, yaw rate and rudder angle of all ships are set as zero. To assess the collision risk, Table 3 lists the initial DCPA between the OS and the TS calculated from the ship's initial position, course and velocity. Based on the data in Table 1, the coordinated ship domain* $R_c = 4.5L_{max} = 0.78$ *nm. The setting of a safe distance is closely related to the visibility conditions at sea, and the safe distance can be set to 1 nm when the visibility is good [33]. Combining the size of the ship domain, set the safe distance as* $d_{safe} = 1.6$ *nm in this simulation. Set different distances to take avoiding action for different encounter situations between the OS and the TS, e.g., as 6 nm for a head-on situation, as 4 nm for a crossing situation and as 3 nm for an overtaking situation. The primary simulation parameters are shown in Table 4.*

Table 3. The initial state of the own ship and target ships in Simulation 1.

	Position	Course (°)	Surge Velocity (kn)	DCPA (nm)
OS	(10,01)	000	15.5	
TS1	(13,05)	270	15.5	0.71
TS2	(10,14)	170	15.5	1.13
TS3	(14,08)	346	5.5	3.06

Table 4. The significant parameters in the simulation.

Description	Notations	Value
Reference distance in a head-on situation (nm)	d_1	2
Reference distance in a crossing situation (nm)	d_2	2
Reference distance in an overtaking situation (nm)	d_3	2
Attractive potential field coefficient	k_{att}	5
Repulsive potential field coefficient	k_{rep}	200
Time of each calculation step (s)	τ	5
Predictive horizon	N_p	10
Control horizon	N_c	8
Steering factor	T_E	2.5

The simulation results are shown in Figures 2–6. Figure 2 shows the complete motion trajectories of four ships, indicating the position and the ship domain at the moment of closest proximity between the own ship and the other target ships. The curve represents the trajectory of the OS, and the other straight lines represent the trajectory of the TS. In addition, it can be observed that there is no collision risk between the target ships. Figure 2a shows the trajectories generated by the scheme based on the IAPF and NMPC methods, considering ship maneuverability. The trajectory in Figure 2b is generated by an existing method, ignoring ship maneuverability. Combined with the variation diagram of the own ship's heading angle in Figure 3, it can be seen that the trajectory and the heading angle of the OS in Figures 2b and 3b are not sufficiently smooth compared with those in Figures 2a and 3a, and this difference is caused by whether the ship maneuverability is considered. In addition, there is an apparent course oscillation in Figure 3b caused by the sudden appearance or disappearance of the repulsive force at the edge of the TS's potential field. NMPC will predict the potential field information for a period of time in the future and choose the path with the gradient of steepest descent in the total potential field to avoid repeated disappearance and appearance of the repulsive force; consequently, there is no course oscillation in Figure 3a. As observed in the figure, there is a crossing situation between the OS and TS1, a head-on situation between the OS and TS2 and an overtaking situation between the OS and TS3. According to the requirement of COLREGs, the OS takes an alternative course of action to avoid TS1, TS2 and TS3 in a sequence. There is no crossing of the ship domains, which shows that the collision avoidance effect is good. However, when the OS avoids TS1, it does not take into account the collision risk with TS2 and TS3, resulting in frequent steering maneuvers.

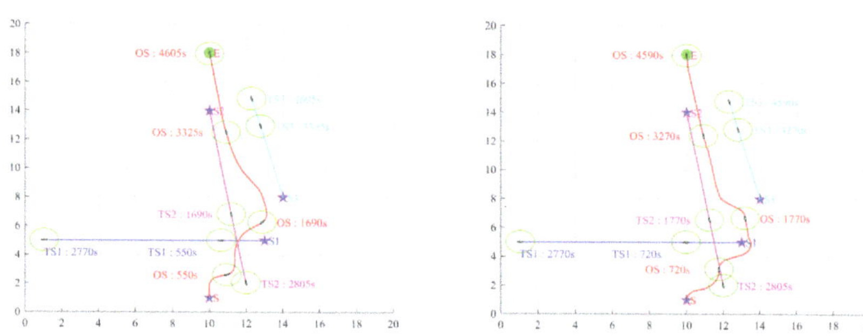

(**a**) (**b**)

Figure 2. Results of multi-ship, multiple encounter scenarios' collision avoidance validation in Simulation 1: (**a**) result of the proposed method; (**b**) result of the existing method.

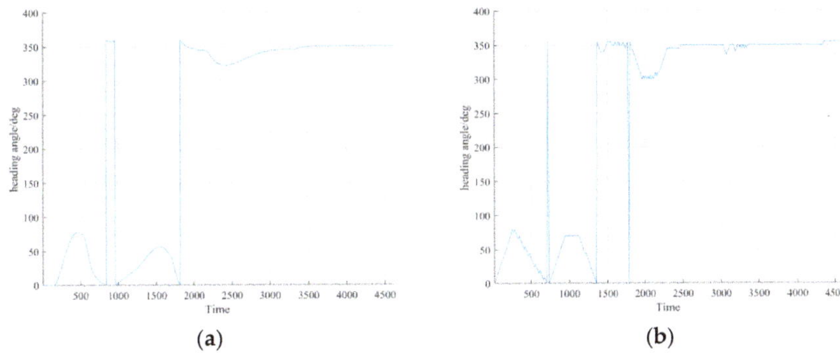

Figure 3. Variation diagram of the own ship's heading angle in Simulation 1: (**a**) result of the proposed method; (**b**) result of the existing method.

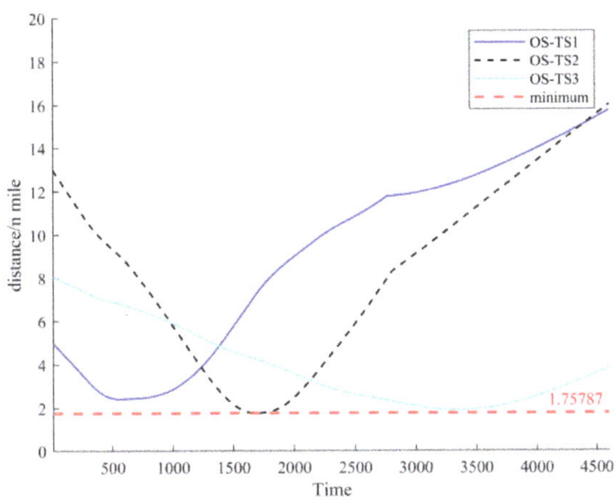

Figure 4. Variation diagram of the distance between the own ship and target ships in Simulation 1.

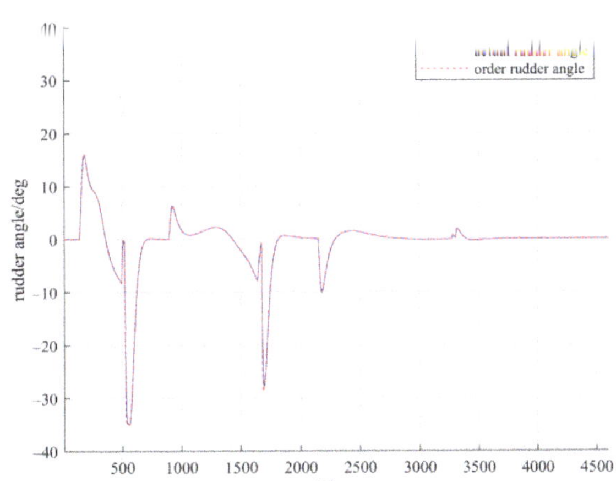

Figure 5. Variation diagram of the rudder angle of the own ship in Simulation 1.

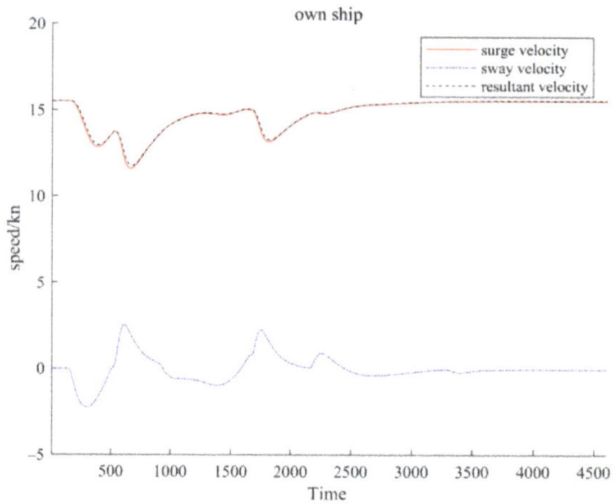

Figure 6. Variation diagram of the velocity of the own ship in Simulation 1.

Figure 4 shows the variation in the distance between the OS and the TS. Obviously, combined with Figure 2, the minimum distance between the OS and the TS is higher than the safe distance d_{safe}, which reflects that the coordinated ship domain is more intuitive for evaluating the effect of collision avoidance. It can be seen from Figure 5 that the change in the rudder angle of the OS is always within the prescribed range in advance, which is also the most intuitive manifestation of ship maneuverability constraints. The variation diagrams of the OS's surge, sway and resultant velocity are drawn in Figure 6, and it can be easily seen that the surge, sway and resultant velocity vary with the change in the rudder angle, which reflects the maneuvering characteristics of the ship.

Simulation 2. *In this simulation, there is a risk of collision between the OS and all other ships, and there is no risk of collision between the target ships. As shown in Table 5, the initial states of the OS and the TS are given, as well as the initial DCPA between the OS and the TS. Here, the distance to take collision avoidance action is enlarged, so that the OS takes collision risk with all target ships into account when taking avoidance action. The repulsive potential field coefficient is set as* $k_{rep} = 800$*, and the other parameters and values are configured identically to those used in Simulation 1.*

Table 5. The initial state of the own ship and target ships in Simulation 2.

	Position	Course (°)	Surge Velocity (kn)	DCPA (nm)
OS	(01,01)	045	15.5	
TS1	(04,05)	045	5.5	0.71
TS2	(15,02)	315	14.5	0.53
TS3	(13,11)	225	15.5	1.41
TS4	(04,16)	135	8.5	1.32

Figure 7 shows the complete motion trajectories of the OS and four target ships and marks the moment and the position of the closest distance between the OS and the TS. Figure 7a shows the trajectories generated by the scheme based on the IAPF and NMPC methods, considering ship maneuverability. The trajectory in Figure 7b is generated by an existing method, ignoring ship maneuverability. The own ship's heading angle changes during collision avoidance using the two methods are described in Figure 8a,b, respectively. The reasons for the differences in the OS trajectory and heading angle in Figures 7 and 8 are consistent with those mentioned in Simulation 1. Combined with the variation in the distance between the OS and the TS depicted in Figure 9, it can be seen that all target ships are successfully avoided by relying only on the OS to take actions. From the perspective of the OS's trajectory, the OS adopts fewer course alterations to avoid multiple target ships. This suggests that increasing the distance of collision avoidance actions can simultaneously consider the collision risk with more target ships, effectively reducing the number of course alterations required by the OS. Meanwhile, as shown in the rudder angle change diagram in Figure 10, the maneuvering amplitude adopted by the OS in the collision avoidance process is also small. Figure 11 displays the variation diagrams of the OS's surge, sway and resultant velocity, which vary with the change in the rudder angle, indicating the ship's maneuvering characteristics.

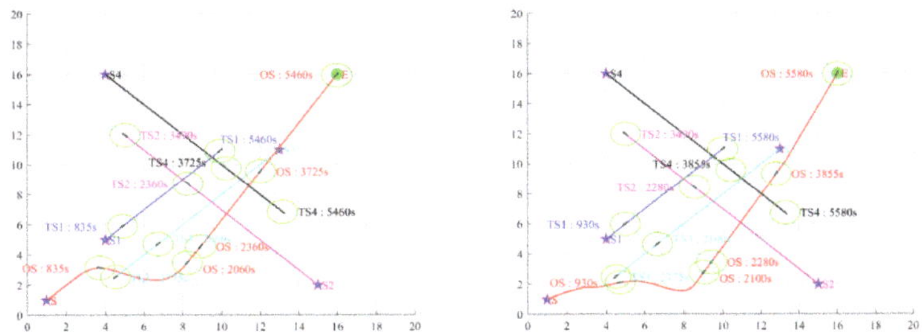

(a) (b)

Figure 7. Results of multi-ship, multiple encounter scenarios' collision avoidance validation in Simulation 2: (**a**) result of the proposed method; (**b**) result of the existing method.

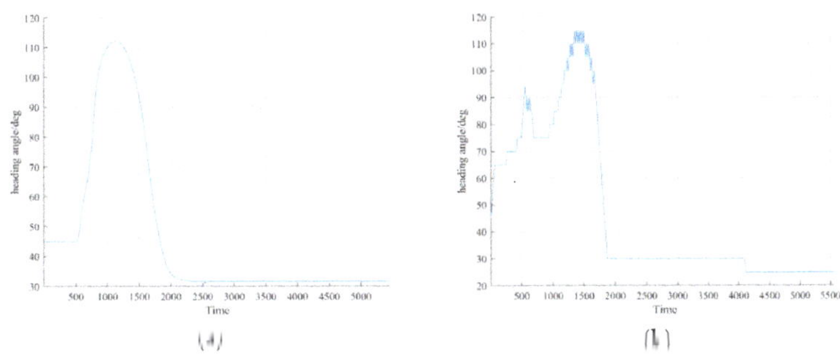

Figure 8. Variation diagram of the own ship's heading angle in Simulation 2: (**a**) result of the proposed method; (**b**) result of the existing method.

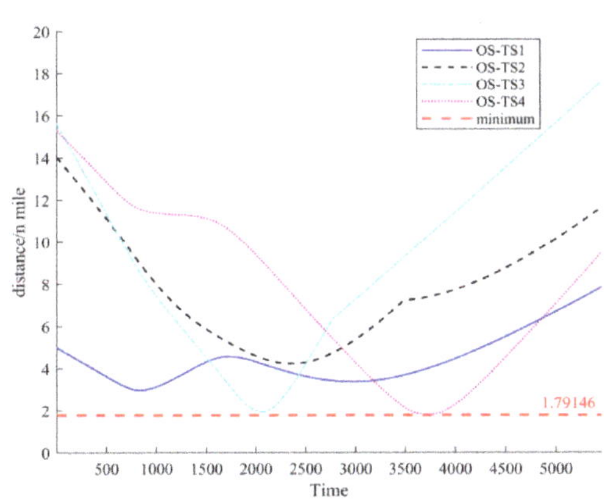

Figure 9. Variation diagram of the distance between the own ship and target ships in Simulation 2.

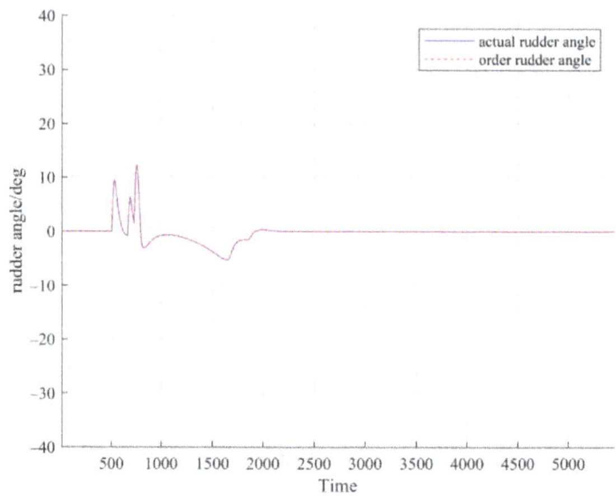

Figure 10. Variation diagram of the rudder angle of the own ship in Simulation 2.

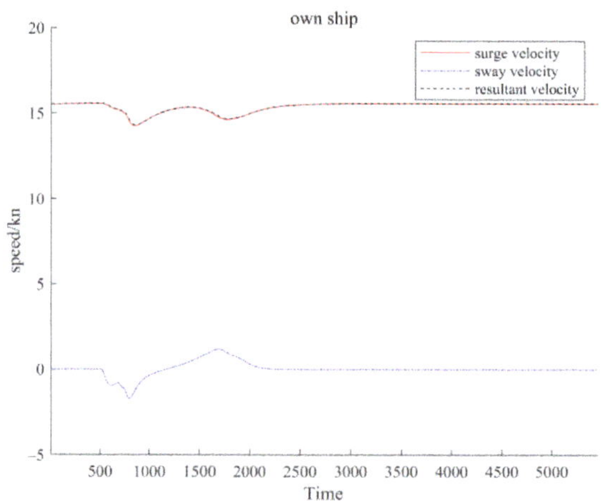

Figure 11. Variation diagram of the velocity of the own ship in Simulation 2.

5. Conclusions

In this paper, a collision avoidance decision-making support scheme based on an improved artificial potential field method and nonlinear model predictive control algorithm is proposed. In order to make the ship's collision avoidance actions conform with the requirements of COLREGs, the repulsive potential fields of the ships in head-on, crossing and overtaking encounter situations are improved, respectively. Meanwhile, the coordinated ship domain is applied to the own ship and target ships simultaneously to ensure passage at a safe distance. The 3 DOF MMG model is utilized and transformed with the Runge–Kutta method to obtain the discrete-time nonlinear system, and then, the collision avoidance decision-making scheme is designed by combining the IAPF and NMPC methods. Moreover, two sets of simulation results show that the collision avoidance path planned by the proposed scheme not only complies with the requirements of COLREGs but also considers the maneuvering motion characteristics of the ship. Compared with the trajectory generated by an existing method, the trajectory obtained by the proposed scheme is smoother, and the proposed scheme overcomes the deficiency of course oscillation in traditional APF.

In addition, this paper does not consider the external disturbances and system uncertainties, which will be further studied to improve the practicability of the planned path in the future.

Author Contributions: Conceptualization, H.L. and X.W.; methodology, H.L. and T.W.; software, H.L.; validation, T.W. and H.L.; formal analysis, H.L. and T.W.; investigation, H.L. and T.W.; resources, X.W. and S.N.; writing—original draft preparation, H.L.; writing—review and editing, H.L., X.W. and S.N.; supervision, X.W. and S.N.; project administration, X.W. and S.N.; funding acquisition, X.W. and S.N. All authors have read and agreed to the published version of the manuscript.

Funding: This research was funded by the National Natural Science Foundation of China, grant numbers 51909022 and 61976033, and by the China Postdoctoral Science Foundation, grant number 2022M710572.

Institutional Review Board Statement: Not applicable.

Informed Consent Statement: Not applicable.

Data Availability Statement: The data used to support the findings of this study are included within the article and are also available from the corresponding authors upon request.

Conflicts of Interest: The authors declare no conflict of interest.

References

1. Chauvin, C.; Lardjane, S.; Morel, G.; Clostermann, J.P.; Langard, B. Human and organisational factors in maritime accidents: Analysis of collisions at sea using the HFACS. *Accid. Anal. Prev.* **2013**, *59*, 26–37. [CrossRef] [PubMed]
2. Huang, Y.; Chen, L.; Chen, P.; Negenborn, R.R.; Van Gelder, P.H.A.J.M. Ship collision avoidance methods: State-of-the-art. *Saf. Sci.* **2020**, *121*, 451–473. [CrossRef]
3. Ma, Y.; Zhao, Y.; Li, Z.; Yan, X.; Bi, H.; Królczyk, G. A new coverage path planning algorithm for unmanned surface mapping vehicle based on A-star based searching. *Appl. Ocean Res.* **2022**, *123*, 103163. [CrossRef]
4. Zhang, J.; Zhang, H.; Liu, J.; Wu, D.; Soares, C.G. A Two-Stage Path Planning Algorithm Based on Rapid-Exploring Random Tree for Ships Navigating in Multi-Obstacle Water Areas Considering COLREGs. *J. Mar. Sci. Eng.* **2022**, *10*, 1441. [CrossRef]
5. Chen, C.; Chen, X.Q.; Ma, F.; Zeng, X.J.; Wang, J. A knowledge-free path planning approach for smart ships based on reinforcement learning. *Ocean Eng.* **2019**, *189*, 106299. [CrossRef]
6. Khatib, O. The potential field approach and operational space formulation in robot control. In *Adaptive and Learning Systems: Theory and Applications*; Springer: Boston, MA, USA, 1986; pp. 367–377.
7. Pan, Z.; Zhang, C.; Xia, Y.; Xiong, H.; Shao, X. An improved artificial potential field method for path planning and formation control of the multi-UAV systems. *IEEE Trans. Circuits Syst. II Express Briefs* **2021**, *69*, 1129–1133. [CrossRef]
8. Huang, Y.; Ding, H.; Zhang, Y.; Wang, H.; Cao, D.; Xu, N.; Hu, C. A motion planning and tracking framework for autonomous vehicles [illegible] *IEEE Trans. Ind. Electron.* **2019**, *67*, 1376–1386. [CrossRef]
9. Lyu, H.; Yin, Y. Ship's trajectory planning for collision avoidance at sea based on modified artificial potential field. In Proceedings of the 2017 2nd International Conference on Robotics and Automation Engineering (ICRAE), Shanghai, China, 29–31 December 2017; pp. 351–357.
10. Lazarowska, A. A discrete artificial potential field for ship trajectory planning. *J. Navig.* **2020**, *73*, 233–251. [CrossRef]
11. Lyu, H.; Yin, Y. COLREGS-constrained real-time path planning for autonomous ships using modified artificial potential fields. *J. Navig.* **2019**, *72*, 588–608. [CrossRef]
12. Han, S.; Wang, L.; Wang, Y. A COLREGs-compliant guidance strategy for an underactuated unmanned surface vehicle combining potential field with grid map. *Ocean Eng.* **2022**, *255*, 111355. [CrossRef]
13. Li, L.; Wu, D.; Huang, Y.; Yuan, Z.M. A path planning strategy unified with a COLREGS collision avoidance function based on deep reinforcement learning and artificial potential field. *Appl. Ocean Res.* **2021**, *113*, 102759. [CrossRef]
14. Ni, S.; Liu, Z.; Huang, D.; Cai, Y.; Wang, X.; Gao, S. An application-orientated anti-collision path planning algorithm for unmanned surface vehicles. *Ocean Eng.* **2021**, *235*, 109298. [CrossRef]
15. Oh, S.R.; Sun, J. Path following of underactuated marine surface vessels using line-of-sight based model predictive control. *Ocean Eng.* **2010**, *37*, 289–295. [CrossRef]
16. Liu, C.; Wang, D.; Zhang, Y.; Meng, X. Model predictive control for path following and roll stabilization of marine vessels based on neurodynamic optimization. *Ocean Eng.* **2020**, *217*, 107524. [CrossRef]
17. Liu, C. Motion Control of Unmanned Surface Vehicles Based on Model Predictive Control. Ph.D. Thesis, Wuhan University of Technology, Wuhan, China, 2017.
18. Abdelaal, M.; Fränzle, M.; Hahn, A. Nonlinear Model Predictive Control for trajectory tracking and collision avoidance of underactuated vessels with disturbances. *Ocean Eng.* **2018**, *160*, 168–180. [CrossRef]
19. Liu, C.; Hu, Q.; Wang, X.; Yin, J. Event-triggered-based nonlinear model predictive control for trajectory tracking of underactuated ship with multi-obstacle avoidance. *Ocean Eng.* **2022**, *253*, 111278. [CrossRef]

20. Fahimi, F. Non-linear model predictive formation control for groups of autonomous surface vessels. *Int. J. Control* **2007**, *80*, 1248–1259. [CrossRef]
21. He, Z.; Chu, X.; Liu, C.; Wu, W. A novel model predictive artificial potential field based ship motion planning method considering COLREGs for complex encounter scenarios. *ISA Trans.* **2023**, *134*, 58–73. [CrossRef]
22. Sutulo, S.; Guedes Soares, C. Mathematical models for simulation of manoeuvring performance of ships. In *Marine Technology and Engineering*; Taylor & Francis Group: London, UK, 2011; pp. 661–698.
23. Fossen, T.I. *Guidance and Control of Ocean Vehicles*; Wiley: Chichester, UK, 1994.
24. Yasukawa, H.; Yoshimura, Y. Introduction of MMG standard method for ship maneuvering predictions. *J. Mar. Sci. Technol.* **2015**, *20*, 37–52. [CrossRef]
25. Wang, X.; Liu, Z.; Li, T. Collision avoidance dynamic support model for ships meeting at a close range. *J. Harbin Eng. Univ.* **2021**, *42*, 1256–1261.
26. Wang, X. The Research on the Safety of Ship Navigation on the Open Sea Based on Ship Maneuverability. Ph.D. Thesis, Dalian Maritime University, Dalian, China, 2017.
27. Hu, L.; Naeem, W.; Rajabally, E.; Watson, G.; Mills, T.; Bhuiyan, Z.; Raeburn, C.; Salter, I.; Pekcan, C. A multiobjective optimization approach for COLREGs-compliant path planning of autonomous surface vehicles verified on networked bridge simulators. *IEEE Trans. Intell. Transp. Syst.* **2019**, *21*, 1167–1179. [CrossRef]
28. International Maritime Organization. International Regulations for Preventing Collisions at Sea (COLREGs). Available online: http://www.imo.org/en/About/Conventions/ListOfConventions/Pages/COLREG.aspx (accessed on 20 October 2022).
29. Fujii, Y.; Tanaka, K. Traffic capacity. *J. Navig.* **1971**, *24*, 543–552. [CrossRef]
30. Goodwin, E.M. A statistical study of ship domains. *J. Navig.* **1975**, *28*, 328–344. [CrossRef]
31. Zhang, W. A Research on Numerical Prediction of Ship Maneuverability in Regular Waves. Ph.D. Thesis, Shanghai Jiao Tong University, Shanghai, China, 2016.
32. Qin, S.J.; Badgwell, T.A. A survey of industrial model predictive control technology. *Control Eng. Pract.* **2003**, *11*, 733–764. [CrossRef]
33. Lisowski, J. Multistage Dynamic Optimization with Different Forms of Neural-State Constraints to Avoid Many Object Collisions Based on Radar Remote Sensing. *Remote Sens.* **2020**, *12*, 1020. [CrossRef]

Journal of
Marine Science and Engineering

MDPI

Article

A Deep Learning Method for Ship Detection and Traffic Monitoring in an Offshore Wind Farm Area

Xintong Liu [1,†], Yutian Hu [1,†], Huiting Ji [1,†], Mingyang Zhang [2] and Qing Yu [1,3,*]

1 Maritime Risk & Behavioral Sciences Lab, School of Navigation, Jimei University, Xiamen 361021, China; 202212861022@jmu.edu.cn (X.L.); 202021011031@jmu.edu.cn (Y.H.); 202021012061@jmu.edu.cn (H.J.)
2 Marine Technology Group, Department of Mechanical Engineering, Aalto University, 00076 Espoo, FI, Finland; mingyang.0.zhang@aalto.fi
3 Hubei Key Laboratory of Inland Shipping Technology, Wuhan University of Technology, Wuhan 430063, China
* Correspondence: qing.yu@jmu.edu.cn
† These authors contributed equally to this work.

Abstract: Newly built offshore wind farms (OWFs) create a collision risk between ships and installations. The paper proposes a real-time traffic monitoring method based on machine vision and deep learning technology to improve the efficiency and accuracy of the traffic monitoring system in the vicinity of offshore wind farms. Specifically, the method employs real automatic identification system (AIS) data to train a machine vision model, which is then used to identify passing ships in OWF waters. Furthermore, the system utilizes stereo vision techniques to track and locate the positions of passing ships. The method was tested in offshore waters in China to validate its reliability. The results prove that the system sensitively detects the dynamic information of the passing ships, such as the distance between ships and OWFs, and ship speed and course. Overall, this study provides a novel approach to enhancing the safety of OWFs, which is increasingly important as the number of such installations continues to grow. By employing advanced machine vision and deep learning techniques, the proposed monitoring system offers an effective means of improving the accuracy and efficiency of ship monitoring in challenging offshore environments.

Keywords: traffic safety; offshore wind farms; YOLOv7; stereo vision; deep learning

Citation: Liu, X.; Hu, Y.; Ji, H.; Zhang, M.; Yu, Q. A Deep Learning Method for Ship Detection and Traffic Monitoring in an Offshore Wind Farm Area. *J. Mar. Sci. Eng.* **2023**, *11*, 1259. https://doi.org/10.3390/jmse11071259

Academic Editor: Eugen Rusu

Received: 14 April 2023
Revised: 2 June 2023
Accepted: 16 June 2023
Published: 21 June 2023

1. Introduction

Motivated by the demands for clean energy in the context of ongoing climate change, the number of offshore wind farms (OWFs) grows rapidly in lots of coastal countries [1]. The present development shows the advantages of offshore wind farms, for example, delivering secure, affordable, and clean energy while fostering industrial development and job creation. Based on the annual report from the Global Wind Energy Council, 2021 becomes the best year in the offshore wind industry, in which 21.1 GW of offshore wind capacity was commissioned, bringing global offshore wind capacity reach to 56 GW, three times more than in 2020. The GWEC Market Intelligence forecasts that 260 GW of new offshore wind capacity could be added during 2022–2030 under the current positive policies, allowing the total global offshore wind installations to rise from 23% growth in 2021 to at least 30% growth by 2031. In the year 2021, China constructed 80% of new offshore installations worldwide becoming the world's largest offshore market [2].

However, this trend of increasing numbers of offshore wind farms puts pressure on local water traffic management. For example, the newly added obstacles increase the difficulties for both the navigation of passing vessels [3] and of Search and Rescue. Once an accident happens, it results in water pollution, significant damage to facilities and other catastrophic casualties and economic losses. For the year 2021, the Global Offshore Wind Health and Safety Organisation (G+) report pointed out that there were a total of

204 high-potential incidents and injuries recorded [4]. Recently, some collision accidents were also reported for the UK, China, and the Netherlands, leading to shipping hull and turbine damage, and electric power loss, especially for construction and fishing ships. For instance, on 31 January 2021, the drifting bulk carrier Julietta D collided with a transformer platform in the Hollandse Kust Zuid wind farm, which was under construction. On 2 July 2022, a dragging accident in southern waters led to 25 casualties and a vessel sank. Therefore, monitoring vessels in the offshore wind farm waters and detecting potential hazards becomes an urgent question for stakeholders of offshore wind farms.

The current monitoring system for water traffic is the Vessel Traffic Services (VTS), which uses the Automatic Identification System (AIS), radar and other detection sensors to show the water traffic situation dynamically. The topics of technology development, information collection, communication, and system design have been studied for the VTS. However, the limitations of its use in offshore wind farm waters are noted due to trespasses and the deliberate turning off of the AIS. Developing a reliable monitoring system can aid the safety of navigation of passing vessels and offshore wind farms. To improve the reliability of using VTS in offshore wind farm water, previous studies proposed several novel methodologies to develop monitoring models. Li et al. reviewed 153 papers related to ship detection and classification and suggested that the current ship detection methods could be divided into two categories, feature design-based methods and Convolutional Neural Network (CNN) based methods [5]. For instance, Zhao et al. developed a coupled CNN model to detect small ships in waterways with a high density of shipping water using a clustered Synthetic Aperture Radar image approach [6]; Zhang et al. proposed a fast region-based convolutional neural network (RCNN) method to detect ships from high-resolution remote sensing images [7]. Using feature design-based methods, Hu et al. employed dilated rate selection and attention guided feature representation strategies for ship detection, in which the dilated convolution selection strategy was applied to a multi-branch extraction network, extracting context information of different receptive fields [8]. Yu et al. employed an advanced ship detection method using Synthetic Aperture Radar images based on the YOLOv5 [9]. The results of this study showed a great improvement in the feature expression ability and sensitivity for target detection after adding a coordinate attention mechanism and a feature fusion network structure. Priakanth et al. proposed a hybrid system by using wireless and IoT technology to avoid boundary invasions [10]. Ouelmokhtar et al. suggested using Unmanned Surface Vehicles (USV) to monitor waters, in which an onboard RADAR is used to detect the targets [11]. Nyman discussed the possibility of using satellites for monitoring, which allows for the visual surveillance of a large area [12]. The relatively low cost of data acquisition makes the use of satellites appealing, however, some prior theory or knowledge is needed to sort through the vast collection of satellite data and images. Although these studies have their advantages, these technologies may typically focus on large water areas, increasing the financial burden for offshore wind farms. Therefore, low-cost, high efficiency and reliable systems are still needed. The low cost of equipment makes video surveillance technologies a possible way to monitor water traffic at a close distance. However, original video surveillance requires human supervision to achieve continuous monitoring, consuming a large number of human resources. In addition, the challenges faced by offshore wind farm operators and managers in the environment and the high workload of their daily tasks may lead to error-prone and tedious situations. Nowadays, automated surveillance systems observing the environment utilizing cameras are widely used. These automatic systems can accomplish a multitude of tasks, which include detection, interpretation, understanding, recording, and creating alarms based on the analysis. They are widely used in different situations, for instance, in road traffic, Pawar and Attar got detection and localization of road accidents from traffic surveillance videos [13]. Vieira et al. and Thallinger et al. quantitatively evaluated the safety of traffic by utilizing security cameras [1,14]. A video surveillance-based system that can detect the pre-events, with an automatic alarm generated in the control room was proposed for improving road safety [15].

Motivated by the above-mentioned difficulties and advantages, this study aimed to pioneer the use of machine vision technologies to aid traffic monitoring in offshore wind farm waters. Specifically, a vision-based monitoring system was developed for OWFs to improve the reliability and efficiency of ship traffic detection and tracking. The system trains a "YOLOv7" machine visual model using automatic identification system (AIS) data. The fine-trained model can identify the passing ship in OWF waters, not only providing the identified target, but also the degree of confidence during the monitoring process. Then a stereo vision algorithm is applied in the model to locate and track the positions of the passing ships. So that the dynamic information (e.g., speed, course, and position) for each target can be provided by the system. In addition, the proposed system was validated by comparing the provided dynamic information with AIS data to ensure the reliability of the results. The contributions of this work are highlighted as follows. The study pioneers the uses of machine vision to aid traffic monitoring in offshore wind farm waters, covers the gap for the current VTS system and provides a novel way for offshore water traffic management. In addition, the model can provide more efficient and reliable dynamic data for individual and overall traffic that can be expanded to providing early warning for ship risk and accident prevention. The main contributions of this work are summarized as follows:

(1) This paper evaluates the performance of using the YOLOv7 model in offshore wind farm waters. It shows that the YOLOv7 model has a high accuracy rate for vessel detection. The results prove that the YOLOv7 models have higher dynamicity for ship detection compared to other ship detection methods.
(2) An optimization strategy for training a visual-based identification model is presented. The study collects hybrid data sources (e.g., AIS data, images) to develop the model training database and validate the model by comparing the positions between AIS data and the detection results. consequently, this ensures the performance of continuous target detection with significant accuracy.
(3) The proposed target identification system was further tested in an offshore case. By using an embedded device, the inference time reached real-time performance (less than 0.1 s) and the overall processing time for one frame was 0.76 s, proving the possibility of implementation of the system in real-time ship traffic monitoring.

The remainder of this paper is organized as follows: Section 2 outlines a review of related research. In Section 3, the framework of the system is introduced in detail. The experimental data and the test results are reported in Section 4. The obtained results are discussed in Section 5. Finally, conclusions are drawn in Section 6.

2. Literature Review

This section can be categorized into three groups based on the method used in the studies: ship monitoring, machine vision in target detections, and machine vision in target tracking, respectively.

2.1. Ship Monitoring Technology

Implemented to promote safe and efficient marine traffic, VTS are typically the most widely used technology for ship supervision. It is a shoreside service within a country's territorial waters that, by detecting and tracking the ship, aims to monitor the traffic, assist the traffic control and manage navigational matters, and provide support and required information for passing ships. The VTS collects dynamic data via two sensors, the radar and AIS; however, both have their limitations. For instance, the radar echo can be disturbed in an environment with the external noise of RF interference and clutter, which creates potentially dangerous situations and decrease VTS functionality. Regarding this matter, Root proposed high-frequency radar ship detection by clutter elimination [16]. Dzvonkovskaya et al. pioneered a new detection algorithm in ship detection and tracking but ignored the external influences position [17]. Another question concerns the limitations of detecting small ships in coastal waters (e.g., offshore wind farm water). From this point of view, Margarit et al. proposed a ship operation monitoring system based on Synthetic Aperture Radar image processing to achieve inferred ship monitoring and classification information, which further improves the Synthetic Aperture Radar image from the new sensor data [18]. Moreover, radar is unable to provide sufficient static information such as ship type and size, which means other systems (e.g., AIS) are needed. The AIS is another important piece of information for ship monitoring, which can compensate for the limitations of radar. It achieves the automatic exchange of ship information and navigation status between ships and shore. As a type of reliable data source, AIS data has been widely used in lots of studies to analyze ship traffic, making it important for water traffic management. For instance, Brekke et al. combined AIS data with satellite Synthetic Aperture Radar images to detect ship dynamic information (e.g., speed, course) [19]. To improve the reliability of radar, Stateczny collected sets of data from AIS and radar, and then applied a numerical model to compare the covariance between the two types of data [20]. Pan and Deng proposed a real-time monitoring system for shore-based monitoring of ship traffic [21]. Although the AIS data are valuable for ship traffic management, several limitations remain. For instance, the AIS can be closed manually. The AIS is not mandatory for some small ships such as dinghies and fishing boats [22].

In coastal waters, the offshore wind farm is a relatively new type of installation that influences the existing ship traffic, not only occupying the navigable waters but also creating blind areas by blocking radio signals, reducing the ability to detect and track small targets. Relevant studies using traditional data sources (e.g., AIS, radar), including Yu et al., used AIS to analyze the characteristics of the ship traffic in the vicinity of offshore wind farms [23] and then developed models to assess the risk for individual ships [24] or for the ship traffic flows [25]. However, as a highly accurate and reliable model for ship detection is still the basis for ship monitoring in offshore waters, traditional ship detection methods (e.g., AIS, radar) were updated to overcome the potential uncertainties, such as small target detection, visual monitoring, and unpredicted invasions.

Consequently, some vision-based technologies have been applied for ship detection, ship tracking, and ship monitoring. For instance, to overcome the difficulty of remote ship control and monitoring in high traffic waters, Liu et al. designed a portable integrated ship monitoring and commanding system [26]. To test the data availability, Shao et al. used image data captured from surveillance cameras to achieve target detection [27]. To improve the function of the target automatic monitoring and tracking, Chen et al. proposed a mean-shift ship monitoring and tracking system [28], which shows the possibility of using machine visual technologies for water traffic monitoring.

2.2. Applications of Machine Vision in Target Detections

Machine vision technology enables a machine with a visual perception system, with the help of hardware (e.g., camera, infrared thermal imager, night vision device) and a software program. It can recognize and manipulate the activities and perform image-based process control and surveillance for traffic monitoring, manufacture inspection, autopilot, and other scene perception usages [29,30]. A widely used application of machine vision is the Tesla driverless system, which was equipped with the hardware needed for Autopilot and the software program to realize Full Self-Driving.

Central to machine vision is the target recognition and detection algorithms. Figure 1 shows the development of recognition and detection algorithms that are used for object detection. The early studies of machine vision came up with Scale Invariant Feature Transform, which involved five steps to match the similarity of two images and to detect targets [31]. Then SIFT was upgraded to the Viola-Jones detection algorithm [32], histogram of oriented gradients (HOG) [33], Data Management Platform (DMP), and so on.

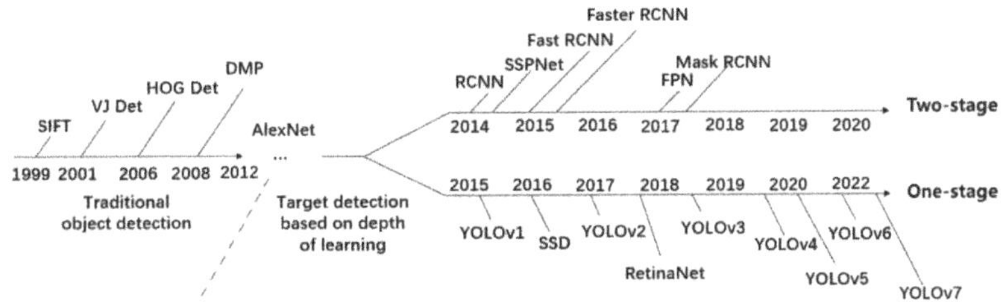

Figure 1. Development of machine vision technologies.

However, the above-mentioned algorithms extract target features manually, and can only perform well when they are guaranteed to extract sufficiently accurate features, so that they are inapplicable where a large number of targets exist. They are replaced by applying deep learning approaches to detect targets. Deep learning-based detection algorithms have the advantage of being able to extract features from complex images. The deep learning-based methods can be grouped into two categories based on the way they extract target features: the anchor-based methods (i.e., Convolutional Neural Network (CNN) methods and You Only Look Once (YOLO) methods) and the anchor-free methods (e.g., adaptively spatial feature fusion methods [34], Corner Net methods [35]). The Anchor-based algorithms are further classified into single-stage detection and two-stage detection. Due to stability and accuracy, anchor-based methods have become more popular in recent years. Typical methods include YOLOv1-v5 [16,36–40], single shot multibox detector [41], and Region CNN etc. For instance, Girshick et al. [42] proposed a novel method of R-CNN for target detection. The method uses image segmentation combining regions and CNNs to improve accuracy; however, it requires a large database to train the detection model, which reduces the detection speed. To improve detection speed, Meng et al. developed an improved Mask R-CNN, which ignores the RolAlign layer in the R-CNN [43]. Zhao et al. suggested enhancing the relationship among non-local features and refining the information on different feature maps to improve the detection performance of R-CNN [44]. Redmon and Farhadi proposed a joint training method to improve the traditional YOLOv1 model [45]. The upgraded YOLOv3 model uses binary cross-entropy loss and scale prediction to improve the accuracy of the model while ensuring the detection speed of detection [16]. The YOLOv3 model was adopted in vision detection studies including Gong et al. [46]. Patel et al. proposed a novel deep learning approach, which combines the capabilities of Graph Neural Networks (GNN) and the You Look Only Once (YOLOv7) deep learning framework; the results show an increased detection accuracy [47], Li et al. [48] and etc., which prove its fast speed in the convergence and detection process. The applications of vision detection have been done in various domains, as well as water traffic management. To design a deep learning-based detector for ship detection, Li et al. applied the Faster-CNN algorithm to train the ship target detection model, which achieves higher accuracy [49]. To address the shortcomings of the region proposal computation, Ren et al. introduced a region proposal network (RPN) by sharing the convolutional features of Fast R-CNN and RPN to further merge the two into one network [50].

2.3. Applications of Machine Vision in Target Position and Tracking

Machine vision methods used for target tracking can be categorized as monocular vision and binocular vision based on the tracking mechanism [51,52]. The monocular vision was first proposed by Davison [53], who used overall decomposition sampling to solve the challenge of real-time feature initialization. The basis of the monocular vision systems is the simultaneous localization and mapping (SLAM) method, which calculates the distance of the target within the camera's field of view. Although monocular visual localization is simple to operate and does not require feature point matching, it is less accurate and only suitable for specific environments. Therefore, it is not suitable for use in complex environments such as maritime target localization and tracking. To solve these problems, binocular vision positioning has been proposed and is widely used in many fields. However, binocular vision pre-localization only works on a flat surface and cannot accurately localize objects, so scholars have extended binocular vision to stereoscopic vision [54,55].

Binocular stereo vision technology can simulate the human eye to perceive the surrounding environment in three dimensions, making it widely used in various fields [56–59]. To reduce errors in the localization part of binocular stereo vision systems due to interference from complex environments, Zou et al. proposed a binocular stereo vision system based on a virtual manipulator and the localization principle [60]. They designed a binocular stereo vision measurement system to achieve an accurate estimation of target object positions. Zuo et al. used binoculars to capture point and line features and selected orthogonality as the minimum parameter for feature extraction, which solved the problem of unreliability of binocular stereo vision in detecting objects [61]. Therefore, compared to monocular vision techniques, binocular stereo vision is a more effective technique for target tracking. It is more accurate, simpler to operate, and suitable for dynamic environments, making binocular vision systems more appropriate for ship supervision in offshore wind farms than monocular vision systems. Video tracking allows for continuous monitoring of the ship, which is very helpful.

Based on the above studies, it can be noted that traditional approaches of using AIS and radar data for ship identification, tracking and localization have their drawbacks. For example, radar has difficulty in detecting small targets and fails to provide some important information (i.e., ship type). The AIS provides dynamic and static information for individual ships but it can be turned off manually, which cause difficulties in ship tracking and invasion detection. Hence, this study suggests a novel way for ship detection and tracking by using a visual-based target detection model which not only overcomes the difficulties of current ship detection and tracking approaches but also provides new ideas to enhance dynamic ship monitoring in coastal waters.

3. Methods

This section proposes a novel framework for ship detection and tracking in the waters near OWFs. The framework uses the vanilla YOLOv7 algorithm as a critical component to detect ships in dynamic situations while applying binocular stereo vision to track the ships. The framework structure is shown in Figure 2.

The framework consists of seven steps. (1) to collect real-time ship video from the waters in the vicinity of the OWFs; (2) to process the collected ship video and picture information; (3) to set the relevant parameters; (4) to construct the training database from the collected video and AIS data; (5) to train the ship detection model using the YOLOv7 approach; (6) to map the ship position from videos into the physical world with the aid of binocular stereo vision and (7) to validate and output the results. The details of those steps are introduced as follows.

Step 1: Cameras placed at wind turbines are used to provide real-time videos of ship traffic in the vicinity of offshore waters. Then, images containing ships are collected to develop the training databases. In the database, images that may have ships that can be overlapped with another image are deleted.

Step 2: This step defines the parameters used in the monitoring system. For the detection model (i.e., YOLOv7), the three parameters learning rate, momentum, and decay are used. The learning rate is a hyperparameter used by the optimizer to control the step size of the parameter update. It can influence the adjustment of the parameter value. Momentum is a regulating hyperparameter in the optimizer, which is used to control the direction and speed of parameter updates. By setting a suitable momentum hyperparameter, the parameter updates can be made more stable and avoid problems such as excessive oscillation or too slow descent during training. Decay or decay rate is a regulating hyperparameter in the optimizer that is used to make the learning rate gradually decrease. It allows for a more detailed search of the parameter space, making it easier for the model to reach a minimum and produce more accurate predictions.

In addition, the camera includes internal and external parameters. The internal parameters are focal, pixel, and resolution. They are mainly used as the basis to transform the ship's position from pixel coordinates to camera coordinates. The external parameters cover camera positions, and high, horizontal, and vertical angles, which convert the ship's position from camera coordinates to world coordinates. By calibrating the left and right cameras, those internal and external camera parameters are crucial for ship positioning.

Step 3: Establish the initial training samples for YOLO model training, which is used to obtain the object in the port videos (i.e., generate a bounding box for each ship in the video). Before training, the ship pictures need to be processed through annotation using an image annotation tool. Each ship in the picture is selected in this study, and the corresponding AIS data are input to develop the database. The database includes the position coordinates of the corners of the ship's box, as well as the width and height of the ship in the picture. In the training process, standard techniques such as multi-scale training, data augmentation, and batch normalization are used to train the ship detector.

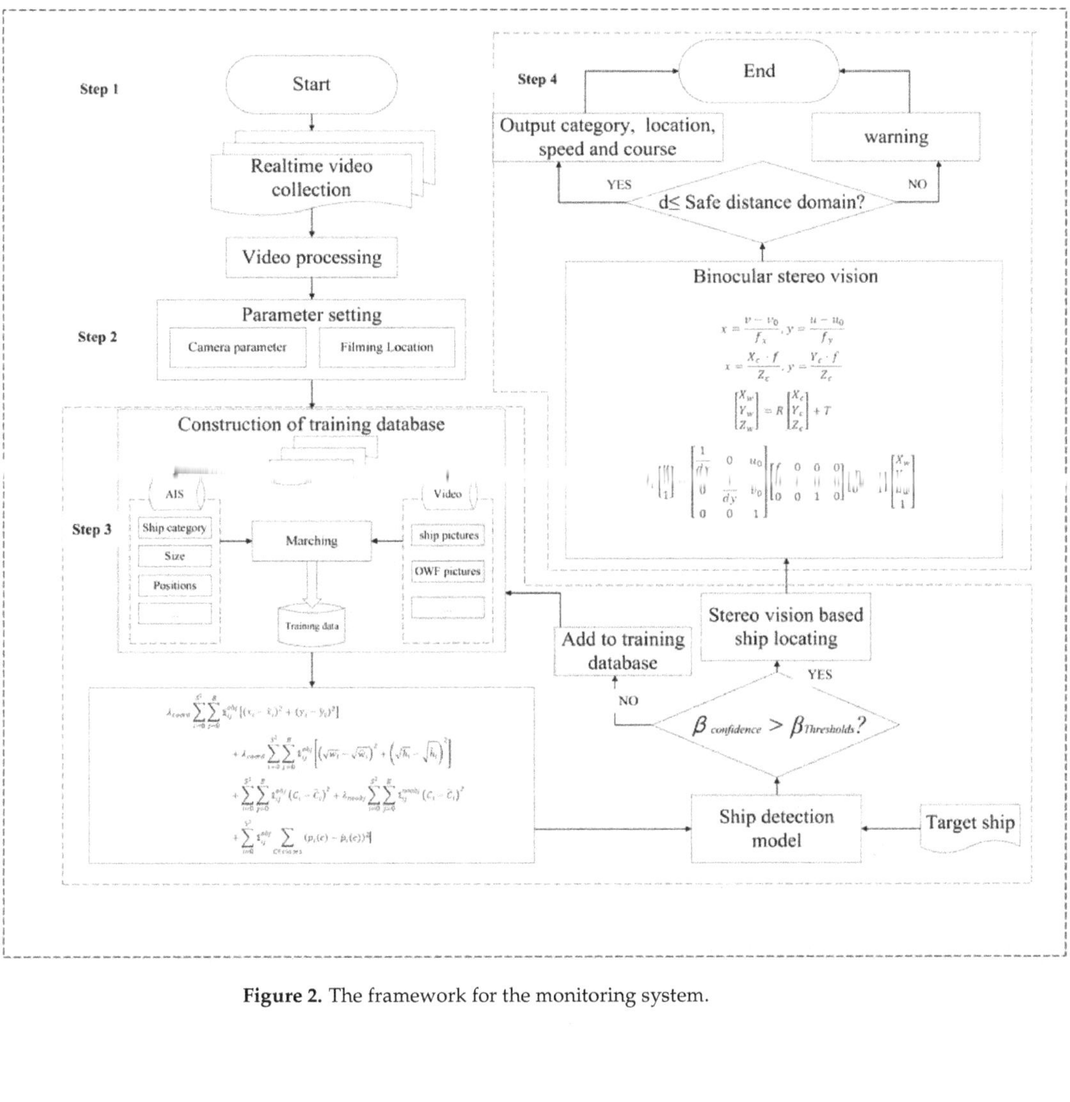

Figure 2. The framework for the monitoring system.

Figure 3 shows the training process of the YOLOv7 model, which consists of the backbone network, a convolutional feature fusion network, and the decoding processing.

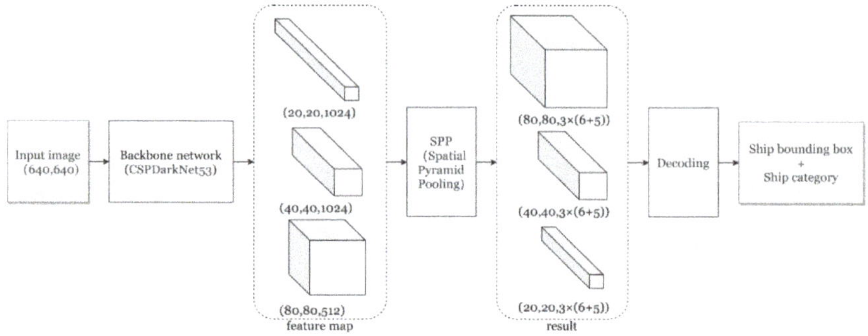

Figure 3. The steps to train the YOLOv7 model.

The YOLOv7 model consists of a backbone, neck, and a head. The input image is scaled to a size of 640 × 640 with a constant aspect ratio, and the blank part of the image is filled with gray. As shown in Figure 3, the input image is grided as 640 × 640 to detect the target. The backbone is a convolutional neural network responsible for extracting features from a given input image; the neck mainly generates a feature pyramid and passes various scales of features to the head, which generates prediction boxes and ship categories as outputs for each object. In addition, sensitivity and accuracy can be affected by the size of the grids. As shown in Figure 4, a scale of 20 × 20 is used for big target detection, a medium scale of 40 × 40 is used to detect the middle target, and a small scale of 80 × 80 for the small target.

(**a**) (**b**) (**c**)

Figure 4. The sizes of the grids to capture ships: (**a**) 20 × 20, large ships; (**b**) 40 × 40, medium ships; (**c**) 80 × 80, small ships.

Step 4: This study uses a 3D reconstruction technique called binocular stereo vision to achieve ship positioning. Stereo vision is a technique that involves detecting objects using two or more cameras. By simultaneously calibrating the left and right cameras, the internal and external parameters of both cameras can be determined. To obtain the position of the ship, the target coordinates are mapped from the video to the physical world using imaging principles, the whole structure is shown in Figure 5.

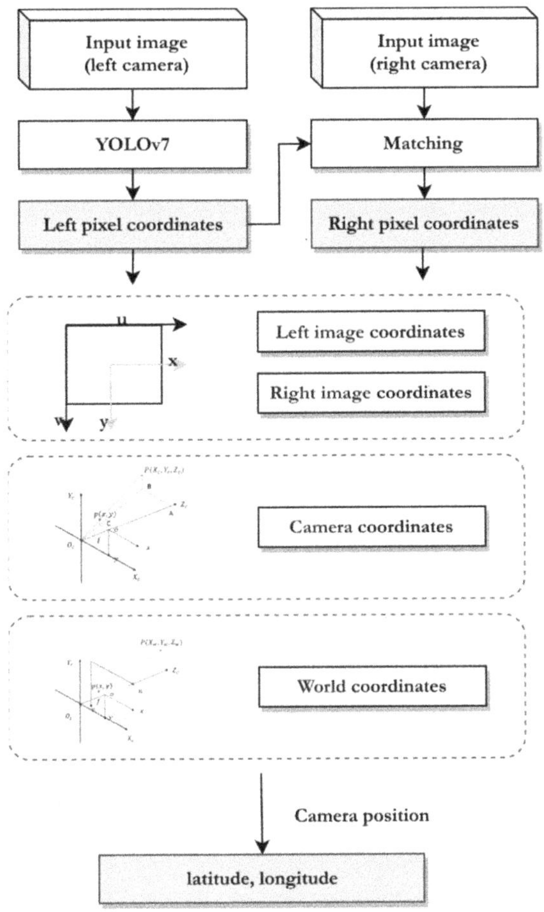

Figure 5. The steps for ship detection and positioning.

In the proposed system, the left pixel coordinate $\left(u_{left}, v_{left}\right)$ is obtained from the input image captured by the left camera into YOLOv7. Meanwhile, the right pixel coordinate $\left(u_{right}, r_{right}\right)$ is obtained after matching the image captured by the right camera and left camera.

The pixel point is denoted by $\boldsymbol{m} = [u, v]^T$. The world point is denoted by $\boldsymbol{M} = [X, Y, Z]^T$. We use $\widetilde{x}$ to denote the augmented vector by adding 1 as the last element: $\widetilde{\boldsymbol{m}} = [u, v, 1]^T$, $\widetilde{\boldsymbol{M}} = [X, Y, Z, 1]^T$. The relationship between the world point $\boldsymbol{M}$ and its pixel projection $\boldsymbol{m}$ is given by:

$$s\widetilde{\boldsymbol{m}} = \boldsymbol{A}[\boldsymbol{R}\ \boldsymbol{T}]\,\widetilde{\boldsymbol{M}}, \text{ with } \boldsymbol{A} = \begin{bmatrix} \alpha & \gamma & u_0 \\ 0 & \beta & v_0 \\ 0 & 0 & 1 \end{bmatrix} \tag{1}$$

where s is an arbitrary scale factor, the extrinsic parameters $(\boldsymbol{R}, \boldsymbol{T})$ is the rotation and translation, which related the world coordinate system to the camera coordinate system, and $\boldsymbol{A}$ is the camera intrinsic matrix, with (u_0, v_0) the coordinates of the principal point, α and β the scale factors in u and v axes, and γ the parameter describing the skew of the two axes.

In order to obtain the relative positions between any two coordinate systems, the rotation $\boldsymbol{R}$ and translation $\boldsymbol{T}$ need to be acquired by calibrating the left and right cameras, simultaneously. This calibration process involves capturing images of a checkerboard pattern at different orientations, as shown in Figure 6. The images are then processed using the "Stereo Camera Calibrator" tool in MATLAB to obtain the camera parameters, including the rotation and translation matrices.

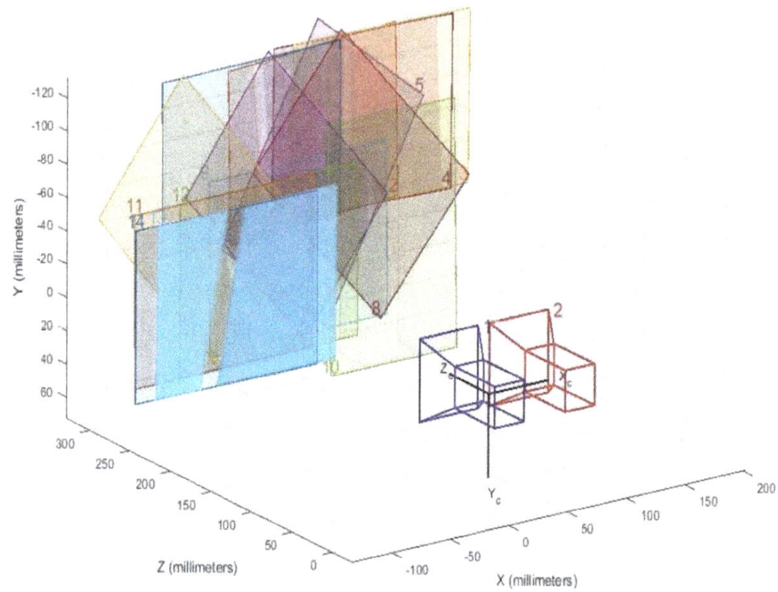

Figure 6. The checkerboard calibration.

As a result, the world coordinate $[X, Y, Z]^T$ of the ship is obtained. The latitude–longitude coordinate of the ship is denoted by $\boldsymbol{P} = \left[P_{latitude}, P_{longitude}\right]^T$ can be formulated as follows:

$$\boldsymbol{P} = \boldsymbol{P_{camera}} + \begin{bmatrix} X \\ Y \end{bmatrix} \tag{2}$$

where $\boldsymbol{P_{camera}} = \left[P^{camera}_{latitude}, P^{camera}_{longitude}\right]^T$ is the latitude–longitude coordinate of the camera.

4. Case Study

4.1. Database and Processing

In this experiment, our team collected a total of 1000 images of marine ships to form the MYSHIP database[1]. It had a resolution of either 1920 × 1080 or 2840 × 2160. Because the training of a convolutional neural network requires a considerable number of samples, we also added the SEASHIP database to our database. The SEASHIP database consists of a total of 7000 images with a resolution of 1920 × 1080. As shown in Figure 7, the database is divided into six categories of ships: ore carriers, bulk cargo carriers, general cargo ships, container ships, fishing boats, and passenger ships.

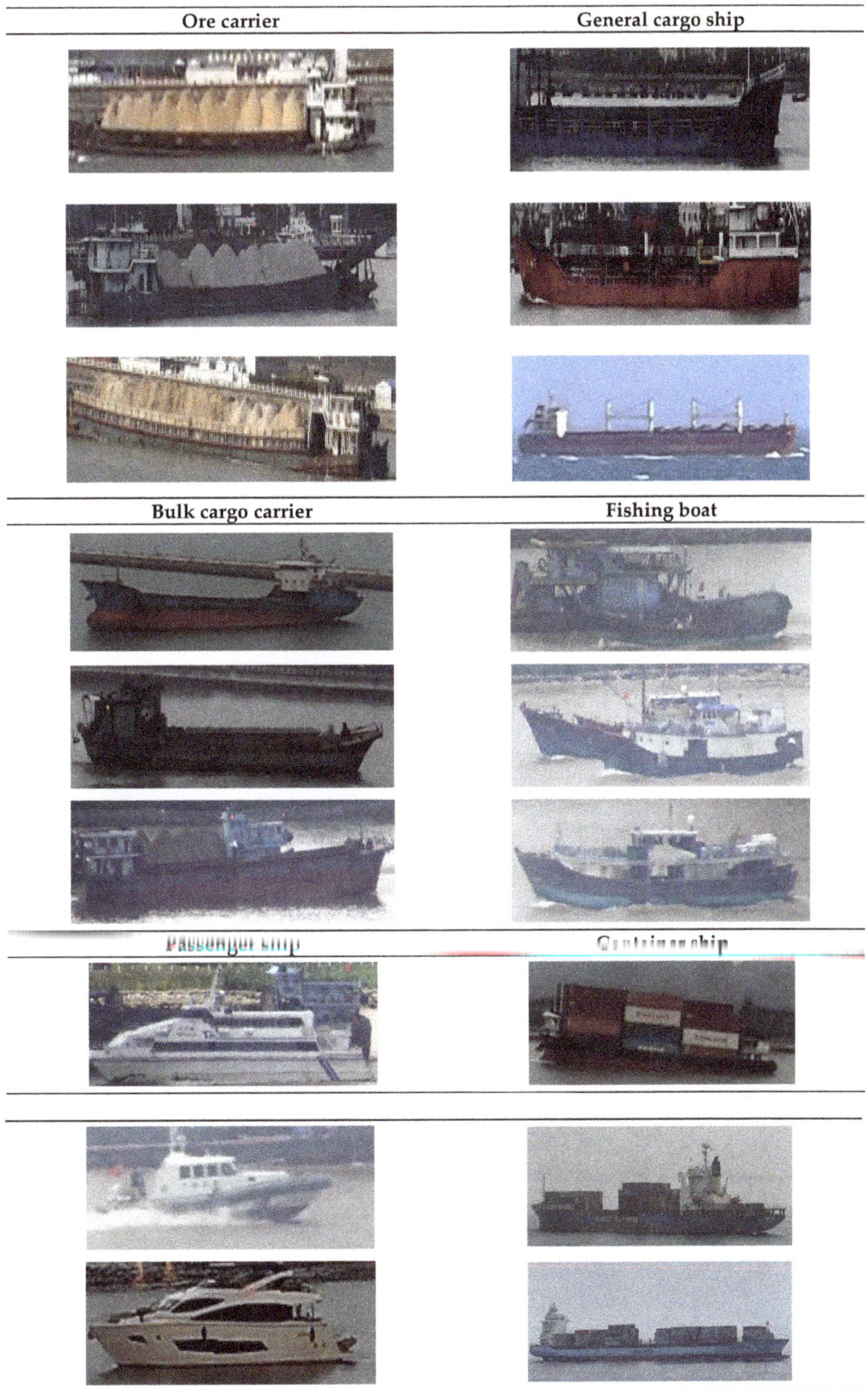

Figure 7. The six types of ships included in the training database.

We divide the database into three parts: the training set, the validation set, and the test set in the proportion 6:2:2. The division of the database is shown in Table 1.

Table 1. The number of images in each database.

Database	Number of Samples		Number of Ships	
	MYSHIP	SEASHIP	MYSHIP	SEASHIP
Training set	600	4200	1334	4934
Validation set	200	1400	444	1610
Test set	200	1400	446	1669
Total	1000	7000	2224	8213

4.2. Parameter Setting

In this experiment, the parameters of the camera used to capture ship video and image data are shown in Table 2.

Table 2. The hardware parameters of the camera.

Parameter Name	Parameter	Unit
Focal length	12.5–775	mm
Pixel	3	mp
Resolution	2048*1536	ppi
Wide dynamic range	120	db

The internal and external parameters of the two cameras are shown in Table 3.

Table 3. The calibration parameters.

Setting	Intrinsic Matrix	Rotation	Translation	Distortion
	Pixel	m	m	m
Left camera	$\begin{bmatrix} 534.31 & 0.00 & 342.64 \\ 0.00 & 534.31 & 234.42 \\ 0.00 & 0.00 & 1.00 \end{bmatrix}$	$\begin{bmatrix} 1.00 & 1.00 & 1.00 \\ 1.00 & 1.00 & 1.00 \\ 1.00 & 1.00 & 1.00 \end{bmatrix}$	$\begin{bmatrix} 1.00 \\ 1.00 \\ 1.00 \end{bmatrix}$	$\begin{bmatrix} -0.29 \\ 0.11 \\ 0.00 \end{bmatrix}, \begin{bmatrix} 0.00 \\ 0.00 \end{bmatrix}$
Right camera	$\begin{bmatrix} 537.39 & 0.00 & 326.62 \\ 0.00 & 537.01 & 250.58 \\ 0.00 & 0.00 & 1.00 \end{bmatrix}$	$\begin{bmatrix} 1.00 & 0.00 & 0.01 \\ -0.00 & 1.00 & 250.58 \\ -0.01 & 0.01 & 1.00 \end{bmatrix}$	$\begin{bmatrix} -99.72 \\ 1.27 \\ 0.05 \end{bmatrix}$	$\begin{bmatrix} -0.29 \\ 0.10 \\ 0.00 \end{bmatrix}, \begin{bmatrix} 0.00 \\ 0.00 \end{bmatrix}$

The experiments were carried out on a computer platform configured with 64G memory, an Intel Core i9-12900kF CPU and a NVIDIA GeForce RTX 3090 Ti GPU for training and testing. The system of the experiment platform was Windows 10. Referring to previous studies [62–64], the training parameters of the model were set as follows: the initial learning rate was 0.001, the attenuation coefficient was 0.0005, and the stochastic gradient descent rate was 0.9.

4.3. Construction of the Training Database

In this paper, we used the image annotation tool labellmg to manually annotate the boxes of each ship in the images (https://github.com/heartexlabs/labelImg (accessed on 18 June 2023)). Labellmg is the most widely used image annotation tool for creating custom databases. Once the images are annotated, a dataset is generated that contains the category of the ship, the position of the corners of the ship's box, as well as the width and height of the ship.

To prevent overfitting and improve target detection accuracy, data augmentation strategies are applied to the images in the database, which increased sample diversity and improved the robustness of the model. In the experiment, several augmentation techniques such as horizontal flipping, vertical flipping, random rotation, Mosaic, and cutout were applied to enrich the training samples.

4.4. Detection

The proposed model was then tested with four scenarios to assess its performance. The details for the video of each scenario is shown in Table 4.

Table 4. The details for the videos.

No.	Frame Rate (fps)	Resolution	Duration(s)	Category	Ship Status
video#1	10	720 × 1280	150	Container ship	Underway
video#2	10	2160 × 3840	60	Bulk cargo carrier	Moored
video#3	10	2160 × 3840	480	Passenger ship	Moored
video#4	10	2160 × 3840	60	Passenger ship Fishing boat	Underway Anchored

Video 1 shows an underway container ship navigating a traffic route. The frame rate of the video is 10 per second (fps), the duration is 150 s, and the resolution is 720 × 1280.

Video 2 tests the performance of the model to detect a static target, therefore a moored bulk cargo ship was selected. The frame rate of the video is 10 per second (fps), the duration is 60 s, and the resolution is 2160 × 3840.

Video 3 is a berth with several moored ships. It tests whether the model could detect the passenger ship within a complex water environment. The frame rate of video is 10 per second (fps), the duration is 480 s, and the resolution is 2160 × 3840.

Video 4 shows a scenario of two ships. One is a passenger ship passing a bridge and another is an anchored fishing boat. It can be considered as a typical condition to test whether the model can detect ships when obstacles exist. The frame rate of the video is 10 per second (fps), the duration is 60 s, and the resolution is 2160 × 3840.

The video clips are shown in Figure 8.

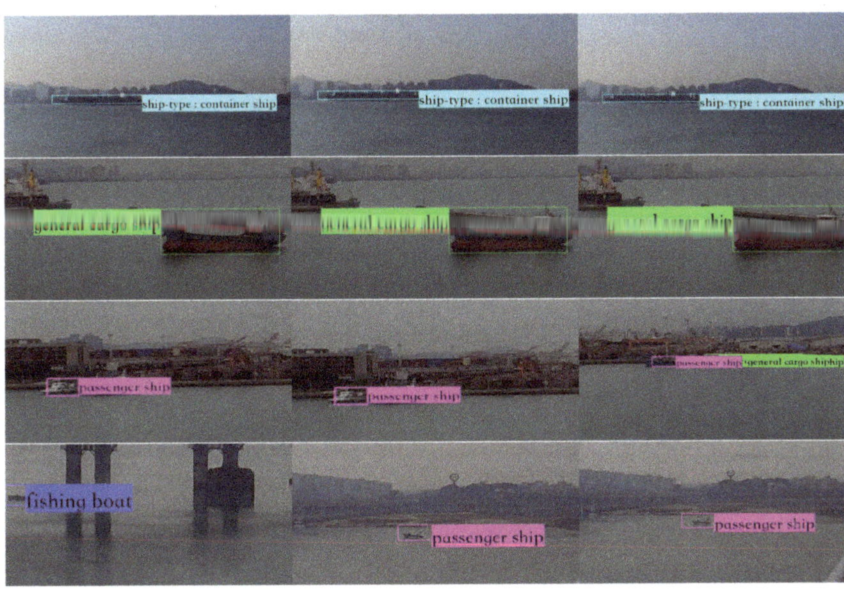

Figure 8. The four scenarios used to test the model: video 1; video 2; video 3 and video 4.

It can be seen from Figure 8, that with use of the proposed model, the targets in the different scenarios can be detected; the associated ship type and confidence rates are given in Table 5.

Table 5. The confidence of detect results for videos.

No.	Minimum Confidence	Maximum Confidence	Average Confidence
Video 1	0.95	1.00	0.97
Video 2	0.99	1.00	0.99
Video 3	0.50	0.94	0.76
Video 4	0.53	0.84	0.72

In Table 5, the minimum confidence, maximum confidence, and average confidence for the target in Video 1 and Video 2 are (0.95, 1.00, 0.97) and (0.99, 1.00, 0.99), respectively, showing high accuracy and reliability of the model to detect ships underway and moored. Based on this, it can be concluded that the proposed system provides satisfactory detection performance and is capable of successfully detecting target objects in a typical offshore wind farm. However, in Video 3 and Video 4, although the target can be detected by the proposed model, the minimum confidence decreased to 0.50 and 0.53, and the average confidence decreased to 0.76 and 0.72. The main reason for this reduced confidence is the impacts from the complex environment conditions and surroundings, which can be overcome by (1) providing more training data to enhance the model detection probability, (2) using additional segmentation methods before detecting the ship, however, the time-consumption will become another problem.

4.5. Position and Tracking Results

Figures 9 and 10 show typical object position and tracking results (i.e., Video 1 and Video 2). The results shown in Figure 9 detect an underway container ship, with a confidence degree of 0.99. After synchronizing images from two cameras and matching them with binocular stereo vision model, the ship's position was obtained as 118.745 E and 24.4749 N. In addition, the ship's dynamic information of speed and course was calculated as well, where the speed was 11.61 knots and the course was 290.01 degrees.

Figure 9. Position and tracking result for Video 1.

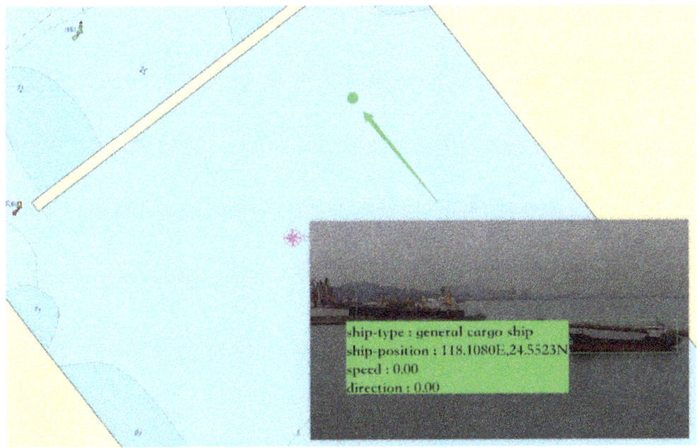

Figure 10. Position and tracking result for Video 2.

In Figure 10, an anchored bulk ship was detected; the model assigns the confidence degree of 0.99 and gives the position of the ship as longitude of 118.1074 E and latitude of 24.5521 N. As it is an anchored ship, the speed and the course for this ship is 0.

5. Validation

5.1. Detection Validation

To validate the proposed model, the two databases MYSHIP and SEASHIP were used. The detection capabilities were evaluated by the following indices: accuracy rate (P), recall rate (R), false alarm rate (F), miss alarm rate (M), and average precision (AP), which were defined as follows:

- Precision (P): proportion of samples that are correctly detected in all test results.
- Recall rate (R): proportion of actual positive samples that are correctly detected.
- False alarm rate (F): proportion of negative samples that are incorrectly detected as positive samples.
- Miss alarm rate (M): proportion of actual positive samples that are incorrectly detected as negative samples.
- Average precision (AP): the integral value of the Precision Rate-Recall rate curve (P-R curve).
- *Average precision*$_{50}$ (AP_{50}): the average accuracy of the test when the IOU threshold is 0.5.
- *Average precision*$_{50:95}$ ($AP_{50:95}$): the average accuracy of the test when the IOU threshold is 0.5–0.95.

The relationship between these indices are mathematically described as function 3:

$$\begin{cases} P = \frac{TP}{TP+FP} \\ R = \frac{TP}{TP+FN} \\ F = \frac{FP}{TP+FP} = 1 - P \\ M = \frac{FN}{TP+FN} = 1 - R \\ AP = \int_0^1 P(R)dR \end{cases} \quad (3)$$

where TP is the counts of the true positive that correctly predicts positive samples, TN is the counts of the true negative that correctly predicts negative samples, FP is the counts of the false positive that incorrectly predicts negative samples and FN is the counts of false negative that incorrectly predicts positive samples.

5.2. Validating the Proposed Models

By using the validation index, the performance of the model is shown in Table 6.

Table 6. The performance of the YOLOv7 model on the database.

Database	P	R	F	M	AP_{50}	$AP_{50:95}$
MYSHIP	74.38%	78.20%	21.80%	25.62%	97.10%	72.61%
SEASHIP	62.49%	70.97%	29.03%	37.52%	96.40%	61.61%

In general, the system has high recognition accuracy and precision for ships. The rates of detection accuracy were relatively high (74.38% for the MYSHIP and 62.46% for SEASHIP datasets). However, by comparing the evaluation results, Table 6 shows the YOLOv7 model demonstrated better performance on the latter database (i.e., SEASHIP database). This is due to the MYSHIP database including more distant and overlapping ships, making identification more challenging.

Then, to validate the accuracy of the model's predictions, the ship's real-time AIS data in the four cases (Video 1–4) was collected. The detected results from the proposed model were compared with information provided in the AIS data, which is considered more accurate and reliable. Table 7 displays the comparison of different detection systems for ships.

Table 7. The results comparing different detections systems for ships.

	Video #1	Video #2	Video #3	Video #4
System result	Container ship	General cargo ship	Passenger ship × 5 General cargo ship × 2	Fishing boat Passenger ship
AIS data	Container ship	General cargo ship	Passenger ship × 2	Passenger ship

Table 7 shows the proposed system can detect ship types with high accuracy, while showing advantages of ship detection that are not provided by AIS data. For instance, in Video 3, only two passenger ships have AIS data but in reality, there were five passenger ships and two general cargo ships but some of them had turned off their AIS or were missing signals. Similarly, Video 4 shows a fishing boat that did not provide its AIS data, but it was detected by the video system.

The accuracy of the positioning results was compared as well. As the AIS transmits ship position with different rates (e.g., 2–10 s when ships are underway and 180 s when ships are moored or anchored), this study linked all the AIS positions with time sequences to obtain the ship trajectories. In this study, the update frequency of the proposed model was set as 2 s each time and all positions were linked following the time sequences to obtain the trajectories. Then, the deviation between the AIS based trajectories and the model detected trajectories in Video 1 were calculated and the results are given in Figure 11 and the deviations are given in Table 8.

Table 8. The comparison between the AIS data and the system output result.

		Sample#1	Sample#2	Sample#3	Sample#4
System output	Longitude	118.0796 E	118.1074 E	118.1081 E	118.1117 E
	Latitude	24.4806 N	24.5521 N	24.5466 N	24.5579 N
AIS data (°)	Longitude	118.07962 E	118.10737 E	118.10814 E	118.11169 E
	Latitude	24.48064 N	24.55212 N	24.54661 N	24.55791 N
Deviation (°)	Longitude	0.00002	−0.00003	0.00004	−0.00001
	Latitude	0.00004	0.00002	0.00001	0.00001

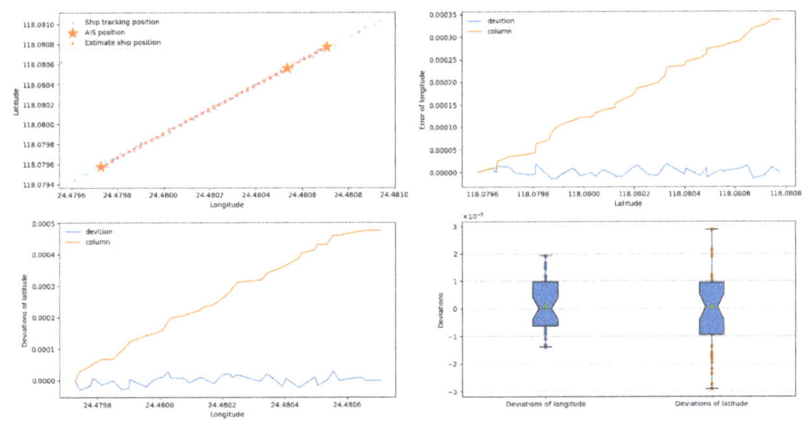

Figure 11. The comparison between the AIS based trajectories and model detected trajectories.

It can be seen in Figure 11 that due to the low update rate, only three AIS records were received for the ship in Video 1. Therefore, three AIS positions were linked and the deviations on the longitude and latitude were calculated (see Table 8). Notably, although the deviation between the two trajectories existed, they were all less than ±0.0001 degree (11 m approximately), which is broadly acceptable.

6. Discussion

6.1. Comparing the Current System Used for Ship Detection and Tracking

Based on the case study results, the advantages of use of visual based technologies for ship detection are as follows:

(1) The proposed approach provides a way for ship monitoring with satisfactory accuracy and reliability. The developed system can obtain necessary information about ship traffic (e.g., speed, position, course), so that it becomes a novel supplement for VTS and OWF managers. In addition, in contrast to AIS and radar, this system uses cameras to collect ship videos, and the CCTVs are already fixed in most of the OWFs worldwide. Therefore, the cost of developing this system in practice can be very low.

(2) As already mentioned, the proposed system has a high frequency of updating the ship information (i.e., 10 fps) and high accuracy in individual ship tracking (the tracking errors less than 15 m/0.0001 degrees). Thus, the developed system reasonably provides ship dynamic data and ensures collision risk assessment for ships in the vicinity of OWFs.

(3) The proposed framework hybridizes several visual technologies. The study not only proves the possibility of using these technologies to aid ship monitoring in offshore wind farm waters but also can be considered as evidence to apply the machine visual model for ship detection and tracking in other similar waters, such as narrow channels, bridges, on a river, etc.

Moreover, the advantages and disadvantages of current ship detection approaches were compared, and the results are listed in Table 9.

Table 9 presents several highlights between the current ship detection approaches. The radar tracking system can accurately measure and track moving or stationary targets, but it has a blind spot, and the echoes are susceptible to environmental factors, which can lead to target loss. The AIS-based ship reporting system can compensate for these shortcomings of radar, but AIS cannot identify targets that are not equipped with AIS or have the AIS turned off, so that leads to missed tracking. Both the CNN-based and the YOLO-based model show high accuracy for target detection. However, The YOLO algorithm is time consuming due to its faster detection speed and frequency.

Table 9. The comparison between the existing system and our system.

Method Name	Advantage	Disadvantage
AIS based ship reporting system	1. Capable of automatically obtaining target position, status, speed, and other information 2. High accuracy of ship dynamic data 3. Strong performance against rain, snow, and wave interference 4. Eliminate the blocking area of some radars	1. Operates only at VHF 2. Cannot identify targets without AIS devices installed 3. The data are not continuous and poor timeliness
Radar tracking system	1. Capable of automatic tracking and automatic warning 2. Accurate measurement and display of distance, bearing, speed and size of the target	1. Blind spots exist 2. The echo is easily affected by weather, sea state and terrain obstruction, etc. 3. In the multi-target dense area is prone to false tracking and target loss phenomenon
YOLO model	1. Has a high recognition speed, and can efficiently identify multiple targets 2. Superiority in real-time target detection 3. Suitable for detecting objects of various shapes and sizes	1. Global information has a good performance, but poor performance on a small range of information 2. Compared to some recognition algorithms, the accuracy rate is slightly worse
RCNN model	1. Powerful adaptability for recognition and classification of complex data 2. Feature classification works well	1. Large number of operations, large sample size and time are required to train the model 2. Lack of annotation resources for specific target samples, high quality for the database

6.2. YOLO Series

In this study, the YOLOv7 model was trained with AIS data and incorporated binocular vision to improve the accuracy of system detection. The proposed system can complete ship detection and target tracking, which significantly increases the model's accuracy and applicability. However, the YOLO series are developing rapidly. To evaluate the performance of detecting targets in maritime domain by using the YOLO series, this study selected the YOLOv5 model as the benchmark. In the comparison, both methods use the same database and hardware platforms. The results of the comparison experiment is shown in Table 10. The original YOLOv5 model achieved 73.80% of detection precision, while the YOLOv7 model achieved 77.48%, which is 3.68% higher than the YOLOv5 model. The recall rates for the YOLOv5 and YOLOv7 models were not very different (73.41% and 73.19%). The false alarm rate of the YOLOv5 model was 3.68%, which was higher than for the YOLOv7 model. For the index of miss alarm rate, the YOLOv5 and YOLOv7 models had similar performance, with 26.6% and 26.81%, respectively. Then the study set the IOU threshold at 0.5 to test the index of average accuracy. The YOLOv5 model obtained an average accuracy of 95.76%, while the YOLOv7 model reached 97.03%. Similarly, setting the IOU threshold in the interval of 0.5–0.95, the results showed the average accuracy for the YOLOv5 model dropped to 70.37% and for the YOLOv7 model it was 71.51%, which is 1.14% higher than the YOLOv5 model. Consequently, based on these comparisons, it can be concluded that the proposed method can be effective for ship detection and real-time traffic monitoring.

Table 10. The results of the comparison experiment between the YOLOv5 and YOLOv7 models.

Methods	P	R	F	M	AP_{50}	$AP_{50:95}$
YOLOv5	73.80%	73.41%	26.20%	26.60%	95.76%	70.37%
YOLOv7	77.48%	73.19%	22.52%	26.81%	97.03%	71.51%

However, the system proposed above also needs to be improved. First, there are still some ships that may not be well detected in the proposed system (current accuracy is 77.48%) due to their position and orientation. When the ship drives independently and laterally, it is easily detected by the system. Moreover, when the ship is longitudinal, and multiple ships overlap, the system may regard these ships as obstacles and ignore them. For example, the ship in Video 3 has the lowest confidence. This is because, among the vessels detected by the system, those with a low confidence level are characterized by facing the camera end on. This orientation makes the ship's characteristics less obvious, so it is easily ignored by the system. Second, the database needs to be expanded to include more training data. Third, the potential effects of critical weather conditions (e.g., snow, heavy fog, and rain) were not tested in this study. Therefore, further studies can be carried out providing more labelled samples, and we will collect more data to enrich the system database and train better classifiers in following research. Meanwhile, the system will be tested under critical environments in practice soon.

7. Conclusions

This paper proposed a ship detection and positioning system based on the YOLOv7 algorithm and stereo vision technologies and introduces the framework and detailed methods used in this system. In addition, this study suggests a novel concept of using AIS data as the training resource for model training, which improves the accuracy of using the YOLOv7 model and stereo vision algorithms in ship detection and tracking. Applying the proposed model in a real ship case study validates the possibility of using the YOLOv7 algorithm to track and identify ships when the stereo vision algorithm is applied to locate ship positions.

The benefit of the proposed system is that it can detect vessels automatically and achieve real-time tracking and positioning. The system not only eases the workload of OWF operators during CCTV monitoring but also provides a possible way for ship traffic management in the water in the vicinity of OWFs. Moreover, the novel system shares the idea of using machine vision technology for ship collision prevention. In addition, the proposed method applies not only to offshore wind farm waters but can be applied in the future to any place of interest. Based on the analysis of the proposed system, further study can investigate applying the proposed system in ships to achieve situation forecasting.

Author Contributions: Conceptualization, validation, writing, original draft, methodology and formal analysis—H.J., X.L. and Y.H.; writing, review and editing, supervision, funding acquisition—Q.Y. and M.Z. All authors have read and agreed to the published version of the manuscript.

Funding: This work is supported by National Natural Science Foundation of China under grant 52201412, Natural Science Foundation of Fujian Province under grant No. 2022J05067 and Fund of Hubei Key Laboratory of Inland Shipping Technology (NO. NHHY2021001).

Institutional Review Board Statement: Not applicable.

Informed Consent Statement: Not applicable.

Data Availability Statement: Not applicable.

Conflicts of Interest: The authors declare no conflict of interest.

Note

1 1000 images of inland ships were captured at Bay Park, Xiamen Bridge, and Gao Qi Wharf in Xiamen City, Fujian Province.

References

1. Vieira, M.; Henriques, E.; Snyder, B.; Reis, L. Insights on the impact of structural health monitoring systems on the operation and maintenance of offshore wind support structures. *Struct. Saf.* **2022**, *94*, 102154. [CrossRef]
2. Musial, W.; Spitsen, P.; Duffy, P.; Beiter, P.; Marquis, M.; Hammond, R.; Shields, M. *Offshore Wind Market Report: 2022 Edition*; National Renewable Energy Lab. (NREL): Golden, CO, USA, 2022.
3. Torres-Rincón, S.; Bastidas-Arteaga, E.; Sánchez-Silva, M. A flexibility-based approach for the design and management of floating offshore wind farms. *Renew. Energy* **2021**, *175*, 910–925. [CrossRef]
4. Brady, R.L. *Offshore Wind Industry Interorganizational Collaboration Strategies in Emergency Management*; Walden University: Minneapolis, MN, USA, 2022.
5. Li, B.; Xie, X.; Wei, X.; Tang, W. Ship detection and classification from optical remote sensing images: A survey. *Chin. J. Aeronaut.* **2021**, *34*, 145–163. [CrossRef]

6. Zhao, J.; Guo, W.; Zhang, Z.; Yu, W. A coupled convolutional neural network for small and densely clustered ship detection in SAR images. *Sci. China Inf. Sci.* **2018**, *62*, 42301. [CrossRef]
7. Zhang, S.; Wu, R.; Xu, K.; Wang, J.; Sun, W. R-CNN-Based Ship Detection from High Resolution Remote Sensing Imagery. *Remote Sens.* **2019**, *11*, 631. [CrossRef]
8. Hu, J.; Zhi, X.; Shi, T.; Yu, L.; Zhang, W. Ship Detection via Dilated Rate Search and Attention-Guided Feature Representation. *Remote Sens.* **2021**, *13*, 4840. [CrossRef]
9. Yu, C.; Shin, Y. SAR ship detection based on improved YOLOv5 and BiFPN. *ICT Express* **2023**. [CrossRef]
10. Priakanth, P.; Thangamani, M.; Ganthimathi, M. *WITHDRAWN: Design and Development of IOT Based Maritime Monitoring Scheme for Fishermen in India*; Elsevier: Amsterdam, The Netherlands, 2021.
11. Ouelmokhtar, H.; Benmoussa, Y.; Benazzouz, D.; Ait-Chikh, M.A.; Lemarchand, L. Energy-based USV maritime monitoring using multi-objective evolutionary algorithms. *Ocean Eng.* **2022**, *253*, 111182. [CrossRef]
12. Nyman, E. Techno-optimism and ocean governance: New trends in maritime monitoring. *Mar. Policy* **2019**, *99*, 30–33. [CrossRef]
13. Pawar, K.; Attar, V. Deep learning based detection and localization of road accidents from traffic surveillance videos. *ICT Express* **2022**, *8*, 379–387. [CrossRef]
14. Thallinger, G.; Krebs, F.; Kolla, E.; Vertal, P.; Kasanický, G.; Neuschmied, H.; Ambrosch, K.-E. Near-miss accidents–classification and automatic detection. In Proceedings of the Intelligent Transport Systems—From Research and Development to the Market Uptake: First International Conference, INTSYS 2017, Hyvinkää, Finland, 29–30 November 2017; Proceedings 1; pp. 144–152.
15. Pramanik, A.; Sarkar, S.; Maiti, J. A real-time video surveillance system for traffic pre-events detection. *Accid. Anal. Prev.* **2021**, *154*, 106019. [CrossRef] [PubMed]
16. Root, B. HF radar ship detection through clutter cancellation. In Proceedings of the 1998 IEEE Radar Conference, RADARCON'98, Challenges in Radar Systems and Solutions, Dallas, TX, USA, 14 May 1998; pp. 281–286.
17. Dzvonkovskaya, A.; Gurgel, K.-W.; Rohling, H.; Schlick, T. Low power high frequency surface wave radar application for ship detection and tracking. In Proceedings of the 2008 International Conference on Radar, Adelaide, SA, Australia, 2–5 September 2008; pp. 627–632.
18. Margarit, G.; Barba Milanés, J.; Tabasco, A. Operational Ship Monitoring System Based on Synthetic Aperture Radar Processing. *Remote Sens.* **2009**, *1*, 375–392. [CrossRef]
19. Brekke, C. *Automatic Ship Detection Based on Satellite SAR*; FFI-rapport: Kjeller, Norway, 2008.
20. Kazimierski, W.; Stateczny, A. Fusion of Data from AIS and Tracking Radar for the Needs of ECDIS. In Proceedings of the 2013 Signal Processing Symposium (SPS), Serock, Poland, 5–7 June 2013; pp. 1–6.
21. Pan, Z.; Deng, S. Vessel Real-Time Monitoring System Based on AIS Temporal Database. In Proceedings of the 2009 International Conference on Information Management, Innovation Management and Industrial Engineering, Kuala Lumpur, Malaysia, 3–5 April 2009; pp. 611–614.
22. Lin, B.; Huang, C.-H. Comparison between Arpa Radar and Ais Characteristics for Vessel Traffic Services. *J. Mar. Sci. Technol.* **2006**, *14*, 7. [CrossRef]
23. Yu, Q.; Liu, K.; Chang, C.-H.; Yang, Z. Realising advanced risk assessment of vessel traffic flows near offshore wind farms. *Reliab. Eng. Syst. Saf.* **2020**, *203*, 107086. [CrossRef]
24. Chang, C.-H.; Kontovas, C.; Yu, Q.; Yang, Z. Risk assessment of the operations of maritime autonomous surface ships. *Reliab. Eng. Syst. Saf.* **2021**, *207*, 107324. [CrossRef]
25. Yu, Q.; Liu, K.; Teixeira, A.; Soares, C.G. Assessment of the influence of offshore wind farms on ship traffic flow based on AIS data. *J. Navig.* **2020**, *73*, 131–148. [CrossRef]
26. Liu, Y.; Su, H.; Zeng, C.; Li, X. A Robust Thermal Infrared Vehicle and Pedestrian Detection Method in Complex Scenes. *Sensor* **2021**, *21*, 1240. [CrossRef] [PubMed]
27. Shao, Z.; Wang, L.; Wang, Z.; Du, W.; Wu, W. Saliency-aware convolution neural network for ship detection in surveillance video. *IEEE Trans. Circuits Syst. Video Technol.* **2019**, *30*, 781–794. [CrossRef]
28. Chen, Z.; Li, B.; Tian, L.F.; Chao, D. Automatic detection and tracking of ship based on mean shift in corrected video sequences. In Proceedings of the 2017 2nd International Conference on Image, Vision and Computing (ICIVC), Chengdu, China, 2–4 June 2017; pp. 449–453.
29. Rahmatov, N.; Paul, A.; Saeed, F.; Hong, W.-H.; Seo, H.; Kim, J. Machine learning–based automated image processing for quality management in industrial Internet of Things. *Int. J. Distrib. Sens. Netw.* **2019**, *15*, 1550147719883551. [CrossRef]
30. Chávez Heras, D.; Blanke, T. On machine vision and photographic imagination. *AI Soc.* **2020**, *36*, 1153–1165. [CrossRef]
31. Cho, S.; Kwon, J. Abnormal event detection by variation matching. *Mach. Vis. Appl.* **2021**, *32*, 80. [CrossRef]
32. Lowe, D.G. Object recognition from local scale-invariant features. In Proceedings of the Seventh IEEE International Conference on Computer Vision, Kerkyra, Greece, 20–27 September 1999; pp. 1150–1157.
33. Orr, G.B.; Müller, K.-R. *Neural Networks: Tricks of the Trade*; Springer: Berlin/Heidelberg, Germany, 1998.
34. Qi, S.; Ma, J.; Lin, J.; Li, Y.; Tian, J. Unsupervised ship detection based on saliency and S-HOG descriptor from optical satellite images. *IEEE Geosci. Remote Sens. Lett.* **2015**, *12*, 1451–1455.
35. Zhu, C.; He, Y.; Savvides, M. Feature selective anchor-free module for single-shot object detection. In Proceedings of the IEEE/CVF Conference on Computer Vision and Pattern Recognition, Long Beach, CA, USA, 15–20 June 2019; pp. 840–849.

36. Law, H.; Deng, J. CornerNet: Detecting objects as paired keypoints. In Proceedings of the European conference on computer vision (ECCV), Munich, Germany, 8–14 September 2018; pp. 734–750.
37. Bochkovskiy, A.; Wang, C.-Y.; Liao, H.-Y.M. Yolov4: Optimal speed and accuracy of object detection. *arXiv* **2020**, arXiv:2004.10934.
38. Chen, Z.; Liu, C.; Filaretov, V.F.; Yukhimets, D.A. Multi-Scale Ship Detection Algorithm Based on YOLOv7 for Complex Scene SAR Images. *Remote Sens.* **2023**, *15*, 2071. [CrossRef]
39. Cen, J.; Feng, H.; Liu, X.; Hu, Y.; Li, H.; Li, H.; Huang, W. An Improved Ship Classification Method Based on YOLOv7 Model with Attention Mechanism. *Wirel. Commun. Mob. Comput.* **2023**, *2023*, 7196323. [CrossRef]
40. Yasir, M.; Zhan, L.; Liu, S.; Wan, J.; Hossain, M.S.; Isiacik Colak, A.T.; Liu, M.; Islam, Q.U.; Raza Mehdi, S.; Yang, Q. Instance segmentation ship detection based on improved Yolov7 using complex background SAR images. *Front. Mar. Sci.* **2023**, *10*. [CrossRef]
41. Liu, W.; Anguelov, D.; Erhan, D.; Szegedy, C.; Reed, S.; Fu, C.-Y.; Berg, A.C. Ssd: Single shot multibox detector. In Proceedings of the Computer Vision–ECCV 2016: 14th European Conference, Amsterdam, The Netherlands, 11–14 October 2016; Proceedings, Part I 14. pp. 21–37.
42. Girshick, R.; Donahue, J.; Darrell, T.; Malik, J. Rich feature hierarchies for accurate object detection and semantic segmentation. In Proceedings of the IEEE Conference on Computer Vision and Pattern Recognition, Portland, OR, USA, 23–28 June 2013; pp. 580–587.
43. Meng, C.; Bao, H.; Ma, Y.; Xu, X.; Li, Y. Visual Meterstick: Preceding Vehicle Ranging Using Monocular Vision Based on the Fitting Method. *Symmetry* **2019**, *11*, 1081. [CrossRef]
44. Zhao, Y.; Zhao, L.; Xiong, B.; Kuang, G. Attention receptive pyramid network for ship detection in SAR images. *IEEE J. Sel. Top. Appl. Earth Obs. Remote Sens.* **2020**, *13*, 2738–2756. [CrossRef]
45. Redmon, J.; Divvala, S.; Girshick, R.; Farhadi, A. You only look once: Unified, real-time object detection. In Proceedings of the IEEE Conference on Computer Vision and Pattern Recognition, Boston, MA, USA, 7–12 June 2015; pp. 779–788.
46. Gong, H.; Li, H.; Xu, K.; Zhang, Y. Object detection based on improved YOLOv3-tiny. In Proceedings of the 2019 Chinese automation congress (CAC), Hangzhou, China, 22–24 November 2019; pp. 3240–3245.
47. Patel, K.; Bhatt, C.; Mazzeo, P.L. Improved Ship Detection Algorithm from Satellite Images Using YOLOv7 and Graph Neural Network. *Algorithms* **2022**, *15*, 473. [CrossRef]
48. Li, J.; Qu, C.; Shao, J. Ship detection in SAR images based on an improved faster R-CNN. In Proceedings of the 2017 SAR in Big Data Era: Models, Methods and Applications (BIGSARDATA), Beijing, China, 13–14 November 2017; pp. 1–6.
49. Li, Y.; Rong, L.; Li, R.; Xu, Y. Fire Object Detection Algorithm Based on Improved YOLOv3-tiny. In Proceedings of the 2022 7th International Conference on Cloud Computing and Big Data Analytics (ICCCBDA), Chengdu, China, 22–24 April 2022; pp. 264–269.
50. Ren, S.; He, K.; Girshick, R.; Sun, J. Faster r-cnn: Towards real-time object detection with region proposal networks. *Adv. Neural Inf. Process. Syst.* **2015**, *28*, 1137–1149. [CrossRef]
51. Campbell, J.; Sukthankar, R.; Nourbakhsh, I.; Pahwa, A. A robust visual odometry and precipice detection system using consumer-grade monocular vision. In Proceedings of the 2005 IEEE International Conference on Robotics and Automation, Barcelona, Spain, 18–22 April 2005; pp. 3421–3427.
52. Matthies, L.; Shafer, S. Error modeling in stereo navigation. *IEEE J. Robot. Autom.* **1987**, *3*, 239–248. [CrossRef]
53. Davison, A.J. Real-time simultaneous localisation and mapping with a single camera. In Proceedings of the Computer Vision, IEEE International Conference, Nice, France, 13–16 October 2003; p. 1403.
54. Schwartzkroin, P.A. Neural Mechanisms: Synaptic Plasticity. Molecular, Cellular, and Functional Aspects. Michel Baudry, Richard F. Thompson, and Joel L. Davis, Eds. MIT Press, Cambridge, MA, 1993. xiv, 263 pp., illus. $50 or£ 44.95. *Science* **1994**, *264*, 1179–1180. [CrossRef] [PubMed]
55. Howard, I.P.; Rogers, B.J. *Binocular Vision and Stereopsis*; Oxford University Press: Oxford, UK, 1995.
56. Xu, Y.; Zhao, Y.; Wu, F.; Yang, K. Error analysis of calibration parameters estimation for binocular stereo vision system. In Proceedings of the 2013 IEEE International Conference on Imaging Systems and Techniques (IST), Beijing, China, 22–23 October 2013; pp. 317–320.
57. Yu, Y.; Tingting, W.; Long, C.; Weiwei, Z. Stereo vision based obstacle avoidance strategy for quadcopter UAV. In Proceedings of the 2018 Chinese Control And Decision Conference (CCDC), Shenyang, China, 9–11 June 2018; pp. 490–494.
58. Sun, X.; Jiang, Y.; Ji, Y.; Fu, W.; Yan, S.; Chen, Q.; Yu, B.; Gan, X. Distance measurement system based on binocular stereo vision. In Proceedings of the IOP Conference Series: Earth and Environmental Science, Hainan, China, 25–26 April 2009; p. 052051.
59. Wang, C.; Zou, X.; Tang, Y.; Luo, L.; Feng, W. Localisation of litchi in an unstructured environment using binocular stereo vision. *Biosyst. Eng.* **2016**, *145*, 39–51. [CrossRef]
60. Zou, X.; Zou, H.; Lu, J. Virtual manipulator-based binocular stereo vision positioning system and errors modelling. *Mach. Vis. Appl.* **2012**, *23*, 43–63. [CrossRef]
61. Zuo, X.; Xie, X.; Liu, Y.; Huang, G. Robust visual SLAM with point and line features. In Proceedings of the 2017 IEEE/RSJ International Conference on Intelligent Robots and Systems (IROS), Vancouver, BC, Canada, 24–28 September 2017; pp. 1775–1782.
62. Hong, Z.; Yang, T.; Tong, X.; Zhang, Y.; Jiang, S.; Zhou, R.; Han, Y.; Wang, J.; Yang, S.; Liu, S. Multi-Scale Ship Detection From SAR and Optical Imagery Via A More Accurate YOLOv3. *IEEE J. Sel. Top. Appl. Earth Obs. Remote Sens.* **2021**, *14*, 6083–6101. [CrossRef]

63. Cao, C.; Wu, J.; Zeng, X.; Feng, Z.; Wang, T.; Yan, X.; Wu, Z.; Wu, Q.; Huang, Z. Research on Airplane and Ship Detection of Aerial Remote Sensing Images Based on Convolutional Neural Network. *Sensor* **2020**, *20*, 4696. [CrossRef]
64. Yang, R.; Wang, R.; Deng, Y.; Jia, X.; Zhang, H. Rethinking the Random Cropping Data Augmentation Method Used in the Training of CNN-Based SAR Image Ship Detector. *Remote Sens.* **2020**, *13*, 34. [CrossRef]

Journal of
Marine Science and Engineering

MDPI

Article

A Ship Route Planning Method under the Sailing Time Constraint

Yuankui Li [1], Jinlong Cui [1,*], Xinyu Zhang [1] and Xuefeng Yang [2]

[1] College of Navigation, Dalian Maritime University, Dalian 116026, China; liyuankui@dlmu.edu.cn (Y.L.); zhangxy@dlmu.edu.cn (X.Z.)
[2] School of Shipping and Naval Architecture, Chongqing Jiaotong University, Chongqing 400074, China; yangxuefeng523@126.com
* Correspondence: fuming97@dlmu.edu.cn

Abstract: This paper realizes the simultaneous optimization of a vessel's course and speed for a whole voyage within the estimated time of arrival (ETA), which can ensure the voyage is safe and energy-saving through proper planning of the route and speed. Firstly, a dynamic sea area model with meteorological and oceanographic data sets is established to delineate the navigable and prohibited areas; secondly, some data are extracted from the records of previous voyages, to train two artificial neural network models to predict fuel consumption rate and revolutions per minute (RPM), which are the keys to route optimization. After that, speed configuration is introduced to the optimization process, and a simultaneous optimization model for the ship's course and speed is proposed. Then, based on a customized version of the A* algorithm, the optimization is solved in simulation. Two simulations of a ship crossing the North Pacific show that the proposed methods can make navigation decisions in advance that ensure the voyage's safety, and compared with a naive route, the optimized navigation program can reduce fuel consumption while retaining an approximately constant time to destination and adapting to variations in oceanic conditions.

Keywords: ship routing; artificial neural network; speed configuration; A* algorithm

Citation: Li, Y.; Cui, J.; Zhang, X.; Yang, X. A Ship Route Planning Method under the Sailing Time Constraint. *J. Mar. Sci. Eng.* **2023**, *11*, 1242. https://doi.org/10.3390/jmse11061242

Academic Editor: Mihalis Golias

Received: 16 May 2023
Revised: 7 June 2023
Accepted: 15 June 2023
Published: 17 June 2023

1. Introduction

With improvements in ship intelligence, big data technology, and navigation-related sensors, Maritime Autonomous Surface Ships (MASS) have attracted significant attention. At the same time, safety and energy-saving issues related to maritime navigation have gradually become the focus of attention in this new low-carbon era. In 2018, the International Maritime Organization adopted the *IMOInitial Strategy for Greenhouse Gas Emission Reduction from Ships*, sending a strong signal to the international community that the shipping industry is changing to a low-carbon industry [1]. Reducing fuel consumption through proper route planning is an important means to respond to the requirement for a low-carbon strategy. However, due to the length of ships' transoceanic voyages, which can range from two weeks to more than a month, weather and sea state forecasts cannot be accurately made, and the accuracy of these forecasts decreases as the time increases. Therefore, it is required that the ship route optimization should fully consider the dynamically meteorological marine environment, and the route needs to be re-evaluated and updated when forecast data are updated [2].

Weather routing can be defined as finding an optimum route in consideration of the ETA, sailing waypoints, sailing speed, and fuel consumption based on the weather forecast data and ship performance [3]. According to this definition, some researchers mainly extract and analyze sailing records to find the association of vessels to routes [4] or summarize a data-driven optimal route in a specific navigation region [5]. Others focus on solving a path-finding problem at the operation level. The main goal of this operational problem is to plan a navigation route from one port to another based on a predefined purpose [6], in which conditions (weather, waves, and other environment variables) are along the route.

Since the ship's navigational task often has complex influences, planning a ship route is often regarded as a pathfinding problem, which assumes that the ship's speed or main engine power is maintained at a fixed value and only the ship's geographical path is considered as a variable. In this way, route optimization has been researched based on various methods or algorithms, such as the modified isochrones method [7], dynamic programming [8–10], Dijkstra's algorithm [11], and the A* algorithm [12], and some evolutionary algorithms such as the genetic algorithm [13,14] and the swarm algorithm [15] have also been used. However, for ocean-going ships, the meteorological and marine environment changes rapidly, and it is not easy to maintain a stable and uniform speed for a long time on the route. The above research ignores many constraints of the meteorological environment and the ship itself. Thus, it is difficult to ensure the result is the globally optimal solution [2].

In order to solve such problems, some 3D algorithms have been developed, in which the time variable is considered as the third dimension. Taking minimal fuel consumption as the goal, some scholars have used genetic algorithms to optimize meteorological routes under the premise of ensuring the ships' safety in adverse sea conditions [16–18]. Evolutionary algorithms, such as genetic algorithms, have the potential to balance local searches and global searches. However, when the ocean voyage is so long that it has many feasible routes, the limited search capability in the optimization process may fail to obtain the optimal global solution. Another research interest is to add a time variable to an existing two-dimensional route searching deterministic algorithm so that it becomes a three-dimensional path searching deterministic algorithm, such as the modified three-dimensional isochrones method with weighting factors [19] and three-dimensional dynamic programming algorithms that include meteorological factors [20–24]. These algorithms mainly use the idea of staged optimization in dynamic programming. In each stage of the optimization process, only a few nodes with the best results can enter the next stage, while the other sub-path nodes are eliminated to reduce the computational effort [6]. Meanwhile, in each stage of the recursion process, the discarded nodes may be associated with the optimal global solution with deterministic constraints. In contrast, the A* algorithm has the advantage of retaining all nodes in each recursion and finding the optimal nodes in stages; thus, it increases the possibility of finding the optimal global solution. However, it also results in an exponential increase in the number of search nodes with each stage of the recursion process, which makes it difficult to add a time-like dimension to the route optimization problem.

With the continuous development of artificial intelligence technology, its use to achieve ship weather route optimization has become a new research interest. The idea is to generate recommended routes after summarizing and analyzing a large amount of ship navigation data. Some scholars process AIS data to recognize key turning regions and connect these turning regions via cluster similarity measuring to generate reasonable routes for different types of ships [25,26]. Others try to use artificial neural networks and machine learning algorithms to predict the fuel consumption of a ship under different sailing conditions to achieve the goal of determining the expected duration or saving energy [27]. For example, artificial neural networks have been used to optimize a ship's speed based on large amounts of ship operation data to reduce fuel consumption [28,29]. Three different statistical models have been used to forecast and optimize the speed of container ship routes with the goal of improving navigational safety and determining expected duration by integrating meteorological information such as wave height, wave period, wind speed, and other information [30]. From the above research, it can be shown that different weather route optimization methods often require different applicable conditions and generally need to assume some ideal conditions, such as a more stable ship navigation state, fewer optimization nodes or dimensions, etc.

For a voyage's safety and economy, ship course and speed are the two key variables in saving energy and controlling arrival time. In an actual navigation situation, distinct course and speed choices may cause a vessel to encounter radically different sea states, so a ship's course and speed should be planned according to the dynamic sea environment, and most commercial software do so, such as the Bon voyage system, Seaware enroute, SPOS, and so on [31]. Normally, some environmental factors, like the resistance of calm water [32], waves [33], and winds [34], can significantly affect a ship's speed. Calculating the values of ship dynamics is extremely complex because of the consideration of hull shape, seakeeping characteristics of the ship, the sea spectrum, and other parameters. Thus, it is too laborious and time-consuming for ship weather routing. In addition, under severe sea conditions, fuel consumption can be reduced exponentially by properly reducing a ship's speed, thus the ship can adjust its speeds in different sea states to reduce its fuel consumption [27].

Therefore, this paper proposes a ship route planning method under a sailing time constraint, which introduces speed as a variable in the ship route optimization process to realize the co-optimization of ship course and speed. In the next section, a three-dimensional sea environment model is constructed, which can load weather data in real time to inform the route optimization process. Section 3 builds an ANN-based model that relates ship parameters to fuel consumption rate and RPM for use in the optimization process. Section 4 introduces a speed variable to the route optimization process to jointly optimize ship course and speed, to design the voyage route, and to plan for deceleration or acceleration to effectively reduce fuel consumption. As a result of adding a variable, it can be understood from the above description of the 3D deterministic algorithms that the number of sub-paths increases exponentially as the algorithm iterates [35,36]. For ship route planning oriented towards transoceanic voyages, this will increase the searching time significantly. Thus, to compute a navigation strategy under the sailing time constraint in an acceptable computational time, Section 5 proposes a customized A* algorithm, which uses ETA to constrain the search space and a suitable heuristic function to guide the direction of the search. Two simulations are performed in Section 6 to test and highlight the main features of the method, followed by conclusions in Section 7.

2. Dynamic Sea Environment Model

2.1. Three-Dimensional Sea Environment Model

In current navigation situations, route optimization should comply with the following restrictions:

1. The ship should navigate in an area with sufficient water depth.
2. The ship should not sail in dangerous wind and wave conditions.
3. The route plan should consider dynamic meteorological and sea conditions for long-distance and long-term navigation.

To satisfy the above constraints, a three-dimensional sea environment model is established for ship route optimization. Most of the forecast data and reanalysis data of the current meteorological and sea state are in "GRIB" format, and this type of data is used to divide the ocean area into a sequence of dense grids. Each grid cell has uniform meteorological and oceanographic properties; hence, grid cells can be taken as the route nodes and discretize ship routes into sequences of waypoints.

The environmental grid data in this paper include meteorological and oceanographic data from the European Centre for Medium-Range Weather Forecasts [37] and ocean bathymetry data from the General Bathymetric Chart of the Oceans [38]. Each two-dimensional grid datum at a single time is read from the environmental data and recorded in the form of a matrix. These multiple time matrices are stored in a 3D matrix, as shown in Figure 1, where $t_1, t_2, \ldots t_n$ represent the meteorological and oceanographic information in the grid cells at discrete times t, and the time interval between them is generally fixed.

Moreover, the environmental data in row x and column y at time t are defined by $E(x, y, t)$. The environmental effect on a ship can be expressed as follows:

$$E(x, y, t) = f(\varphi, \lambda, v, t_0) \quad (1)$$

Here, (φ, λ) is the ship's location, v is the ship's speed, and t_0 is the ship's departure time. After the calculation of the waypoints sequence and the corresponding speed configuration from the departure time t_0, the goal grid (x, y) will be found to arrive at time t. In addition, the index (x, y) is also related to the range and accuracy of the environmental data.

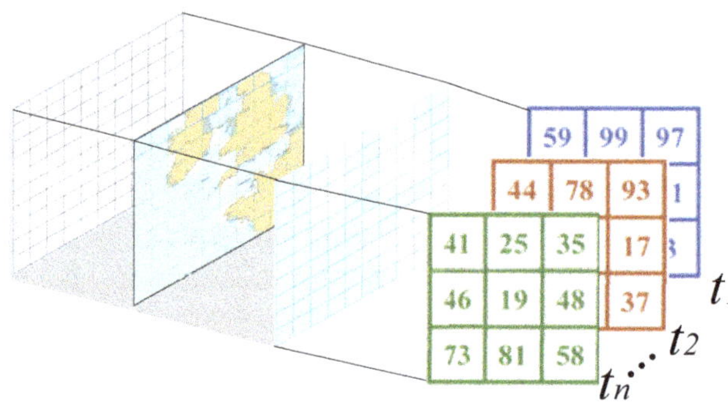

Figure 1. Three-dimensional marine environmental model.

2.2. Binarization of Navigation Area

After reading and storing the environmental data, the data representation is shown in Figure 2. The white contour lines in the figure are contours of wave height, the direction of the arrows indicates the wind direction, their length indicates the value of the wind speed, and the elevation is represented by the isosurface.

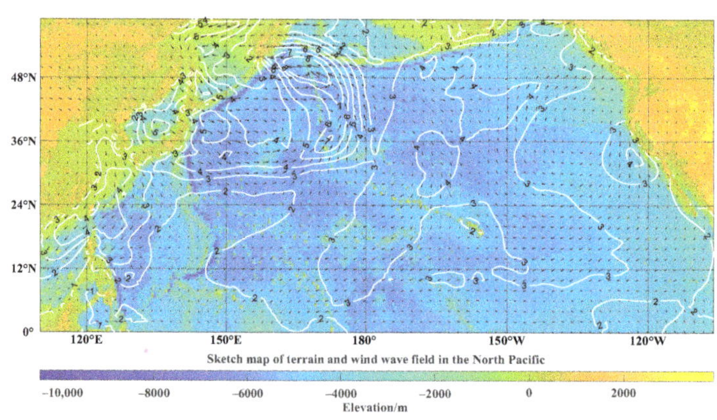

Figure 2. Marine environment visualization.

As can be seen in Figure 2, there are obstructive conditions in the map, including static obstacles such as islands and shoals in the marine environment, and dynamic restricted navigation zones corresponding to areas of strong winds and high waves (e.g., the contour-dense area in Figure 2 has wave heights above 6 m). Grid cells with obstructive conditions should be marked as unnavigable cells through which the route is forbidden to pass, while others marked as navigable, in which the ship can sail through unconstrained. Significantly, the classification and marking of the cells should be pre-processed considering the ship's characteristics and performance factors.

A binary method is used in grid pre-processing to divide the navigation area, and the navigable cells are set to 1; unnavigable cells are set to 0, which are also called restricted grids. The grid pre-processing is shown in the Formula (2). Combining the ship draft and elevation information, when the water depth of the grid (x, y) is not sufficient to ensure the safety of shipping, it is set as unnavigable and expressed as $Navi(x, y) = 0$. Also, when the wind and wave of the grid (x, y) exceeds the ship's anti-wave ability at time t, it is set as unnavigable and expressed as $Navi(x, y, t) = 0$.

$$Navi(x, y, t) = \begin{cases} 0, & insufficient\ water\ depth, \\ 0, & wind\ and\ waves\ over\ the\ threshold, \\ 1, & meet\ the\ ship's\ navigation\ conditions. \end{cases} \quad (2)$$

In this way, the marine environment in Figure 2 can be binarized, and the restricted grids are filled with black while the free grids are filled with white. Thus, the restricted and free grids can be distinguished. And the accuracy of grids is often the same as the acquired environment data, which is set to $0.5° \times 0.5°$ in this paper, as shown in Figure 3. In this figure, a sea area with wave heights over 6 m is set as a unnavigable area, as framed in the red box.

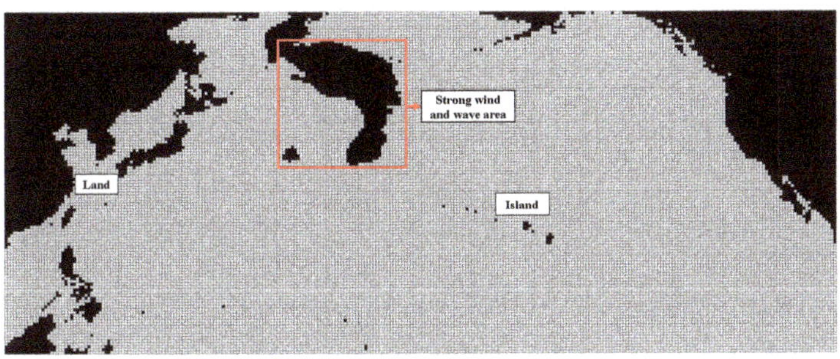

Figure 3. Binary image of the raster marine environment.

3. Predictive Model for Ship Parameters

Since the sea status during the voyage is not always static, the relevant parameters of the ship will constantly change (e.g., fuel consumption rate, speed, course, etc.), which affects the corresponding sailing decisions. Therefore, establishing a reliable and effective model to predict these effects is necessary for accurate route optimization. Artificial neural networks can be used for modeling complex systems. They are highly adaptable, robust, fault-tolerant, and surface-fitting and are well-suited to predictive analysis. Since the prediction model needs to be continuously invoked in route planning, a prediction model with a relatively simple structure and high prediction accuracy is required, and an artificial neural network model with a single hidden layer can meet such needs [39].

3.1. Pre-Processing of Data

To accurately predict the ship's sailing state under various external factors, a large amount of sailing data and an artificial neural network-based ship parameter prediction model are needed. In this work, the training and testing data are from actual navigation records of one container ship's voyages from 1 January to 28 February 2021. The details of that container ship are shown in Table 1.

Table 1. Ship parameters.

Parameter	Value
Ship type	Container ship
Displacement (t)	169,700
Length (m)	348
Beam (m)	51.2
Draft (m)	13.5
Pitch (mm)	8668

A total of 36,440 consecutive records were collected during the voyages. Each record has required data types, such as date, latitude, longitude, speed over ground (SOG), course, draft, fuel consumption, and RPM, etc. Due to errors in equipment and instruments, network signals, and other factors, there are noisy and erroneous data in the records. The predictions of the model may have large errors compared with the actual values if all records are directly used for model training, making it difficult to reflect the actual fuel consumption under many influencing factors. So, a data cleaning method proposed by ISO based on the idea of removing track anomalies was designed to filter the data and eliminate useless data [40]; the major steps are shown below:

1. Remove anomalous records in which data are incomplete, such as missing latitude and longitude data.
2. To reduce the complexity of the artificial neural network mentioned later, the absolute directions of weather are converted to the directions to the ship reference frame:

$$dir_{wr} = |180° - |(dir_c - dir_{wa}) \bmod 360°|| \tag{3}$$

   where dir_{wr} represents directions of weather components (wind and wave) to the ship's reference frame, dir_c is the ship's course, and dir_{wa} is the absolute directions of weather components.
3. According to this method [40], extract sequential records of approximately 10 min and combine them into one segment.
4. Calculate the mean value μ of each type of data d_i in a segment.

   For data that are not measured as angles, the mean μ for the N data points in a segment with values d_i is computed by

$$\mu = \frac{1}{N}\sum_{i=1}^{N} d_i \tag{4}$$

   For data that are measured as angles, the mean μ is computed by

$$\mu = \arctan\left(\frac{\sum_{i=1}^{N}\sin d_i}{N} \Big/ \frac{\sum_{i=1}^{N}\cos d_i}{N}\right) \tag{5}$$

5. Calculate the standard error of the mean σ of each type of data in a segment. The standard error of the mean σ is computed by

$$\sigma = \sqrt{\frac{1}{N}\sum_{i}^{N}\Delta_i^2} \tag{6}$$

   For data not measured as angles, the difference Δ_i is computed by

$$\Delta_i = |d_i - \mu| \tag{7}$$

For data measured as angles, the difference Δ_i is computed by

$$\begin{cases} \Delta_i = 360^\circ - r_i, & r_i = \mathrm{mod}\left(|d_i - \mu|, 360^\circ\right) > 180^\circ \\ \Delta_i = r_i, & r_i = \mathrm{mod}\left(|d_i - \mu|, 360^\circ\right) \leq 180^\circ \end{cases} \quad (8)$$

6. Two thresholds are used to delete segments in which the standard error of the mean of certain data is too large. In particular, if the σ of RPM is greater than 3 min^{-1} or the σ of SOG is greater than 0.5 kt, then all records in that segment are deleted. After this operation, several uniform sailing segments are obtained.
7. We use Chauvenet's criterion to determine if data are anomalous in each uniform sailing segment. Once a data point is identified as an outlier based on the criterion, it is removed. A new mean and standard error of the mean can be calculated based on the remaining values and the new sample size.
The probability for the occurrence of any d_i is computed by

$$P(d_i) = \mathrm{erfc}\left(\frac{\Delta_i}{\sigma \cdot \sqrt{2}}\right) \quad (9)$$

where $P(d_i)$ is the probability of occurrence of d_i and erfc is the complementary error function.
A datum is considered an outlier if Formula (10) is fulfilled.

$$P(d_i) \times N < 0.5 \quad (10)$$

This method cleans the original voyages' records to obtain several navigation segments with a duration of about 10 min and a uniform speed. The mean values of these segments' data can be used for subsequent training and prediction of the model.
8. The different types of input data have different units. It would reduce the performance and convergence of the predictive model if the filtered data of the segments were input into the neural network directly. Therefore, the z-score standardization method is adopted to standardize the data:

$$d_{ty}^* = \frac{d_{ty} - \mu_{ty}}{\sigma_{ty}} \quad (11)$$

where d_{ty}^* is one type of data in a segment after the standardization, d_{ty} is this type of data unprocessed in the segment, and μ_{ty} and σ_{ty} are the mean and standard error value of this type of data in all segments.

3.2. *Predictive Model for Ship Parameters*

An artificial neural network (ANN) has good nonlinear mapping capability, adaptive learning capability, and parallel information processing capability. Compared with traditional system identification methods, ANNs do not require knowledge of the physical causal relationships between the observed system variables, providing a way to model the system without information about the internal state of the system. In this paper, an ANN with a single hidden layer is used as a prediction tool for fuel consumption rate and RPM because of its simplicity, robustness, and ideal prediction effect [39,41]. In addition, the ship's sailing performance also changes as time passes, and the ANN can be reconstructed quickly based on more recent sailing data when necessary.

3.2.1. Structure of the Predictive Model

An artificial neural network consists of several connection layers, which can be divided into an input layer, hidden layer, and output layer according to their position and function. All neurons in the network are connected to adjacent neurons located in different hierarchies. The structure of an ANN with a single hidden layer is shown in Figure 4.

Here, w_{ij} is the connection weight from the ith neuron in the input layer to the jth neuron in the hidden layer. v_{jl} is the connection weight from the jth neuron in the hidden layer to the lth neuron in the output layer. θ_j is the bias of the jth neuron in the hidden layer. γ_l is the bias of the lth neuron in the output layer. f is the activation function of the hidden layer, and g is the activation function of the output layer.

An ANN establishes a mapping from N-dimensional to Q-dimensional data through the above process, and the functional relationships can be expressed by the following formulas:

$$b_j = f\left(\sum_{i=1}^{n} w_{ij}a_i + \theta_j\right) \tag{12}$$

$$c_l = g\left(\sum_{j=1}^{p} v_{jl}b_j + \gamma_l\right) \tag{13}$$

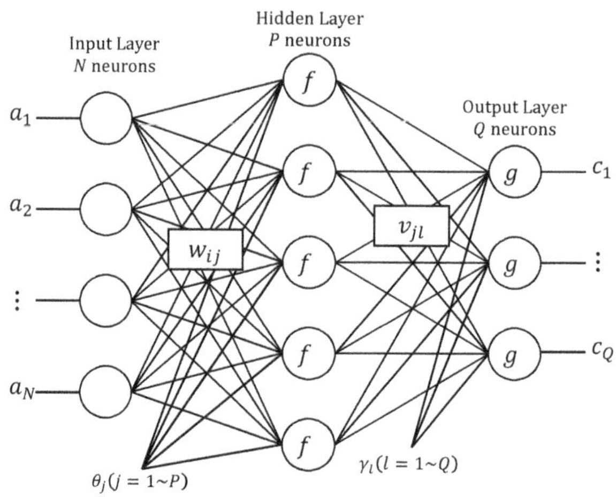

Figure 4. Multiple-layer ANN with a single hidden layer.

The activation function of the hidden layer $f(x)$ must be a bonded nonlinear function that is continuous, smooth, and monotonically increasing. So, the Relu activation function is used for its advantages of simple computation, simple derivatives, fast convergence, one-sided inhibition, and broader boundaries.The activation function of the output layer $g(x)$ does not necessarily need to be nonlinear, so a linear function is used. The relevant activation functions are shown as follows:

$$f(x) = \max(0, x) \tag{14}$$

$$g(x) = x \tag{15}$$

3.2.2. Error Back-Propagation Algorithm

The error back-propagation algorithm [42] is a method that uses the difference between the actual output and the desired output to correct the weight parameters layer by layer from back to front, which applies to supervised learning. When learning samples are provided to the network, if there is a gap between the desired output and the output based on current weights and biases, the error function of the current output and the desired output is calculated, and then the error signal is back-propagated along the original connection path to update the neuron weight parameters of each layer.

Assuming that M learning samples are provided to the network, the error function of the network for each learning sample is constructed as follows.

The goal of learning is to minimize a given error function, and network training is generally stopped when the error function value terminates a downward trend and starts to rise:

$$E^k = \frac{1}{2}\sum_{l=1}^{P}\left(y_l^k - c_l^k\right)^2 \tag{16}$$

where k is the learning sample number ($k = 1 \sim M$), l is the dimension of the output layer ($l = 1 \sim Q$), E^k is the error function of the kth learning sample, y_l^k is the expected output in the lth dimension of the kth learning sample, and c_l^k is the actual output of the lth dimension of the kth learning sample.

The network calculates the global error after obtaining all M samples and updates the parameters uniformly based on the global error E. The global error E of the network is the sum of all the learning sample errors:

$$E = \sum_{k=1}^{M} E^k \tag{17}$$

This work uses the Levenberg–Marquardt (LM) algorithm to update the parameters. The LM algorithm combines the advantages of the gradient descent method and the Newton method, as its error function decreases faster at the beginning, and provides the ideal search direction when the error function is near the optimal value [43]. The parameter updating formula of the LM algorithm is as follows:

$$\omega(i+1) = \omega(i) - (H + \lambda I)^{-1}\frac{\partial E}{\partial \omega} \tag{18}$$

Here, ω represents the neural network parameters, including connection weights between different layers and the offsets of each layer; i is the number of learning times; H is the Hessian Matrix of the error function; and I is the unit matrix.

The parameter λ is updated by taking larger values in the early stages of learning so that the parameters are updated along the inverse direction of the gradient, while the direction of iteration of the step is shifted to the direction of the Newton method as λ gradually decreases to zero in the later stages [43]. Compared with other iterative methods, the LM algorithm has a relatively good convergence rate for training networks of medium size [44].

3.3. Selection of Model Variables

The two essential criteria for evaluating the utility of a sailing route are sailing time and fuel consumption. Except for voyage distance, the variables most closely related to these two criteria are the speed over ground (SOG) and fuel consumption rate. SOG is affected by the navigation environment in a natural sea state, and differs from the calm water speed due to added resistance from waves, winds, and currents, as well as the reduction in propulsive efficiency caused by waves and increased resistance, which is often known as ship speed loss. Ship speed loss is a critical parameter in route planning for voyage time and fuel consumption evaluation, and it is calculated based on the calm water speed and the actual sea state in the general route optimization process; then, the ship's actual SOG is obtained. But in voyage records, it can be noted that the expected calm water speed is not recorded, so the traditional way to calculate speed loss using calm water speed and the SOG in natural sea state is no longer practical. Fortunately, there is the real-time RPM data in the voyage record, which is a measurement closely related to the expected calm water speed.

So, the actual SOG is taken as one optimization variable in this work, and once the SOG is assigned, the sailing time can easily be calculated, and the recommended value of RPM can be predicted through the ANN model, as shown in Figure 4, which can make the ship sail at the assigned SOG. In this way, the output variables of the ANN model are RPM and fuel consumption rate, and they are used to calculate the total sailing time and fuel consumption in the ship route optimization procedure.

To meet the route optimization task, the calculation and prediction of ship-related parameters, which generally need to be updated frequently to serve the route planning algorithm, should not occupy too much running time. Thus, the input variables of the ANN should have the following characteristics:

1. They can represent the current navigation state or navigation environment of the ship.
2. They can be obtained or predicted in some way in the route planning process.
3. How the data are obtained or predicted cannot be too complex or take up too much running time.

Thus, eight input variables which meet the above characteristics have been selected to establish an ANN-based model for fuel consumption rate and RPM prediction, including ship data such as the SOG, ship course, fore draft, aft draft, and sea environment data such as relative wind direction, relative wave direction, wind speed, and wave height. The relevant variable mapping relationship is shown in Figure 5.

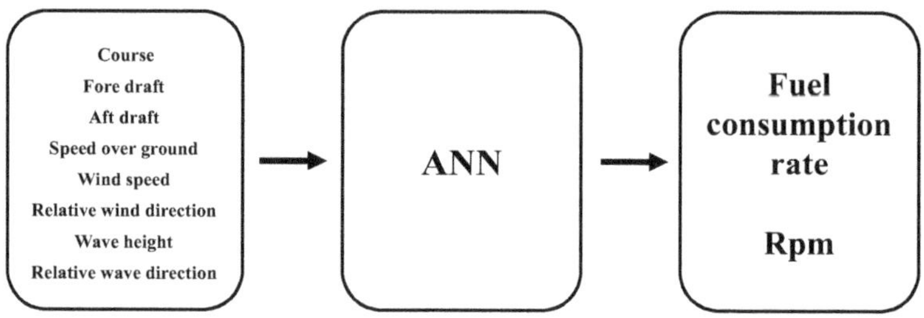

Figure 5. Variable mapping.

4. Co-Optimization of Ship Course and Speed

The route between the starting node and end node consists of several rhumb lines, and each rhumb line connects two adjacent nodes, as shown in Figure 6. Each node includes three variables: position, time, and speed. Then, a sequence of n nodes can represent the whole voyage, including the speed along each leg.

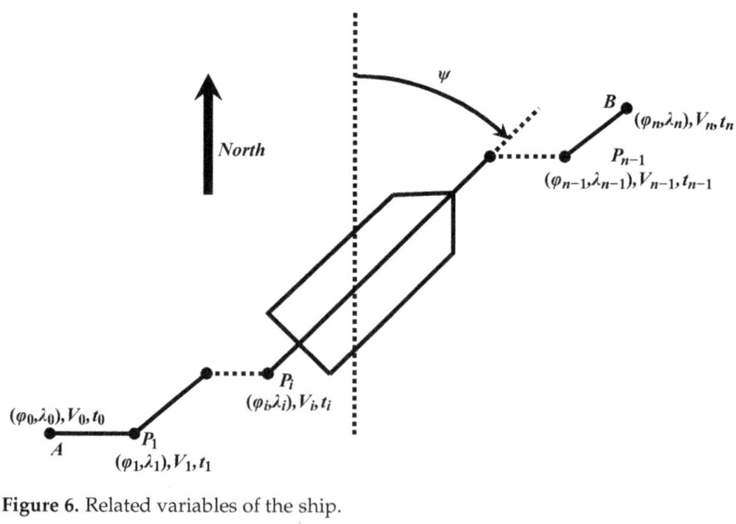

Figure 6. Related variables of the ship.

The course is represented by the orientation of the rhumb line connecting two adjacent nodes; by adjusting the position of the nodes, the course is changed. The time when the ship arrives at the next node is determined by the course and speed, and the meteorological and sea conditions that the ship encounters vary as time goes by. So, the route and its speed are the keys to ship route decision, and by adjusting the position of nodes and the speed between them, the whole route can be optimized.

Figure 7 shows the process of route generation. The centers of grids marked with 83, 67, 45 indicate nodes where the ship is located; the surrounding arrows refer to alternative courses from one node to the next; and t_1, t_2, t_3 are the times when the ship reaches these nodes in sequence. When the ship chooses different courses and sails with different speeds, the nodes and their arrival times are different too, so the ship will encounter different wind and wave states, and the sailing time and fuel consumption may differ accordingly. The wind and wave state the ship encounters can be obtained using the 3D sea environment model in Section 2.1, while the fuel consumption can be calculated using the predictive model in Section 3.

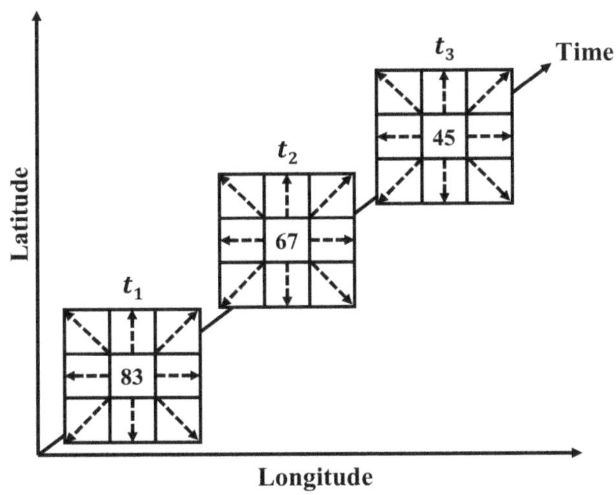

Figure 7. The process of route generation.

Thus, node selection and speed configuration are the two key methods in ship voyage optimization, and optimizing ship course as well as speed is an efficient way to plan a voyage, so a method [illegible] of a ship's course and speed is proposed in this work.

4.1. Speed Configuration

Usually, the speed during the voyage will be maintained in a range according to the ship's performance. To ensure safety and economic efficiency, the ship will sail at a suitable speed in a certain sea area, and the set of permissible speeds is presented as follows:

$$v = [v_{min}, v_{min} + v_{cd}, v_{min} + 2v_{cd}, \ldots, v_{max}] \tag{19}$$

Here, v is a row matrix vector including n permissible speeds, its interval is v_{cd}, v_{min} is the minimum value in v, and v_{max} is the maximum value in v.

Accordingly, since there are multiple permissible speeds in each node, a leg from one node to an adjacent node is expanded from 1 to n. Alternatively, one node i is extended to n "ext-nodes" i_w, described as follows:

$$i_w = \{P(i, v_w) | v_w \in v\} \tag{20}$$

Here, i_w means the ship sails from node i to the next adjacent node with fixed speed v_w, and P contains all navigation states at node i. In addition, one sailing node is expanded into a row matrix. Thus, this method can be used to achieve parallel computation and speed up the algorithm operation.

Therefore, a speed configuration can be a part of an optimized route, as shown in Figure 8. The expression form of the optimized route from the starting node S to the end node T evolves from the original $S - A - B - C - D - E - T$ to $S_{13} - A_{12} - B_{13} - C_{14} - D_{12} - E_{15} - T$. That is, the ship sails from the origin node S to node A at 13 kt, from node A to node B at 12 kt, and continues sailing until from node E to the end node T at 15 kt. With this method, the optimized route not only contains the optimized route (Figure 9) but also includes the optimal speed in each segment (Figure 10) to realize the synergistic optimization of the ship's course and speed.

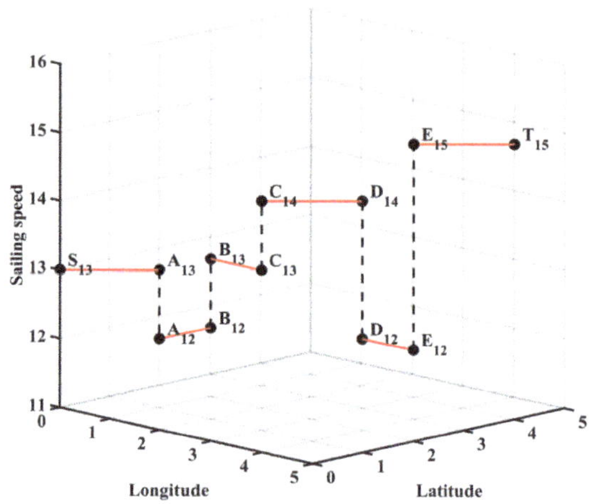

Figure 8. Diagram of the nodes after adding sailing speed.

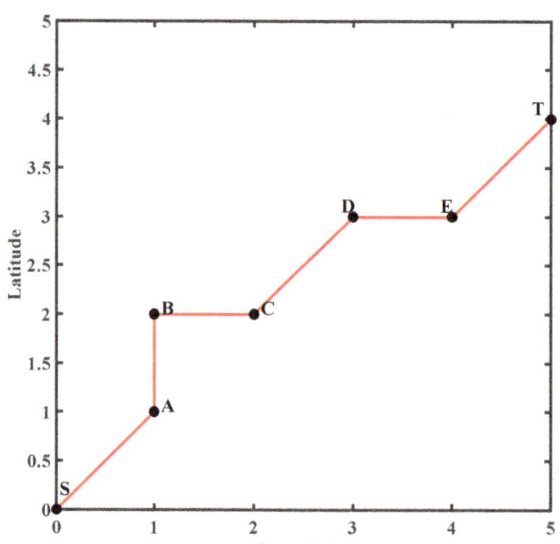

Figure 9. Route formed by the projection.

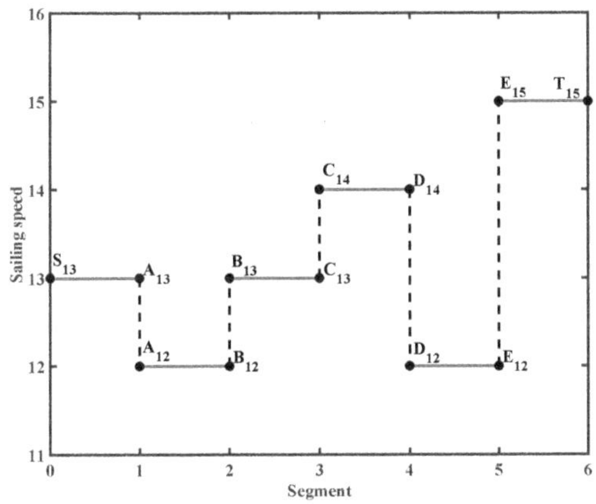

Figure 10. Sailing speed formed by the projection.

5. Weather Route Optimization for Ships with Sailing Time Constraints

5.1. A Algorithm*

The A* algorithm is based on Dijkstra's algorithm, with an added heuristic function for predicting the future cost of movement to influence the search direction and narrow the search space, so that it can find the shortest path in a certain sense in a short time.

The main process of the A* algorithm is to build an "*Openlist*" node set to store nodes to be checked and a "*Closedlist*" node set to store checked nodes, and iterate continuously to select the node with the lowest movement cost to obtain the shortest path. The basic form of the function to evaluate the movement cost is shown in the following formula:

$$F(i) = G(i) + H(i) \tag{21}$$

Here, i is the current node. F is the estimated total movement cost from the starting node to the end node. G is the minimum movement cost recorded from the starting node to the current node, and the heuristic function H is the estimated movement cost from the current node to the end node.

The A* algorithm calculates and iterates the node selection according to this evaluation function. Its basic process is as follows: [illegible] of surrounding nodes is calculated from the starting node. The node with the minimum movement cost is selected in each iteration until the end node is found. The process is shown in Figure 11.

The A* algorithm has characteristics of fast calculation speed to an optimal solution, but there are several problems that need to be solved for route optimization tasks, as follows:

1. The two-dimensional A* algorithm takes distance as the evaluation criterion. At the same time, ship fuel consumption and navigation time are the main factors to evaluate the navigation cost of a route, and taking distance as the evaluation criterion cannot accurately quantify the navigation cost.
2. The two-dimensional A* algorithm does not consider speed. In contrast, the ship does not sail at a fixed speed over the whole voyage; dynamically adjusting speed according to the sea state is desirable.
3. The two-dimensional A* algorithm does not consider dynamic unnavigable areas with heavy winds and waves, which will affect the safety and efficiency of the voyage.

Thus, the A* algorithm needs to be modified to meet the actual needs of ship navigation.

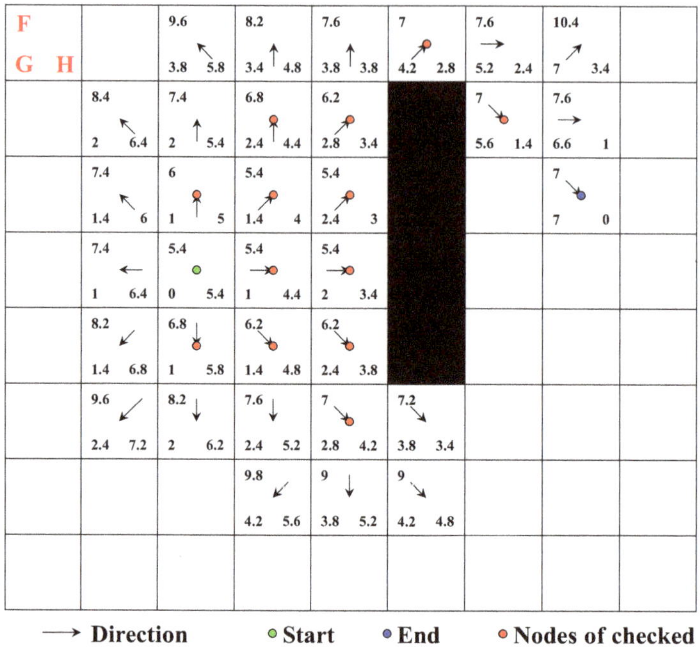

Figure 11. The search process of the A* algorithm.

5.2. Evaluation Functions Concerning Time and Fuel Consumption

The Estimated Time of Arrival (ETA) is the time when a ship or vessel is expected to arrive at a specific destination. An accurate ETA can make entire supply chains more efficient and reliable, and it also helps to determine the expected duration of a vessel's route.

Generally, as the two most important factors in evaluating a route, voyage time and fuel consumption need to be fully considered in the evaluation function. In fact, the voyage schedule is the crucial constraint for merchant vessels, and in the scope of reducing gas emissions, the effect of an increased sailing speed could be enhanced by the waiting time for a free berth, therefore accounting for even more emissions. The ship must arrive at the specific destination at a time close to the ETA, and thus the expected duration is taken as a constraint in the route design process, and minimum fuel consumption is the optimization objective.

In order to meet the above description, two evaluation functions for sailing time and fuel consumption are proposed:

$$FT_{iw} = GT_{iw} + HT_{iw} \tag{22}$$

$$FF_{iw} = GF_{iw} + HF_{iw} \tag{23}$$

Here, GT_{iw} and GF_{iw} are the actual time and fuel consumption of the route from the starting node to the current ext-node i_w, described as follows:

$$GT_{iw} = \sum_{i=1}^{n} \frac{d_i}{v_w} \tag{24}$$

$$GF_{iw} = \sum_{i=1}^{n} \frac{d_i}{v_w} \times q_{iw} \tag{25}$$

Let h_w and i_w describe two adjacent ext-nodes and h_w be the previous node; then, d_i is the distance between h_w and i_w, and $\frac{d_i}{v_w}$ refers to the sailing time from h_w to i_w at the speed v_w. q_{iw} refers to the fuel consumption rate when the ship sails from h_w to i_w. The value of q_{iw} can be calculated using the fuel consumption rate model in Section 4.1.

In addition, it is well-known that since the heuristic function H affects the search efficiency of the A* algorithm, a suitable H function can guarantee a faster search speed and the quality of the route [45]. To ensure the solution quality with a short running time, heuristic functions with suitable search efficiency, named HT and HF, are proposed as follows:

$$HT_{iw} = dE_i / v_w \tag{26}$$

$$HF_{iw} = \frac{dE_i}{v_{eco}} \cdot f_q(v_{eco}) \tag{27}$$

where dE_i is the distance from node i to the end node; HT_{iw} is the estimated time when the ship reaches the end node from the current ext-node i_w at the speed v_w; HF_{iw} is the estimated fuel consumption from the current ext-node i_w to the end node at speed v_{eco}; v_{eco} is the economic speed of the ship, usually artificially set; and f_q is a function of the fuel consumption rate and speed.

The relationship of fuel consumption rate with speed is generally a cubic relationship, as shown in the Formula (28):

$$f_q(v) = av^3 + bv^2 + cv + d \tag{28}$$

where a, b, c, and d are fitting coefficients that vary with the ship's characteristics.

5.3. Customized A* Algorithm with Time Constraint

After reconstructing the evaluation function, the process of the A* algorithm is also customized accordingly. The process is as follows, which is shown in Figure 12:

1. Initialize *Openlist* (the set of ext-nodes to be checked) and *Closedlist* (the set of ext-nodes checked).
2. All ext-nodes belonging to the start node S are added to the *Openlist*, and their GT and GF are set to 0, while the HT and HF are calculated and recorded.
3. If it is judged that the *Openlist* is not empty, the operation goes to Step 4; otherwise, there is no solution to this problem, which means no available [illegible] and T, and the procedure is terminated.
4. The total navigation time FT of each ext-node in the *Openlist* is calculated. If the total navigation time FT of one ext-node is greater than the pre-set ETA, the node is retained in the *Openlist* but will not be executed in the rest of this step. Then, the ext-node i_w with the minimum fuel consumption of the whole voyage is found in these nodes which meet the time limit. And the ext-node i_w will be removed from the *Openlist* and added into the *Closedlist*, setting the SOG record to v_w.
5. Traverse all the adjacent nodes of the navigation node i. There are two situations for the ext-nodes of these adjacent nodes:
    - If there is no ext-node of the adjacent node j in *Openlist* and *Closedlist*, all ext-nodes of node j are added to *Openlist*, the G and H value of these ext-nodes are calculated and recorded, and the parent node of these ext-nodes is set as node i_w.
    - If there is an ext-node j_w of the adjacent node j in the *Openlist*, and the GT and GF values of j_w calculated in this step are both smaller than those stored in the *Openlist*, the G and H values of j_w in the *Openlist* will be updated correspondingly in this step. Also, the ext-node j_w's parent node is set as node i_w. Conversely, all information of the ext-node j_w stored in *Openlist* will remain unchanged.

6. Check whether there is any ext-node in *Openlist* that is subordinate to the end node *T*. If not, repeat Step 3; if it already exists, a path composed of ext-nodes is found by querying the parent node of the current ext-node step by step until the starting node. The sailing route and speed configuration are obtained after mapping like in Section 4.1.
7. The weather forecast data and navigation data at each ext-node on the route are extracted based on the waypoint sequence and speed configuration. Then, the RPM prediction model is called, and the above data are modified as the input of the prediction model to calculate the recommended RPM of each ext-node on the route. Then, the RPM recommendation scheme is output for the whole voyage.

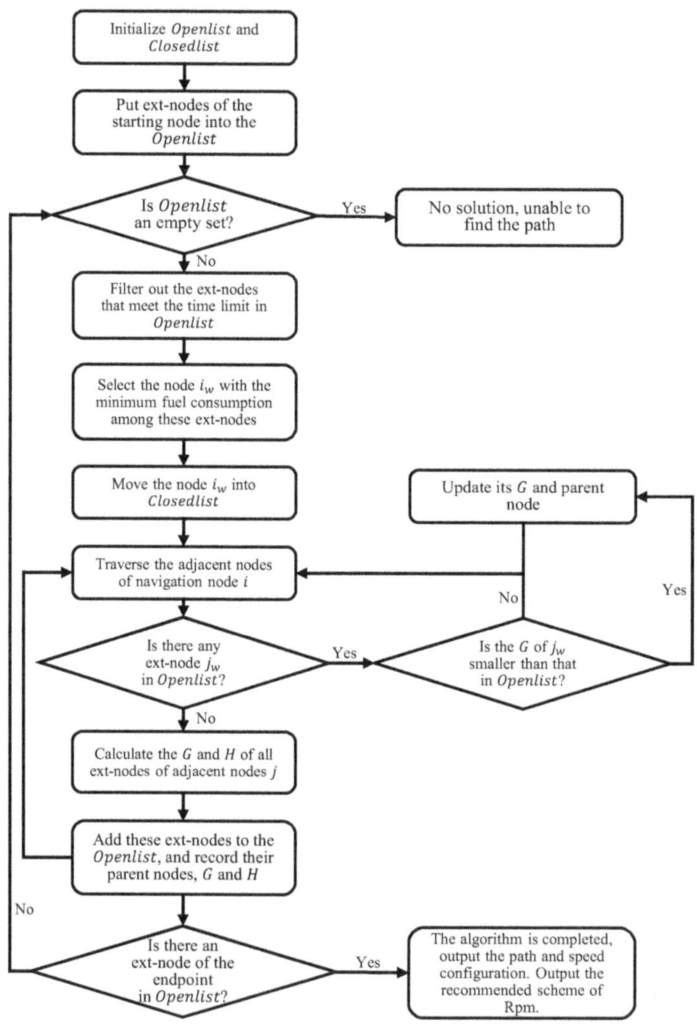

Figure 12. Flow chart of the customized A* algorithm.

5.4. Overall Optimization Flow

The weather route optimization method proposed in this paper consists of four parts: ship historical data processing, the ship parameter prediction model, ship navigation data processing, and ship route optimization. Briefly, the work flow is as follows, which is shown in the Figure 13:

1. The historical data processing part uses historical ship voyage records and corresponding meteorological historical data, etc. These data are standardized by pre-processing data methods, and the pre-processed data are used as input for the ship parameter prediction models.
2. The ship parameter prediction model part is used to build two artificial neural network models with the same frame. The input contains eight variables: SOG, fore draft, aft draft, course, wind speed, wind direction, wave height, and wave direction; the output is fuel consumption rate or RPM. The trained models are used for calculating fuel consumption rate and RPM in the voyage afterward.
3. The ship navigation data processing part obtains and divides the ship's characteristics and the weather forecast data into eight input variables which meet the requirements of the ship parameter prediction model and are used in the route planning process afterward.
4. The ship route optimization part is based on the customized A* algorithm, using the ship fuel consumption rate prediction model and the current sailing data to calculate the fuel consumption for all nodes to be checked in order to find the node with the minimum fuel consumption that satisfies the ETA constraint and thus obtain the optimal ship route. Finally, the RPM recommendation scheme for the whole range is obtained based on the sailing data and the generated route.

Notably, the ship route can be frequently updated during the voyage. The frequency of calculation during the sailing voyage would depend on the fleet's will and the updating frequency of the forecast weather data. If the voyage is too long to acquire the forecast data at the later part of the voyage, this method would find the optimum route based on the current weather information. In addition, the updated route may have great difference with the prior one. But both routes are planned based on the assurance of sailing safety and economy, which can be identified as the optimal sailing strategy at their respective states.

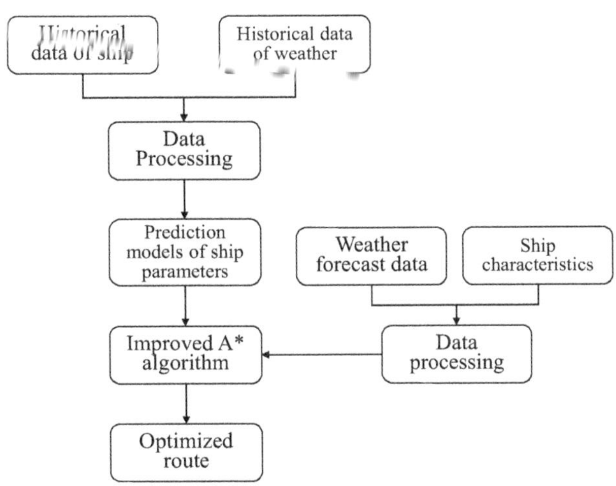

Figure 13. Diagram of the overall optimization process.

6. Simulation Experiment and Results Analysis

6.1. Performance Analysis of Ship Parameter Prediction Models

In order to evaluate the prediction effect of the artificial neural network model designed in this paper, the data set is divided into a training set and a test set. The training set is used to train the model, and the test set is used to test the generalization ability of the model. Each ten-minute segment (extracted in Section 3.1) is taken as a data point by calculating its average value. A total of 1976 data points are obtained from the data set, of which 1580 data points are randomly picked and used as the training set, and the remaining 396 data points are used as the test set. The parameters of fuel consumption rate model and RPM model are shown in Table 2. The prediction effect of the two models' test sets are shown in Figures 14 and 15, and the blue dots are the actual value of the test set.

Table 2. ANN parameters.

Parameters	Value
Number of neurons in hidden layer of fuel consumption rate model	45
Number of neurons in hidden layer of RPM model	140
Activation function of the hidden layer	Relu
Optimizer	Levenberg–Marquardt

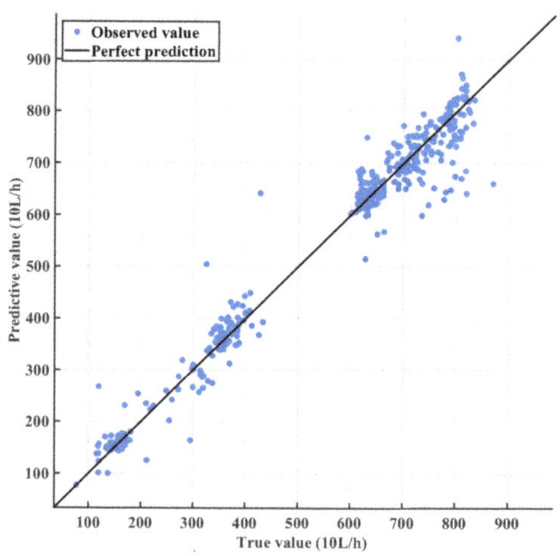

Figure 14. Effect of the fuel consumption rate model.

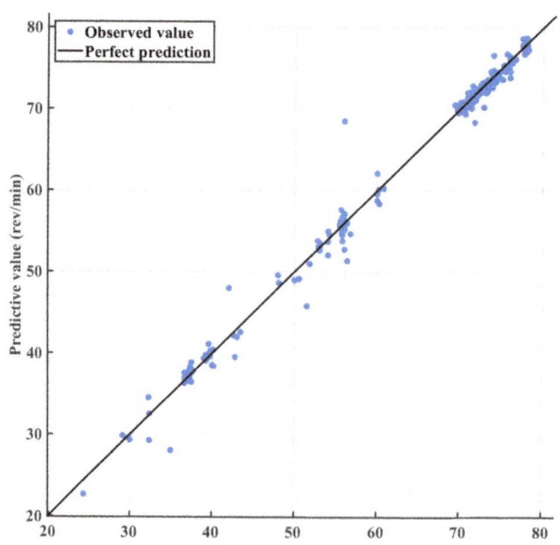

Figure 15. Effect of the RPM model.

A perfect predictive model should have a predicted value equal to the true value, which means the points would lie on the black diagonal line. The vertical distance from the line to any point is the prediction error for that point. A good model has minor errors, as Figures 14 and 15 show; all the blue dots surround the black lines, and the average relative error of the fuel consumption rate model and RPM model are 0.0594 and 0.0101, respectively. Thus, the above models can meet the navigation requirements in practice and can be used in the ship route optimization process to calculate fuel consumption and RPM.

6.2. Simulation Experiment of Ship Weather Route Optimization System

The proposed ship route optimization method under the sailing time constraint, described in detail in Section 5, has been validated and verified in a series of computer simulations based on actual voyage records. Two scenarios have been selected with different passages (voyage departure and destination points) and times (different weather conditions, having a direct impact on the optimization process). The simulations are programmed using MATLAB, and the accuracy of the relevant meteorological data is $0.5° \times 0.5°$, as mentioned in Section 2.

6.2.1. Scenario 1: Voyage from Osaka to Los Angeles on 3 February 2021

In this scenario, a voyage from Osaka with a departure on 3 February 2021, at 18:00 UTC with a destination of Los Angeles has been simulated; its expected duration is 370 h, and its economic speed is set to 13.5 kt. The primary purpose of this simulation is to verify the ability to avoid wind and wave areas.

The experimental parameters are listed in Table 3, in which the threshold values are set according to the actual demand of the ship's voyage.

Table 3. Experimental parameters.

Parameters	Value
Start	35.5° N, 141° E
End	34° N, 120° W
SOG range (knot)	[8, 19]
Maximum allowable wind speed (m/s)	20
Maximum allowable wave height (m)	6
Minimum draught (m)	14

The ship's course and speed were optimized using the methods proposed in this work, and an optimal route was generated. As shown in Figure 16, the optimal route (red curve) is compared with a great circle (GC, green curve) with SOG of 13.5 kt for the whole voyage. For Figure 16, the wave heights are represented by the isosurface and the wind speeds are represented by contour lines. The four charts describe the wind and wave conditions and the routes at different times (80 h, 165 h, 230 h, and 367 h, referring to the time after the ship departs).

From Table 4, it can be concluded that when the ship sails along the great circle route, there are 26.26 h of sailing time in heavy wind and wave areas, accounting for about 7.27% of the total sailing time, which is very unfavorable to the ship's navigational safety, while the optimized route generated by the proposed method in this work avoids the dangerous sea area with heavy wind and waves.

Table 4. Comparison of routes.

	Time (h)	Danger Time (h)
Optimized route	367.15	0
Great circle route	361.38	26.26

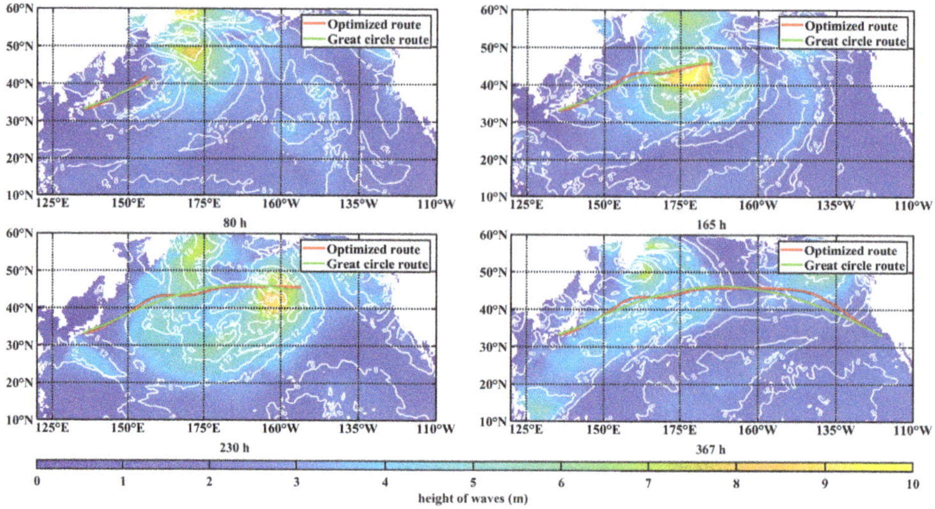

Figure 16. Comparison of ship routes at different moments.

The distribution of and variation in the wind and waves during the voyage are demonstrated in Figure 16, where "80 h" refers to the situation that the optimized route chooses to sail to higher latitudes in advance based on the analysis of meteorological information; at about 165 h and 230 h, there are heavy winds and waves on the great circle route, and the optimized route avoids this area in advance; "367 h" is the comparison of the two routes after reaching the destination point.

The speed distribution is shown in Figure 17, and the wave heights on the two routes are shown in Figure 18. Define the route between each two adjacent nodes as a "segment", and the segment numbers from the starting node to the end node increase in sequence.

It can be seen from Figure 18 that the optimized route encounters a significantly lower incidence of high significant wave heights compared with the great circle route, and the maximum wave height is also smaller. Therefore, the algorithm can handle the upcoming navigational risks in advance. The speed configuration is also more reasonable than the case of uniform speed for the whole voyage.

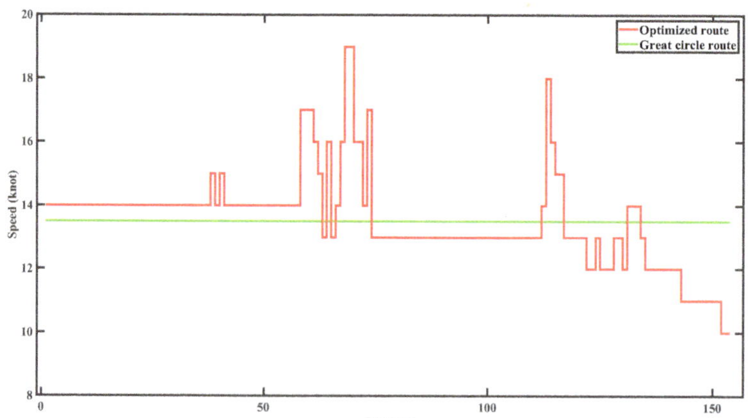

Figure 17. Ship speed configuration for the whole range.

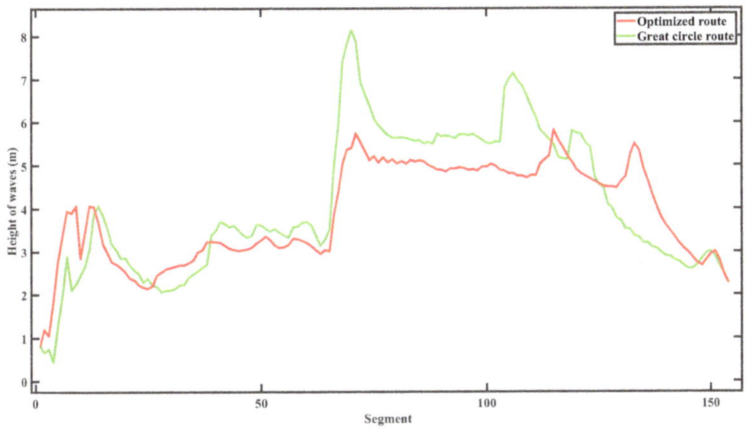

Figure 18. Wave height on the routes.

6.2.2. Scenario 2: Voyage from Tainan to Los Angeles on 17 January 2021

In this scenario, a voyage from Tainan (23° N, 120° E) with a departure on 17 January 2021 at 07:00 UTC with a destination of Los Angeles (34° N, 120° W) was simulated, and its expected duration is 410 h. The primary purpose is to compare the optimized route with the historical route to evaluate the effectiveness of the algorithm. Some experimental parameters of the ship are the same as in Scenario 1.

As shown in Figure 19, the optimized route (red curve) is compared with the historical route (green curve). The economic speed of the optimal route is set to 13.5 kt. It is known that the ship's safety was not compromised on the historical route, so the comparison focuses on the sailing time and fuel consumption (FOC).

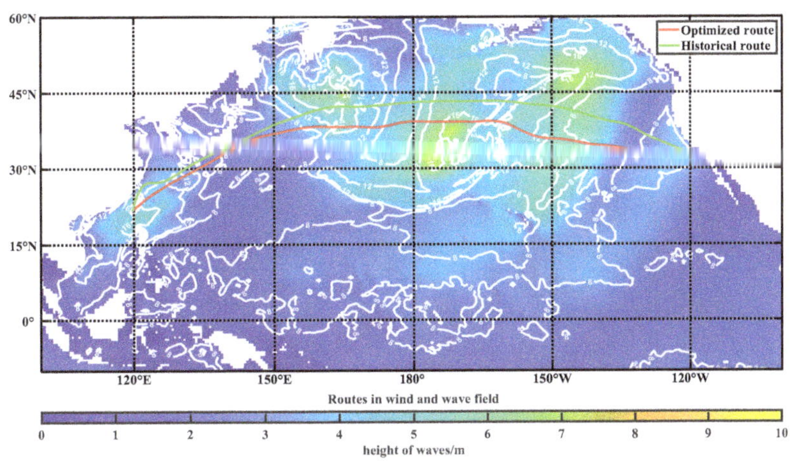

Figure 19. Comparison of routes under wind and wave fields.

As seen from Table 5, compared with the historical route, the optimized route saves about 130.93 tons of fuel consumption (7.25%) within the specified sailing time. And the increase in voyage time is by 4.35 h (1.08%), which is tolerable for the route planning. Thus, it achieves the purpose of saving energy by designing the optimal route and reasonably configuring speeds. In addition, the speed configuration (Figure 20) and required RPM (Figure 21) for the whole voyage are also generated. Moreover, the output RPMs of the whole voyage will also guide the ship to reach the planned SOGs, which provides more assistance for real-time ship navigation.

Table 5. Comparison of routes.

	Time (h)	FOC (t)
Optimized route	408.69	1674.27
Historical route	404.34	1805.20

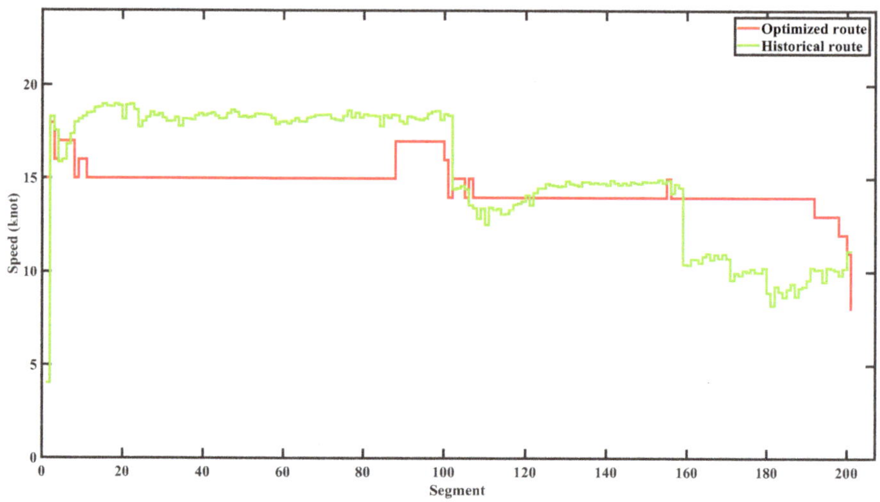

Figure 20. Ship speed configuration for the whole range.

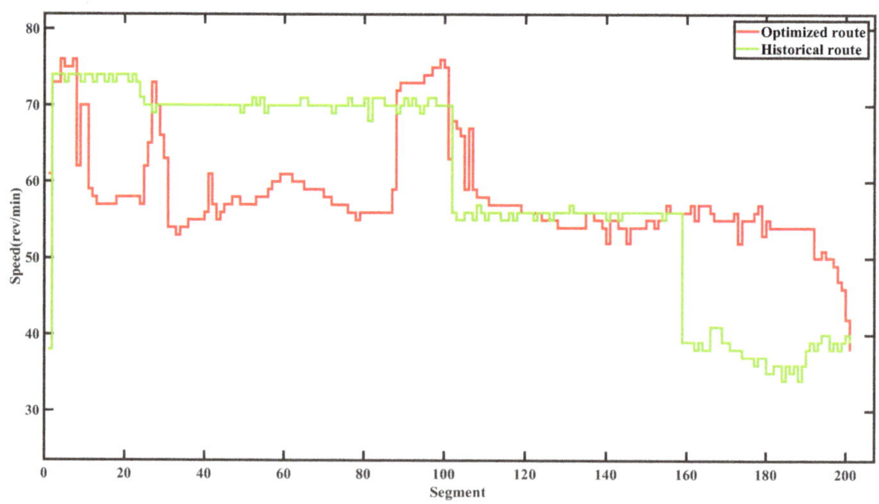

Figure 21. RPM configuration for the whole range.

6.2.3. Computational Complexity of This Method

This paper adds one dimension about the SOG to the route planning method, which may cause a exponential increase in the computational cost. But with the rational design of the customized A* algorithm's evaluation function and operation process, the computational cost is significantly reduced, as Table 6 shows. When the optimal route is output, the computation times for the two scenarios are 991.917 s and 731.639 s, which is tolerable in the practical application. In addition, the storage sizes of the data (includes *Openlist* and *Closedlist*) are 1.10 MB and 1.04 MB. Given all the above, the computational complexity of this method is acceptable.

Table 6. Computational cost of two scenarios.

	Computational Time (s)	Data Size (MB)
Scenario 1	991.917	1.10
Scenario 2	731.639	1.04

7. Conclusions

As the main influencing factor of navigation efficiency in the process of ship route optimization, speed control gives ample space for optimization and should be fully considered in ship route design. Based on this, this paper proposes a ship route planning method under a sailing time constraint that designs the ship course and speed simultaneously to realize an optimal navigation strategy, including route and speed configurations for the whole voyage. This navigation strategy can ensure that a ship arrives at its destination close to its ETA. Such an optimized navigation program can not only avoid the navigational risks brought by a heavy sea state but can also generate an optimal route with minimized fuel consumption while arriving at the required ETA.

In this paper, a rasterized dynamic sea area model was constructed by processing meteorological information, and speed configuration was added to the route optimization; then, single nodes were extended to collections of "ext-nodes" for optimization. Based on the actual voyages' records, an artificial network model was modeled to predict the fuel consumption rate and RPM to assist the route optimization. Then, the evaluation function and operation process of an A* algorithm were designed to achieve the simultaneous optimization of vessel course and speed based on the premise of arriving at ETA. The results of two optimization simulations of the winter route in the North Pacific Ocean showed that the algorithm proposed in this paper could not only guarantee navigational safety but could also reduce fuel consumption by 130.93 tons or about 7.25% compared with the historical route. Therefore, the methods in this paper can help ships reduce their fuel consumption while avoiding windy and stormy areas and still arrive at their ETA, which has particular practical significance in accomplishing complex transoceanic tasks.

In the future, we will consider more specific complexities of the sailing environment (such as currents, ice-covered waters, etc.) by obtaining and analyzing data from more voyages or different types of ships. In addition, while an ANN with a single hidden layer has great performance, there is space to optimize the prediction of the fuel consumption rate, which can be further investigated. We have up to now only used the three-dimensional A* algorithm to plan a low-fuel-consumption route under a time constraint, while the needs of ship navigation may be very diverse, and future attempts will be made to use such three-dimensional algorithms to solve multi-objective ship navigation tasks and further improve the computational speed.

Author Contributions: Conceptualization, Y.L. and J.C.; methodology, Y.L. and J.C.; software, J.C. and X.Y.; validation, Y.L., J.C. and X.Z.; formal analysis, J.C.; data curation, X.Y.; writing—original draft preparation, Y.L.; writing—review and editing, Y.L. and J.C.; supervision, X.Z. All authors have read and agreed to the published version of the manuscript.

Funding: This work was funded by the Fundamental Research Funds for the Central Universities (NO. 3132023150); the Natural Science Foundation Project of Chongqing, Chongqing Science and Technology Commission (NO. cstc2019jcyj-msxmX0729); and Fund of the National Engineering Laboratory of Transport Safety and Emergency Informatics (NO. YW170301-05).

Institutional Review Board Statement: Not applicable.

Informed Consent Statement: Informed consent was obtained from all subjects involved in the study.

Data Availability Statement: The data that support the findings of this study are available from the corresponding author, Jinlong Cui, upon reasonable request.

Conflicts of Interest: The authors declare no conflict of interest.

References

1. Chircop, A. The IMO Initial Strategy for the Reduction of GHGs from International Shipping: A Commentary. *Int. J. Mar. Coast. Law* **2019**, *34*, 482–512. [CrossRef]
2. Wang, H.; Mao, W.; Eriksson, L. Benchmark Study of Five Optimization Algorithms for Weather Routing. In *International Conference on Offshore Mechanics and Arctic Engineering*; American Society of Mechanical Engineers: New York, NY, USA, 2017; Volume 57748, p. V07BT06A023. [CrossRef]
3. Simonsen, M.H.; Larsson, E.; Mao, W.; Ringsberg, J.W. State-of-the-Art within Ship Weather Routing. In Proceedings of the International Conference on Offshore Mechanics and Arctic Engineering. American Society of Mechanical Engineers Digital Collection, St John's, NL, Canada, 31 May–5 June 2015. [CrossRef]
4. Iphar, C.; Jousselme, A.L. A geometry-based fuzzy approach for long-term association of vessels to maritime routes. *Ocean Eng.* **2023**, *281*, 114755. [CrossRef]
5. Pallotta, G.; Vespe, M.; Bryan, K. Vessel Pattern Knowledge Discovery from AIS Data: A Framework for Anomaly Detection and Route Prediction. *Entropy* **2013**, *15*, 2218–2245. [CrossRef]
6. Wang, H.; Mao, W.; Eriksson, L. A Three-Dimensional Dijkstra's Algorithm for Multi-Objective Ship Voyage Optimization. *Ocean Eng.* **2019**, *186*, 106131. [CrossRef]
7. Hagiwara, H. Weather Routing of (Sail-Assisted) Motor Vessels. Ph.D. Thesis, Technical University of Delft, Delft, The Netherlands, 1989.
8. Bijlsma, S.J. On Minimal-Time Ship Routing. Ph.D. Thesis, Delft University of Technology, Staatsdrukkerij Den Haag, The Netherlands, 1975.
9. Bijlsma, S. On the Applications of Optimal Control Theory and Dynamic Programming in Ship Routing. *Navigation* **2002**, *49*, 71–80. [CrossRef]
10. De Wit, C. Proposal for Low Cost Ocean Weather Routeing. *J. Navig.* **1990**, *43*, 428–439. [CrossRef]
11. Padhy, C.P.; Sen, D.; Bhaskaran, P.K. Application of Wave Model for Weather Routing of Ships in the North Indian Ocean. *Nat. Hazards* **2008**, *44*, 373–385. [CrossRef]
12. Liu, Y.; Wang, T.; Xu, H. PE-A* Algorithm for Ship Route Planning Based on Field Theory. *IEEE Access* **2022**, *10*, 36490–36504. [CrossRef]
13. Wang, L.; Zhang, Z.; Zhu, Q.; Ma, S. Ship Route Planning Based on Double-Cycling Genetic Algorithm Considering Ship Maneuverability Constraint. *IEEE Access* **2020**, *8*, 190746–190759. [CrossRef]
14. Pan, C.; Zhang, Z.; Sun, W.; Shi, J.; Wang, H. Development of Ship Weather Routing System with Higher Accuracy Using SPSS and an Improved Genetic Algorithm. *J. Mar. Sci. Technol.* **2021**, *26*, 1324–1339. [CrossRef]
15. Zhao, W.; Wang, H.; Geng, J.; Hu, W.; Zhang, Z.; Zhang, G. Multi-Objective Weather Routing Algorithm for Ships Based on Hybrid Particle Swarm Optimization. *J. Ocean Univ. China* **2022**, *21*, 28–38. [CrossRef]
16. Marie, S.; Courteille, E. Multi-Objective Optimization of Motor Vessel Route. *TransNav Int. J. Mar. Navig. Saf. Sea Transp.* **2009**, *3*, 133–141.
17. Hinnenthal, J.; Clauss, G. Robust Pareto-optimum Routing of Ships Utilising Deterministic and Ensemble Weather Forecasts. *Ships Offshore Struct.* **2010**, *5*, 105–114. [CrossRef]
18. Maki, A.; Akimoto, Y.; Nagata, Y.; Kobayashi, S.; Kobayashi, E.; Shiotani, S.; Ohsawa, T.; Umeda, N. A New Weather-Routing System That Accounts for Ship Stability Based on a Real-Coded Genetic Algorithm. *J. Mar. Sci. Technol.* **2011**, *16*, 311–322. [CrossRef]
19. Lin, Y.H.; Fang, M.C.; Yeung, R.W. The Optimization of Ship Weather-Routing Algorithm Based on the Composite Influence of Multi-Dynamic Elements. *Appl. Ocean Res.* **2013**, *43*, 184–194. [CrossRef]
20. Shao, W. Development of an Intelligent Tool for Energy Efficient and Low Environment Impact Shipping. Ph.D. Thesis, University of Strathclyde, Glasgow, Scotland, 2013.
21. Skoglund, L.; Kuttenkeuler, J.; Rosén, A.; Ovegård, E. A Comparative Study of Deterministic and Ensemble Weather Forecasts for Weather Routing. *J. Mar. Sci. Technol.* **2015**, *20*, 429–441. [CrossRef]
22. Zaccone, R.; Figari, M. Energy Efficient Ship Voyage Planning by 3d Dynamic Programming. *J. Ocean Technol.* **2017**, *12*, 49–71.
23. Zaccone, R.; Ottaviani, E.; Figari, M.; Altosole, M. Ship Voyage Optimization for Safe and Energy-Efficient Navigation: A Dynamic Programming Approach. *Ocean Eng.* **2018**, *153*, 215–224. [CrossRef]
24. Du, W.; Li, Y.; Zhang, G.; Wang, C.; Zhu, B.; Qiao, J. Energy Saving Method for Ship Weather Routing Optimization. *Ocean Eng.* **2022**, *258*, 111771. [CrossRef]
25. Wen, Y.; Sui, Z.; Zhou, C.; Xiao, C.; Chen, Q.; Han, D.; Zhang, Y. Automatic Ship Route Design between Two Ports: A Data-Driven Method. *Appl. Ocean Res.* **2020**, *96*, 102049. [CrossRef]

26. Vouros, G.A.; Vlachou, A.; Santipantakis, G.; Doulkeridis, C.; Pelekis, N.; Georgiou, H.; Theodoridis, Y.; Patroumpas, K.; Alevizos, E.; Artikis, A.; et al. Increasing Maritime Situation Awareness via Trajectory Detection, Enrichment and Recognition of Events. In Proceedings of the Web and Wireless Geographical Information Systems, A Coruna, Spain, 21–22 May 2018; R. Luaces, M., Karimipour, F., Eds.; Springer International Publishing: Cham, Switzerland, 2018. Lecture Notes in Computer Science. pp. 130–140. [CrossRef]
27. Zis, T.P.; Psaraftis, H.N.; Ding, L. Ship Weather Routing: A Taxonomy and Survey. *Ocean Eng.* **2020**, *213*, 107697. [CrossRef]
28. Du, Y.; Meng, Q.; Wang, S.; Kuang, H. Two-Phase Optimal Solutions for Ship Speed and Trim Optimization over a Voyage Using Voyage Report Data. *Transp. Res. Part B Methodol.* **2019**, *122*, 88–114. [CrossRef]
29. Beşikçi, E.B.; Arslan, O.; Turan, O.; Ölçer, A.I. An Artificial Neural Network Based Decision Support System for Energy Efficient Ship Operations. *Comput. Oper. Res.* **2016**, *66*, 393–401. [CrossRef]
30. Mao, W.; Rychlik, I.; Wallin, J.; Storhaug, G. Statistical Models for the Speed Prediction of a Container Ship. *Ocean Eng.* **2016**, *126*, 152–162. [CrossRef]
31. Psaraftis, H.; Morales Llamas, J.; Ding, L.; Nehammer, J. *BlueSIROS Project WP3, Proof of Concept*; Technical Report, BlueSIROS Project Technical Report; Technical University of Denmark: Lyngby, Denmark, 2017.
32. Newman, J.N. *Marine Hydrodynamics*; The MIT Press: Cambridge, MA, USA, 2018.
33. Grifoll, M.; Martorell, L.; Castells, M.; de Osés, F.X.M. Ship weather routing using pathfinding algorithms: The case of Barcelona – Palma de Mallorca. *Transp. Res. Procedia* **2018**, *33*, 299–306. [CrossRef]
34. Windeck, V. Environmental Routing. In *A Liner Shipping Network Design: Routing and Scheduling Considering Environmental Influences*; Windeck, V., Ed.; Produktion und Logistik, Springer Fachmedien: Wiesbaden, Germany, 2013; pp. 39–78. [CrossRef]
35. Du, W.; Li, Y.; Zhang, G.; Wang, C.; Zhu, B.; Qiao, J. Ship weather routing optimization based on improved fractional order particle swarm optimization. *Ocean Eng.* **2022**, *248*, 110680. [CrossRef]
36. Wang, H.; Lang, X.; Mao, W.; Zhang, D.; Storhaug, G. Effectiveness of 2D optimization algorithms considering voluntary speed reduction under uncertain metocean conditions. *Ocean Eng.* **2020**, *200*, 107063. [CrossRef]
37. Hersbach, H.; Bell, B.; Berrisford, P.; Biavati, G.; Horányi, A.; Muñoz Sabater, J.; Nicolas, J.; Peubey, C.; Radu, R.; Rozum, I.; et al. ERA5 Hourly Data on Single Levels from 1940 to Present. Available online: https://cds.climate.copernicus.eu/cdsapp#!/dataset/reanalysis-era5-single-levels?tab=overview (accessed on 21 January 2022).
38. GEBCO Compilation Group. GEBCO Gridded Bathymetry Data. Available online: https://www.gebco.net/data_and_products/gridded_bathymetry_data (accessed on 21 January 2022).
39. Gkerekos, C.; Lazakis, I.; Theotokatos, G. Machine learning models for predicting ship main engine Fuel Oil Consumption: A comparative study. *Ocean Eng.* **2019**, *188*, 106282. [CrossRef]
40. *ISO 19030-2:2016*; Ships and Marine Technology—Measurement of Changes in Hull and Propeller Performance — Part 2: Default Method. International Organization for Standardization: Geneva, Switzerland, 2016.
41. Hu, Z.; Zhou, T.; Zhen, R.; Jin, Y.; Li, X.; Osman, M.T. A two-step strategy for fuel consumption prediction and optimization of ocean-going ships. *Ocean Eng.* **2022**, *249*, 110904. [CrossRef]
42. Rumelhart, D.E.; Hinton, G.E.; Williams, R.J. Learning representations by back-propagating errors. *Nature* **1986**, *323*, 533–536. [CrossRef]
43. Hagan, M.T.; Menhaj, M.B. Training Feedforward Networks with the Marquardt Algorithm. *IEEE Trans. Neural Netw.* **1994**, *5*, 989–993. [CrossRef] [PubMed]
44. Hagan, M.T.; Demuth, H.B.; Beale, M.H. *Neural Network ToolboxTM 6 User's Guide*; MathWorks: Natick, MA, USA, 2015.
45. Shin, Y.W.; Abebe, M.; Noh, Y.; Lee, S.; Lee, I.; Kim, D.; Bae, J.; Kim, K.C. Near-Optimal Weather Routing by Using Improved A* Algorithm. *Appl. Sci.* **2020**, *10*, 6010. [CrossRef]

Journal of
Marine Science and Engineering

MDPI

Article

Parameter Prediction of the Non-Linear Nomoto Model for Different Ship Loading Conditions Using Support Vector Regression

Jiafen Lan [1,2,3], Mao Zheng [2,4,*], Xiumin Chu [2,3,4] and Shigan Ding [1,2]

1 School of Transportation and Logistics Engineering, Wuhan University of Technology, Wuhan 430063, China; lanjiafen@whut.edu.cn (J.L.); dingshigan1998@whut.edu.cn (S.D.)
2 State Key Laboratory of Maritime Technology and Safety, Wuhan University of Technology, Wuhan 430063, China; chuxm@whut.edu.cn
3 School of Physics and Electronic Information Engineering, Minjiang University, Fuzhou 350108, China
4 National Engineering Research Center for Water Transport Safety, Wuhan University of Technology, Wuhan 430063, China
* Correspondence: zhengmao@whut.edu.cn; Tel.: +86-189-8608-8286

Abstract: Significant changes in the load of cargo ships make it difficult to simulate and control their motion. In this work, a parameter prediction method for a ship maneuvering motion model is developed based on parameter identification and support vector regression (SVR). First, the effects of least-squares (LS) and multi-innovation least-squares (MILS) parameter identification methods for the non-linear Nomoto model are investigated. The MILS method is then used to identify the parameters of the non-linear Nomoto model under various load conditions, and model training datasets are established. On this basis, SVR is used to predict the parameters of the non-linear Nomoto model. The results reveal that the MILS method converges faster than the LS method. The SVR method achieves lower accuracy than the MILS method, but exhibits reasonable prediction accuracy for zigzag motions, and the maneuvering motion model can be predicted as navigation conditions change.

Keywords: non-linear Nomoto model; parameter identification; multi-innovation least-squares; support vector regression

Citation: Lan, J.; Zheng, M.; Chu, X.; Ding, S. Parameter Prediction of the Non-Linear Nomoto Model for Different Ship Loading Conditions Using Support Vector Regression. *J. Mar. Sci. Eng.* **2023**, *11*, 903. https://doi.org/10.3390/jmse11050903

Academic Editor: Rafael Morales

Received: 12 March 2023
Revised: 17 April 2023
Accepted: 20 April 2023
Published: 23 April 2023

1. Introduction

In recent years, the impact of economies of scale has led to an increase in ship size, traffic intensity, and related risks [1]. Accurate models for the maneuvering of cargo ships are critical for maneuvering simulations and control applications [2]. Currently, there are three major types of ship maneuvering motion models, the ship hydrodynamics model [3,4], the Maneuvering Modeling Group (MMG) model [5], and the Nomoto model [6]. Although the ship hydrodynamics model and MMG model are of high accuracy, many hydrodynamic parameters are difficult to obtain. The Nomoto model is simple and easy to use, with hydrodynamic parameters that can typically be obtained from captive model tests, numerical calculations based on computational fluid dynamics (CFD), and the system identification method. Captive model tests require expensive towing tanks and numerical calculations consume significant computational resources. The system identification method is practical and effective, and requires little experimental time or cost. This method only needs state information and an inertia term, and does not involve force measurements. As a result, it takes several minutes for a free running test, compared with several weeks for CFD simulations, depending on the propeller modeling and the grid resolution [7]. Thus, it is easier to apply the ship maneuvering model through the system identification method. System identification theory has broad application prospects in ship motion modeling and control, combined with scaled free-running ship models or full-scale ship tests [8]. This theory has been applied in frequency domain identification, Kalman filters (KF), Gaussian process modeling, the maximum likelihood method, neural networks, support vector machines (SVM), least-squares (LS) methods, and hybrid methods. Among these, the LS method is one of the most practical approaches [9]. As an improved recursive form of the LS method, the recursive least-squares (RLS) method has the high identification accuracy and ease-of-use [10], and is widely applied in the identification of ship model parameters.

However, the present work mainly uses a single standard maneuver to generate the training data. For example, Luo et al. [11] used 25°/5° zigzag maneuver data as the training set, and predicted the zigzag maneuver motions of 25°/5° and 35°/5° by SVM and particle swarm optimization (PSO) algorithms. Zhu et al. [12] identified the model parameters by 20°/20° zigzag maneuver data using the RLS method based on SVM. Xue et al. [13] used the empirical Bayesian method to clean simulated polluted responses from a 20°/20° zigzag test and identify hydrodynamic parameters. However, the zigzag test datasets for different rudder angles corresponds to different dynamic characteristics. For this reason, scholars such as He et al. [14] used multiple standard maneuver datasets as training data, which ensures a better ability to predict maneuvering motions under a greater range of rudder angles. Wang et al. [15] used the 20°/20°, 15°/15°, and 10°/10° zigzag test data to cover as many dynamic features as possible. They investigated the fidelity of the model under different levels of perturbation and verified that SVM can achieve better generalization compared to traditional neural networks, but this approach has been validated on constant parameters and cannot be applied to time-varying coefficients. In additional, datasets with different rudder angles are generally used for training without exploring the effect of the rudder angle on the maneuverability parameters, and factors such as ship engine speed, load, and trim are not considered.

For most cargo ships, the load, trim, draft, and engine speed change greatly during daily voyages, making the dynamics characteristics highly variable and leading to uncertainty in maneuverability parameters [16]. More research is required on their effect on maneuverability, which is important for the ease of ship control and safe navigation [17]. If the ship maneuvering motion model cannot be accurately obtained in real-time, the ship's automatic motion control and autonomous collision avoidance control will be compromised. Ship collisions may lead to devastating consequences, such as a ship capsizing/sinking, resulting in oil spills and fatalities [18]. To solve this problem, Zhang et al. [19] proposed a multi-innovation least-squares (MILS) method for identifying the ship maneuvering model parameters. The MILS method can achieve higher identification accuracy and faster convergence than the RLS method. Wang et al. [20] used non-linear Gaussian filtering algorithm to solve the problem of on-line parameter identification in ship autonomous navigation control. However, during the actual navigation of cargo ships, the ship's trajectory is mostly straight without larger rudder angles (larger than 10°), resulting in insufficient parameter excitation and significant random noise in real-time training data. Thus, the real-time parameter identification method is difficult to generate reasonable results during the actual navigation process that the system identification method is almost useless for cargo ships in a daily voyage. To guarantee a strong excitation of the training data, large yaw-amplitude motions, such as zigzag tests, should be conducted. Unfortunately, it is almost impossible to do so for full-scale ships as the cost is too high. It is a reasonable approach to predict a ship's motion model in real-time based on groups of identified parameters under several typical operating conditions.

In this study, a scaled free-running ship model is used to perform zigzag tests under different loads, trims, speeds, and rudder angles to obtain the maneuvering motion model of the cargo ship. The non-linear Nomoto model parameters of the scaled free-running ship model under each navigation condition are obtained using the MILS method, and the effects of rudder angle, engine speed, bow trim, stern trim, and load are investigated. SVR is then used to predict the parameters of the non-linear Nomoto model. The prediction accuracy of the Nomoto model parameters is evaluated by comparing the ship motion simulation data with the scaled free-running ship test data. This study makes the following contributions:

(1) We propose a parameter prediction method for ship maneuvering motion models based on SVR. A training dataset of non-linear Nomoto model parameters under various navigation conditions is established. The proposed algorithm has good prediction accuracy and can quickly obtain the maneuvering model parameters.
(2) We use the MILS method to identify the parameters of the non-linear Nomoto model under various navigation conditions. Additionally, the effects of rudder angle, engine speed, trim, and load on the maneuvering parameters are analyzed, providing an optimal direction for ship maneuvering and control.
(3) Based on the parameter identification of the Nomoto model, we include the engine speed, bow and stern draft, and test rudder angle into the training set. The predicted maneuvering motion model can change with navigation conditions, matching the dynamics of the cargo ship in the daily voyage.

2. Parameter Identification of Maneuvering Motion Model

2.1. Ship Maneuvering Motion Model

Considering the balance between model accuracy and calculation efficiency, ship maneuvering motion models are established based on the first-order non-linear Nomoto models. As shown in Figure 1, we take due east as the X-axis and due north as the Y-axis to establish the X–O–Y Earth fixed coordinate system. We take the bow direction as the x_0-axis and the starboard direction as the y_0-axis to establish the ship's fixed coordinate system x_0–o–y_0.

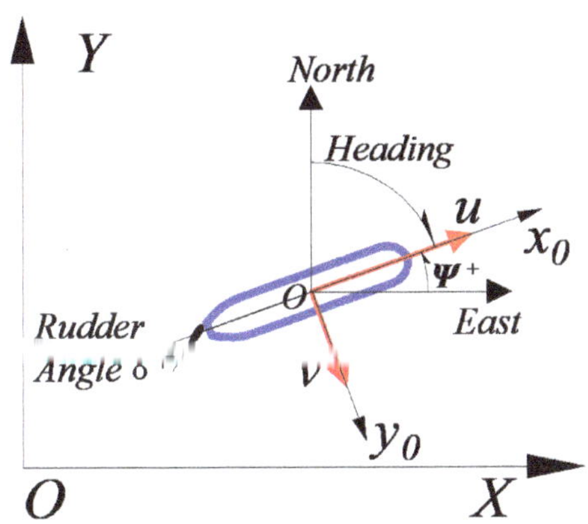

Figure 1. Coordinate system of ship maneuvering motion.

As depicted in Figure 1, Ψ is the ship's heading angle and r is the yaw angular velocity. Thus, $r = \dot{\psi}$. The first-order non-linear Nomoto model [21] are, respectively, expressed as

$$T\dot{r} + r + \alpha r^3 = K(\delta + \delta_r), \quad (1)$$

where K is the turning index; T are time coefficients; δ_r is the effective neutral rudder angle; α is the coefficient of the non-linear term; and δ is the actual rudder angle. The ship speed in the x_0- and y_0- directions are u and v, respectively. The ship speed can be expressed as a vector $\mathbf{v} = [u, v, r]$ and converted to the Earth fixed coordinate system as follows

$$\dot{\eta} = \mathbf{R}(\psi)\mathbf{v}, \quad (2)$$

$$\mathbf{R}(\psi) = \begin{bmatrix} \sin(\psi) & \cos(\psi) & 0 \\ -\cos(\psi) & \sin(\psi) & 0 \\ 0 & 0 & 1 \end{bmatrix}, \tag{3}$$

where $\boldsymbol{\eta}$ is the position vector and $\mathbf{R}(\psi)$ is a rotation matrix that maps vectors from a body fixed frame to an inertial frame.

2.2. Model Parameter Identification

2.2.1. Model Discretization

Assuming that the data samples collected from the experiments are discrete with a time step of h, let $y(t) = \psi(t) - \psi(t-1)$ and $\dot{\psi}(t) = \frac{\psi(t+1)-\psi(t)}{h}$. Then, $\ddot{\psi}(t) = \frac{y(t+2)-y(t+1)}{h^2}$. With the forward difference quotient used to replace the derivative, Equation (1) can be rewritten as:

$$T\ddot{\psi}(t) = K(\delta(t) + \delta_r) - \dot{\psi}(t) - \alpha\dot{\psi}^3(t), \tag{4}$$

$$y(t+2) - y(t+1) = \frac{K}{T}h^2\delta(t) + \frac{K\delta_r}{T}h^2 - \frac{1}{T}hy(t+1) - \frac{\alpha}{T}\frac{y^3(t+1)}{h}, \tag{5}$$

and

$$y(t) - y(t-1) = \frac{K}{T}h^2\delta(t-2) + \frac{K\delta_r}{T}h^2 - \frac{1}{T}hy(t-1) - \frac{\alpha}{T}\frac{y^3(t-1)}{h}. \tag{6}$$

Let all parameters be identified as $\mathbf{A} = \begin{bmatrix} \frac{K}{T} & \frac{K\delta_r}{T} & -\frac{1}{T} & -\frac{\alpha}{T} \end{bmatrix}^T$ and all known variables set as $\mathbf{x}(t) = \begin{bmatrix} h^2\delta(t-2) & h^2 & hy(t-1) & \frac{y^3(t-1)}{h} \end{bmatrix}^T$. Then, Equation (6) can be written as:

$$Y(t) = y(t) - y(t-1) = \mathbf{x}^T(t) \cdot \mathbf{A}. \tag{7}$$

The left-hand side of Equation (7) is the differences in heading angles. $\mathbf{x}(t)$ is the information vector and $\mathbf{A}$ is the vector to be identified.

The zigzag maneuver, also known as the Kempf overshoot or "Z" maneuver, is one of the standard test methods for measuring the maneuverability of ships [22]. It is convenient for observation and comparison. Through zigzag tests of the scaled free-running ship model, the heading angle sequences $\boldsymbol{\psi}(t) = [\psi(1), \psi(2), \ldots, \psi(t)]$ and rudder angle sequences $\delta(t) = [\delta(1), \delta(2), \ldots, \delta(t)]$ are obtained. These test data are utilized for identification based on the LS and MILS methods, respectively. The maneuvering motion model parameters (K, T, α, and δ_r) can then be determined.

2.2.2. Identification Based on LS

It is assumed that the input and output data of the model are sampled from $t = 1$ to $t = n$, and then the output data $\{Y(1), Y(2), \ldots, Y(n)\}$ and information data $\{\mathbf{x}(1), \mathbf{x}(2), \ldots, \mathbf{x}(n)\}$ are obtained. The expanded matrix equations are as follows:

$$\begin{cases} \mathbf{Y} = \mathbf{X}^T\mathbf{A} \\ \mathbf{Y} = [\ Y(n) \quad Y(n-1) \quad \ldots \quad Y(1)\ ]^T \\ \mathbf{X} = [\ \mathbf{x}(n) \quad \mathbf{x}(n-1) \quad \ldots \quad \mathbf{x}(1)\ ] \end{cases}. \tag{8}$$

The LS loss function is expressed as:

$$V_{LS} = \frac{1}{n}\sum_{t=1}^{n}(Y(t) - \mathbf{x}^T(t)\mathbf{A})^2. \tag{9}$$

Minimizing the LS loss function gives $\hat{\mathbf{A}}$, which is the estimated value of $\mathbf{A}$:

$$\hat{\mathbf{A}} = [\mathbf{X}\mathbf{X}^T]^{-1}\mathbf{X}\mathbf{Y}. \tag{10}$$

2.2.3. Identification Based on MILS

The recursive equations of the RLS method [23] are

$$\begin{cases} \mathbf{A}(t) = \mathbf{A}(t-1) + \mathbf{P}(t)\mathbf{x}(t)e(t) \\ e(t) = Y(t) - \mathbf{x}^T(t)\mathbf{A}(t-1) \\ \mathbf{P}^{-1}(t) = \mathbf{P}^{-1}(t-1) + \mathbf{x}(t)\mathbf{x}^T(t) \end{cases}, \tag{11}$$

where $\mathbf{P}(t)$ is the covariance matrix and $e(t)$ is the innovation scalar for each iteration. The feature of RLS is that the estimation of parameters in each iteration mainly depends on the current updated information, and past information is generally not used. In real complex navigation conditions, environmental interference often has a negative impact on traditional identification methods. To solve this problem, a multi-innovation identification theory based on an information window with a certain length is proposed [24]. This solves small-sample estimation problems based on the innovation vector. To improve the recognition accuracy by reusing the past state and measurement information, the multi-innovation length p is introduced and the innovation scalar $e(t)$ is extended to the multi-innovation vector $\mathbf{E}(p,t)$:

$$\mathbf{E}(p,t) = \begin{bmatrix} e(t) \\ e(t-1) \\ \vdots \\ e(t-p+1) \end{bmatrix} = \begin{bmatrix} Y(t) - \mathbf{x}^T(t)\mathbf{A}(t-1) \\ Y(t-1) - \mathbf{x}^T(t)\mathbf{A}(t-1) \\ \vdots \\ Y(t-p+1) - \mathbf{x}^T(t-p+1)\mathbf{A}(t-1) \end{bmatrix}. \tag{12}$$

Similarly, the output vector and information vector of the MILS method are:

$$\begin{cases} \mathbf{Y}(p,t) = \begin{bmatrix} Y(t) & Y(t-1) & \cdots & Y(t-p+1) \end{bmatrix}^T \\ \mathbf{X}(p,t) = \begin{bmatrix} \mathbf{x}(t) & \mathbf{x}(t-1) & \cdots & \mathbf{x}(t-p+1) \end{bmatrix} \end{cases}. \tag{13}$$

Thus, the recursive equations of the MILS method can be written as:

$$\begin{cases} \mathbf{A}(t) = \mathbf{A}(t-1) + \mathbf{P}(t)\mathbf{X}(p,t)\mathbf{E}(p,t) \\ \mathbf{E}(p,t) = \mathbf{Y}(p,t) - \mathbf{X}^T(p,t)\mathbf{A}(t-1) \\ \mathbf{P}^{-1}(t) = \mathbf{P}^{-1}(t-1) + \mathbf{X}(p,t)\mathbf{X}^T(p,t) \end{cases}. \tag{14}$$

3. Parameter Prediction of Maneuvering Motion Model

3.1. Analysis of Navigation Parameters

For cargo ships, the characteristics of the maneuvering motion models will be affected by navigation parameters, such as the ship speed, load, trim, and rudder angle. Considering the detectability, these parameters are simplified as the following.

(1) Engine speed R_T

This study considers a engine that is used to drive the propeller directly, and the engine speed can be accurately measured by the Hall effect element. Thus, the engine speed is indirectly used to reflect the ship speed.

(2) Bow draft D_F and stern draft D_A

The bow and stern draft are used to describe the load and trim of the ship. These quantities can be read from the bow and stern draft scales.

(3) Test rudder angle δ_T

We consider $10°/10°$, $20°/20°$, and $30°/30°$ rudder angles in the zigzag tests. The rudder angle can be measured by an incremental encoder.

Based on the parameter identification of the Nomoto model, the engine speed, bow and stern draft, and test rudder angle are included in the training set for training the SVR model.

3.2. Parameter Training Based on SVR

It is necessary to establish a Nomoto model that can automatically adapt to the navigation conditions. Considering the sample set size, a SVR-based parameter-learning method for the maneuvering motion model is proposed. Let the training sample set be $(\mathbf{x}_i, y_i), i = 1, 2, \ldots, N$, $\mathbf{x}_i \in \mathbf{R}^d$, where $\mathbf{x}_i$ and y_i denote the input and output spaces, respectively, d is the sample dimension, and N is the number of samples. Then, $\mathbf{R}^d$ represents the d-dimensional vector space in which the samples are located. We establish a non-linear mapping from the input space to the output space, $\varphi(\mathbf{x})$:$\mathbf{R}^d \to \mathbf{H}$, and use linear regression to analyze the sample data in the high-dimensional feature space $\mathbf{H}$. The SVR model $f(\mathbf{x}) = \boldsymbol{\omega} \cdot \varphi(\mathbf{x}) + b$ can then be established, where $\boldsymbol{\omega}$ is the weight vector and b is the offset. Introducing the slack variables ξ_i and ξ_i^*, the regularized risk function can be transformed into the dual optimization problem,

$$\min J = \frac{1}{2}\|\boldsymbol{\omega}\|^2 + C\sum_{i=1}^{l}(\xi_i + \xi_i^*), \tag{15}$$

$$\text{subject to}: \begin{cases} y_i - \boldsymbol{\omega} \cdot \varphi(\mathbf{x}_i) - b \leq \varepsilon + \xi_i \\ -y_i + \boldsymbol{\omega} \cdot \varphi(\mathbf{x}_i) + b \leq \varepsilon + \xi_i^* \quad i = 1, 2, \cdots, l. \\ \xi_i \cdot \xi_i^* \geq 0 \end{cases} \tag{16}$$

where J is the regularized risk, C is the penalty factor, ε is the insensitivity factor, and l is the dimension of the Euclidean space. According to the principle of structural risk minimization, $f(x)$ should minimize $\frac{N}{2}\|\boldsymbol{\omega}\|^2$. The weight vector can be expressed as

$$\boldsymbol{\omega} = \sum_{i=1}^{N}(\lambda_i - \lambda_i{}^*)\mathbf{x}_i, \tag{17}$$

where λ is the Lagrange multiplier.

Therefore, the SVR machine can be expressed as:

$$f(\mathbf{x}) = \sum_{i=1}^{N}(\lambda_i - \lambda_i{}^*)\langle \mathbf{x}_i \cdot \mathbf{x} \rangle + b. \tag{18}$$

Based on the non-linear characteristics of ship maneuvering motion parameters, the non-linear radial basis function (RBF) kernel is selected as

$$K(\mathbf{x}, \mathbf{x}_i) = \exp\left\{-\frac{\|\mathbf{x} - \mathbf{x}_i\|^2}{2\sigma^2}\right\}, \tag{19}$$

where σ is the parameter of the kernel function.

Although traditional SVR can effectively deal with high-dimensional, non-linear, and small-sample datasets, it is vulnerable to outliers and noise. This is because all samples are treated equally, resulting in a large deviation in the classification interface. A weighted training set based on the sample error is now presented. Based on the parameter identification results of the maneuvering motion model under condition i (described in Section 3.1), the zigzag simulation motion is deduced by using the fourth-order Runge–Kutta (R-K) algorithm to obtain a series of heading angle data. The *RMSE* between the heading angle simulation and test data are calculated under condition i as:

$$RMSE_i(\hat{\psi}, \psi) = \sqrt{\frac{1}{T-1}\sum_{t=0}^{T}\left[\hat{\psi}_i(t) - \psi_i(t)\right]^2}. \tag{20}$$

In order to improve the accuracy of RBF-SVR, the influence of training samples with larger identification errors should be weaken. The SVR kernel function is improved by introducing sample weights into the RBF-kernel function.

$$K(\mathbf{x},\mathbf{x}_i) = \exp\left\{-\frac{\|\mathbf{x}-\mathbf{x}_i\|^2}{(\omega_i+\omega_j)\sigma^2}\right\}, \tag{21}$$

where ω_i is the i-th training sample weight, and ω_j is the j-th test sample weight. These weights are defined as functions of the root mean square error (*RMSE*) of the training and test samples, as follows:

$$\omega_i = \left\|\frac{RMSE_i - RMSE_{max} - RMSE_{min}}{RMSE_{max}}\right\|. \tag{22}$$

As the main parameters of SVR, ε, C, and σ have a significant impact on the regression accuracy (i.e., generalization ability) [25]. The leave-one-out (LOO) method is used to investigate the generalization ability of SVR. In the LOO method, one sample is selected as the test set $h(\mathbf{x}_j)$, and the remaining samples are used as the training set. SVR is used to obtain the estimated value $\hat{f}(\mathbf{x}_j)$ of the test set. The *RMSE* of SVR is then used as the precision index:

$$RMSE(\hat{f},h) = \frac{1}{N-1}\sum_{i=1}^{N-1}\sqrt{\left[\frac{1}{N}\sum_{j=1 \text{ and } j\neq i}^{N}\left(\hat{f}(\mathbf{x}_j)-h(\mathbf{x}_j)\right)^2\right]}. \tag{23}$$

Taking the minimum *RMSE* of regression estimation as the optimization goal, the parameters ε, C, σ are optimized by a genetic algorithm. The fitness value is as follows:

$$fitness = sort\left[1\big/RMSE\left(\hat{f},h,\varepsilon,C,\sigma\right)\right]. \tag{24}$$

3.3. Evaluation Indicators

The *RMSE* and Pearson's product–moment correlation coefficient (PCC) are used to evaluate the accuracy of the proposed method. The *RMSE* measures the deviation between the predicted value and the true value, but cannot measure the correlation. Therefore, PCC is used to calculate the correlation between the predicted value and the true value as a supplementary evaluation index. PCC is often denoted as R and is calculated as

$$\bar{h} = \frac{1}{N}\sum_{i=1}^{N} h(\mathbf{x}_j), \tag{25}$$

$$\bar{f} = \frac{1}{N}\sum_{i=1}^{N} \hat{f}(\mathbf{x}_j), \tag{26}$$

and

$$R = \frac{\sum_{i=1}^{N}\left\{\left(h(\mathbf{x}_j)-\bar{h}\right)\left(\hat{f}(\mathbf{x}_j)-\bar{f}\right)\right\}}{\sqrt{\sum_{i=1}^{N}\left(h(\mathbf{x}_j)-\bar{h}\right)^2}\sqrt{\sum_{i=1}^{N}\left(\hat{f}(\mathbf{x}_j)-\bar{f}\right)^2}}. \tag{27}$$

4. Test Verification

4.1. Scaled Free-Running Ship Model Test System

The test system is built using the scaled free-running ship model of KVLCC2, as displayed in Figure 2. It is equipped with a main control board, gyrocompass, differential GPS (D-GPS) module (Real-time kinematic, RTK system), rudder angle sensor, and speed sensor, allowing ship motion data to be collected in real-time. The main parameters of the scaled free-running ship model are listed in Table 1, and more details of KVLCC2 can be found on SIMMAN 2014 [26].

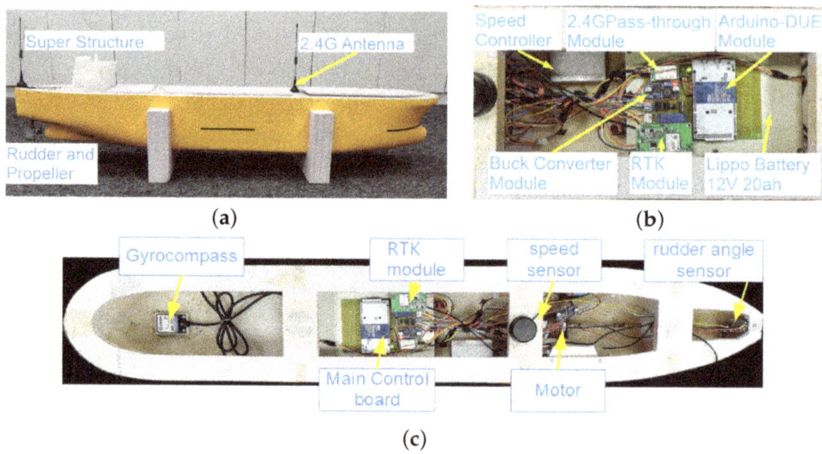

Figure 2. Ship maneuverability test platform. (**a**) Scaled free-running ship model. (**b**) Shipboard control terminal. (**c**) Major sensors and controllers on board.

Table 1. Parameters of test models.

Parameter	Full-Scale Ship	Scaled Ship Model
Scale ratio	1:1	1:266
Length between perpendiculars (L_{pp})	320 m	1.200 m
Maximum beam of waterline	58 m	0.217 m
Depth	30 m	0.112 m
Draft	20.8 m	0.078 m
Displacement	312,622 t	16.42 kg
Propeller diameter	9.86 m	37 mm
Number of propeller blades	4	4
Propeller area ratio	0.431	0.431
Rudder area	135.9 m^2	0.00192 m^2
Rudder turning rate	2.34°/s	38.24°/s

4.2. Test Conditions

To obtain the navigation data of the ship model, the zigzag test conditions are set in terms of the engine speed, test rudder angle, and bow and stern draft, as depicted in Table 2. It should be noted that the light load mentioned is relative to the full load. Bow and stern trim are a lighter condition than the light load. To ensure that the rotational inertia moment of the ship model remains unchanged in each repeated test, the bow and stern draft are adjusted by uniformly pasting ballast lead blocks at fixed positions along the length of the ship. We set the rudder turning rate of the full-scale ship as ω_S, the scale ratio of the model as R_S, and the rudder turning rate of the scaled model as $\omega_M = \omega_S \cdot \sqrt{R_S}$. This ensures that the angular velocity of the rudder is similar to that of the full-scale ship.

Table 2. Zigzag test conditions.

Navigation Condition	Engine Speed (rpm)	Bow Draft (cm)	Stern Draft (cm)	Test Rudder Angle (°)
Light load	2000	7	7	10°/10° 20°/20° 30°/30°
	3000	7	7	
Full load	2000	9	9	
	3000	9	9	
Bow trim	2000	7	5	
	3000	7	5	
Stern trim	2000	5	7	
	3000	5	7	

4.3. Test Data Analysis

4.3.1. Parameter Identification

The free-running model tests were carried out in a lake, as displayed in Figure 3. The heading angle $\psi(t)$ and rudder angle $\delta(t)$ of the scaled free-running ship model were recorded with a frequency of 10 Hz to form a preliminary training set $\mathbf{H}(t) = [\psi(t) \quad \delta(t)]$. To study the convergence of the identification parameters, the preliminary training set was split into a series of training samples according to the time length, and parameter identification based on the LS method was carried out. Meanwhile, considering data convergence and computational efficiency [23], the innovation length was set to 4, and parameter identification based on the MILS method was performed. Taking the navigation condition of a 2000 rpm engine speed, 20° rudder angle, 5 cm bow draft, and 7 cm stern draft as an example, Figure 4 reveals the parameter identification results of the first-order non-linear Nomoto models.

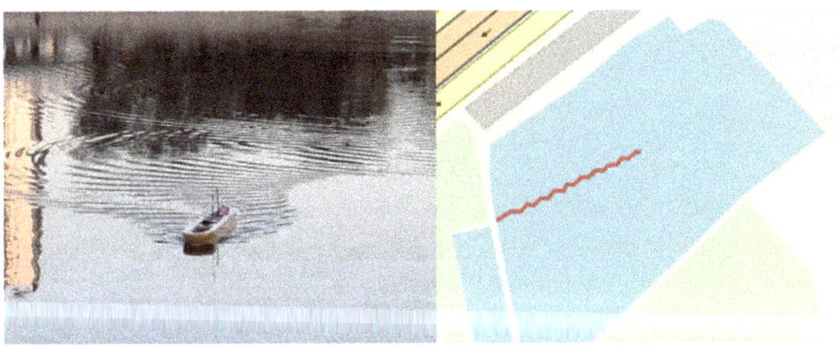

Figure 3. Zigzag tests.

As depicted in Figure 4a–d, for the first-order non-linear Nomoto model, the MILS method converges faster than the LS method. The parameters K, T, and α converge after approximately 500 iterations (i.e., 50 s). The effective neutral rudder angle δ_r converges relatively slowly, requiring around 1500 iterations. This is because low-frequency random disturbances (e.g., randomness of ship motion) influence the scaled free-running ship model as it sails. The fourth-order R-K integration method was used to simulate the motion of the ship model, and the heading angle and yaw angular velocity are shown in Figure 4e. From Figure 4e, the simulation heading angles are very similar to the experimental data. In the remainder of this study, the MILS method is used to identify the parameters of the first-order non-linear Nomoto model. This allows us to analyze the law of the maneuvering motion model parameters under different navigation conditions.

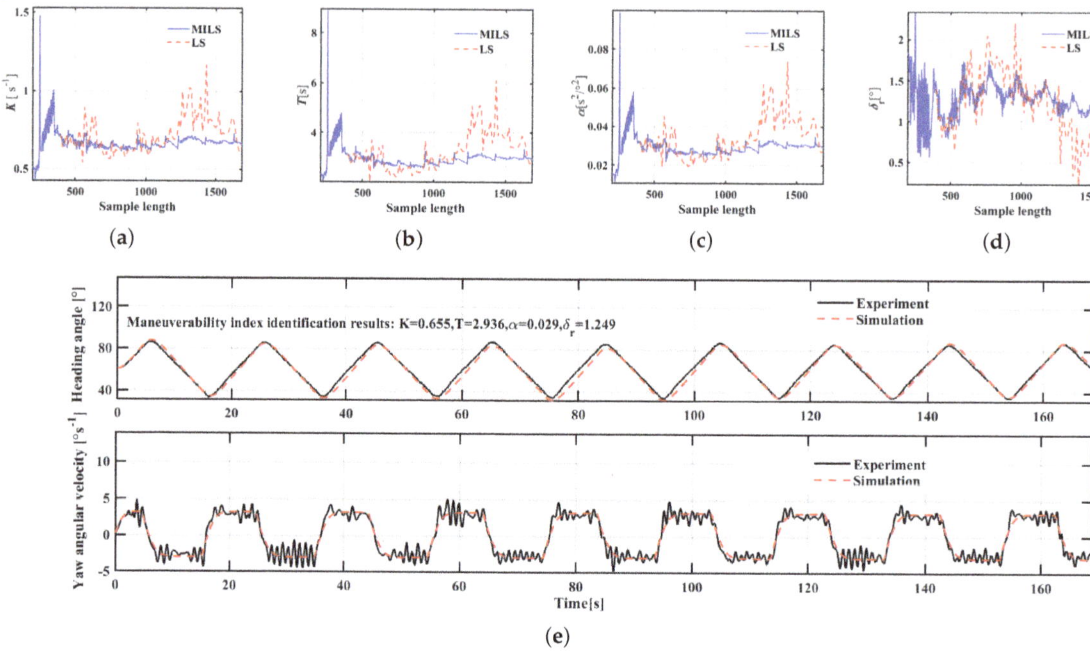

Figure 4. Results of the first-order non-linear Nomoto model. (**a**) Convergence process of K. (**b**) Convergence process of T. (**c**) Convergence process of α. (**d**) Convergence process of δ_r. (**e**) Time history of heading angle and yaw angular velocity for experimental and simulated data.

4.3.2. Parameter Training

A series of zigzag tests were carried out under the navigation conditions listed in Table 2. The first-order non-linear Nomoto model parameters were identified by the MILS method. At least five zigzag tests should be carried out under each navigation condition. Due to the deviation of the results obtained in each identification, the parameters of the maneuvering motion model are shown as box diagrams in Figures 5 and 6, where 7-5 represents the bow trim condition, 5-7 represents the stern trim condition, 7-7 represents the light load condition, and 9-9 represents the full load condition. Under four test conditions (1: rudder angle 30°, bow trim, engine speed 2000 rpm; 2: rudder angle 10°, bow trim, engine speed 3000 rpm; 3: rudder angle 30°, bow trim, engine speed 3000 rpm; and 4: rudder angle 30°, full load, engine speed 3000 rpm), many experimental data are invalid and unstable due to the random environmental disturbances, non-linear ship motion, the data collection process, and other factors. Thus, only one grouping of valid data are identified.

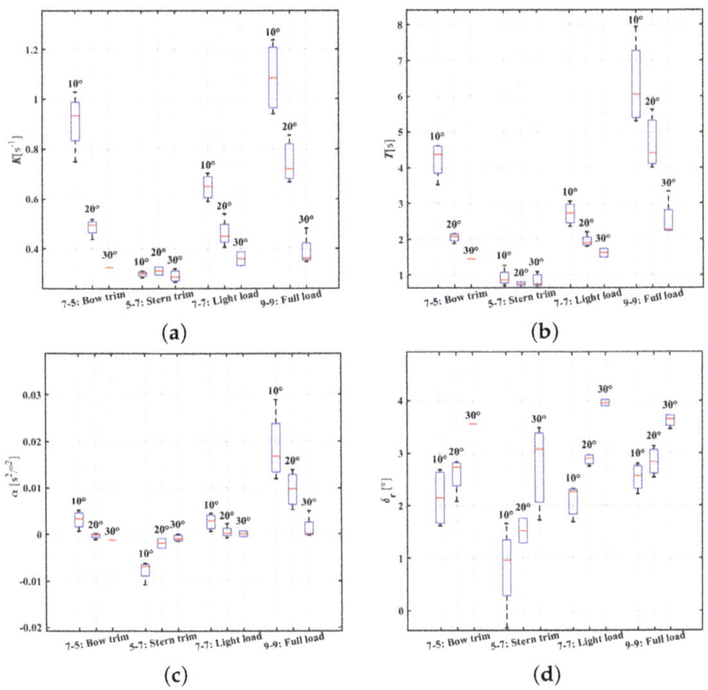

Figure 5. Identification results at 2000 rpm engine speed. (**a**) Identification results of K. (**b**) Identification results of T. (**c**) Identification results of α. (**d**) Identification results of δ_r.

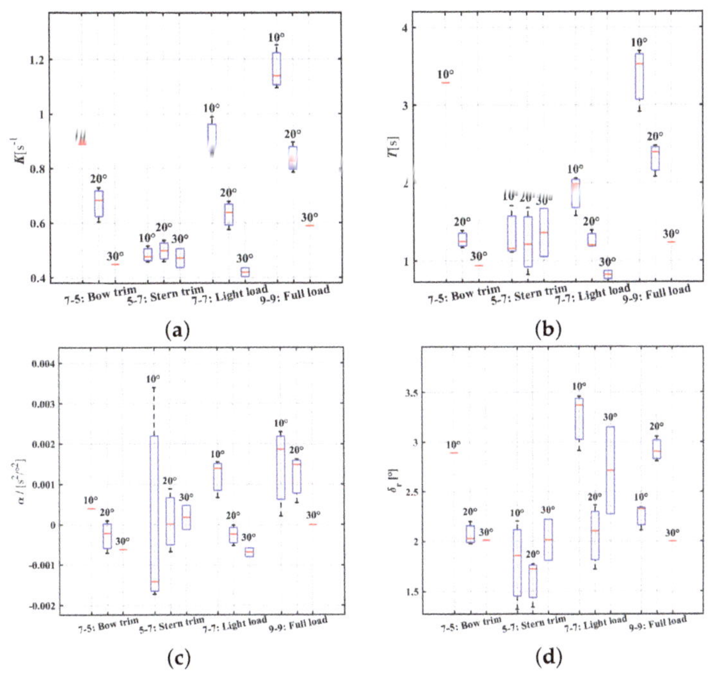

Figure 6. Identification results at 3000 rpm engine speed. (**a**) Identification results of K. (**b**) Identification results of T. (**c**) Identification results of α. (**d**) Identification results of δ_r.

The test rudder angle δ_T, rudder force F_δ, rudder force arm A_δ, ship turning inertia I_{zz}, hydrodynamic force F_W, and hydrodynamic action area of the underwater hull S_h have significant impacts on the maneuvering performance. As shown in Figures 5 and 6, under the same engine speed and rudder angle conditions, K and T are greater under the bow trim condition than under the stern trim condition, and are greater under the full load condition than under the light load condition. In general, I_{zz} and F_δ are very similar in the bow trim (7-5) and stern trim (5-7) conditions; however, A_δ is larger in the bow trim condition, and so K and T are greater in the bow trim condition than in the stern trim condition. Compared with the light load condition (7-7), S_h is larger in the full load condition (9-9), resulting in a larger value of I_{zz}. Thus, K is larger in the full load condition, which means that the turning angular velocity is higher. Because of the larger I_{zz}, the value of T in the full load condition is higher than in the light load condition, meaning that it takes longer to enter the stable turning state. The values of K and T basically decrease with increasing rudder angle. A change in rudder angle usually causes the ship to turn and change its course, which will affect the speed and turning rate of the ship. Within a certain range of rudder angles, the ship's steering response accelerates as the rudder angle increases, but this also enhances the ship's lateral resistance, reducing the driving force and steering ability of the ship; thus, K and T decrease accordingly. Note that an excessive rudder angle not only increases the ship's navigation resistance, but also aggravates the ship's sway and affects the ship's stability. Comparing Figure 5a with Figure 6a and Figure 5b with Figure 6b, it is clear that a higher speed produces a larger value of K and a lower value of T. The reason may be that the higher speed increases the rudder force and turning torque, enhancing the turning rate.

From Figure 5c, α approaches zero as the rudder angle increases, meaning that a larger rudder angle decreases the non-linear motion. This may be because the hydrodynamic forces and moments are sufficiently large at greater rudder angles, while the random disturbances are not significant; however, this phenomenon does not occur at an engine speed of 3000 rpm. The effective neutral rudder angle δ_r exhibits strong randomness in different conditions, making it difficult to determine the change rule and hydrodynamic action mechanism. δ_r is the angle of a ship's rudder necessary to maintain its heading with no lateral force applied. It is affected by the ship speed, draft, and environmental interference. There is no obvious regularity in the change of δ_r. Fortunately, the influence of δ_r is limited.

The first-order non-linear Nomoto model parameters identified by MILS and the engine speeds, bow and stern drafts, and test rudder angles of the scaled free-running ship model were non-dimensionalized and normalized to create a complete training set: $\mathbf{G}_i = [R_{Ti}, D_{Fi}, D_{Ai}, \delta_{Ti}, K_i, T_i, \alpha_i, \delta_{ri}]$. SVR training was used to obtain rapid regression predictions of the ship maneuvering motion parameters under any condition. The specific process is illustrated in Figure 7. Additionally, Table 3 shows the optimal kernel function parameters.

Table 3. The optimal kernel function parameters.

Nomoto Model Parameters	Kernel Function Parameters of SVR		
	ε	C	σ
K (s^{-1})	0.000002	0.86	1.175
T (s)	0.001	2.561	0.922
α (s/°^{2})	0.000001	0.156	0.021
δ_r (°)	0.00002	4.0103	1.041

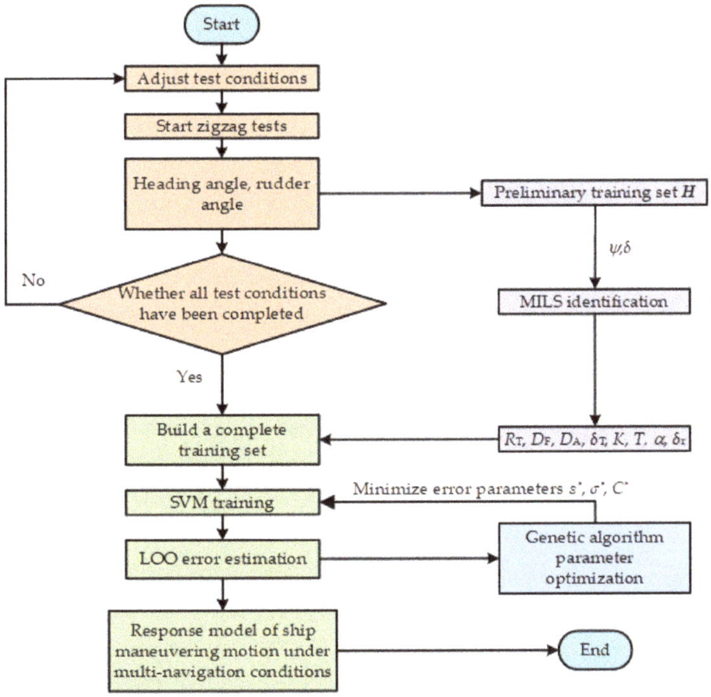

Figure 7. Flowchart of ship maneuvering motion model under multi-navigation conditions.

4.4. Model Accuracy Verification

We used the MILS method to directly identify the parameters of the first-order non-linear Nomoto model, and compared them with the SVR prediction results, as illustrated in Table 4. Under the two verification conditions, the values of K and T predicted by the SVR are close to those directly identified by the MILS, although the effective neutral rudder angle δ_r is slightly different. The variation of the non-linear coefficient α is relatively large. In addition, it can be seen from the analysis in Section 4.3.2 that K and T decrease as the rudder angle increases. K and T decrease in stern trim conditions. Additionally, K and T increase with increasing load, while a higher engine speed lead to a higher K value and lower T value. The coupling effect of rudder angle, load, stern trim, and engine speed makes the reduction in K less than the reduction in T.

The parameters obtained by the MILS and SVR methods were used for dynamic simulations. To further validate the method proposed in this study, the computational results of the MMG parameters of KVLCC2 by Yasukawa et al. [27] were cited and compared with the simulation results by numerical integration of the differential equation, as shown Table 5. Where the primary figure input to the symbol means a non-dimensionalized value, force and moment are non-dimensionalized by $1/2\rho L_{pp}dU^2$ and $1/2\rho L_{pp}^2 dU^2$, respectively. More details can be found in the references [27,28]. It is worth noting that the shape of the blades was modified due to the difficulty of machining the prototype propeller. The simulated and experimental results of the zigzag motions are shown in Figure 8. The SVR and MILS results in condition 2 are close to the experimental data, while the MMG results have a phase lead compared with the experimental data. In condition 1, the SVR and MMG results have a phase lag compared with the experimental data and MILS results. Table 6 compares the overshoot angles in the zigzag maneuvers. The overshoot angles of the SVR and MILS simulation results are smaller than the experimental values. The overshoot angle of MMG simulation results is less than the experimental value in condition 1 and greater than the experimental value in condition 2. In condition 1, the MMG method is more accurate in predicting overshoot angles, while, in condition 2, it has comparable accuracy to MILS. It is difficult to predict the overshoot angle to within a few degrees [27]. The *RMSE* and PCC between the simulated and experimental heading angles are listed in Table 7. The errors and correlation coefficients of the MILS simulation results are better than SVR and MMG. This is because the MILS method is a direct identification, using information from the current navigation condition for identification. The method based on SVR is an indirect identification; it does not use data from the current navigation condition, but trains the MILS identification results of other navigation conditions to predict the current value. The hydrodynamic derivatives of MMG were obtained based on captive tests of a 2.902 m model ship. There is a scale effect with the ship model in this study. In general, however, the *RMSE* and PCC values are reasonable under the verification conditions, indicating that the SVR-based prediction method proposed in this study has good generalization ability.

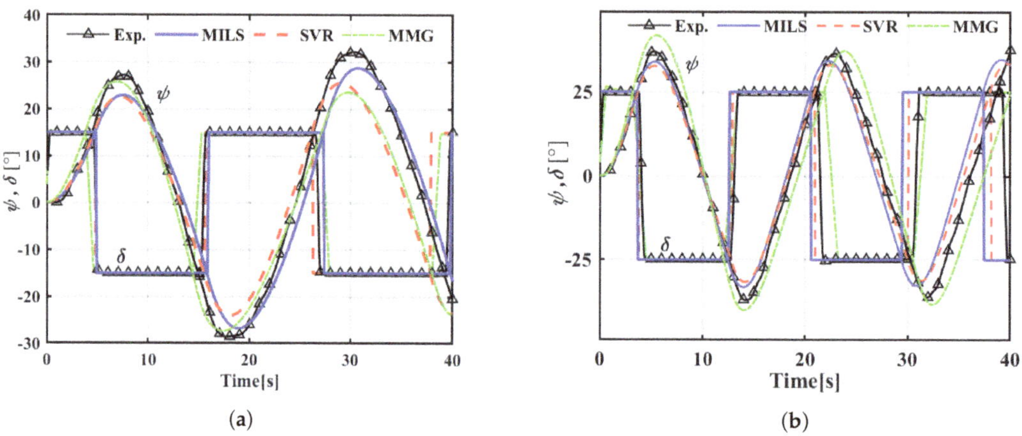

Figure 8. Simulated and experimental results of zigzag motions. (**a**) Comparison of results under verification condition 1. (**b**) Comparison of results under verification condition 2.

Table 4. Comparison of identified results.

Verification Condition	Navigation Condition Parameters: Engine Speed (rpm)	Rudder Angle (°)	Bow Draft (cm)	Stern Draft (cm)	K (s^{-1}) MILS	K (s^{-1}) SVR	T (s) MILS	T (s) SVR	a (s/°^{2}) MILS	a (s/°^{2}) SVR	δ_r (°) MILS	δ_r (°) SVR
1	2000	15	8	8	0.559	0.566	4.282	4.009	−0.004	0.0094	0.492	0.620
2	2500	25	6	8	0.415	0.430	1.228	1.303	0.023	0.0076	2.013	2.419

Table 5. Hydrodynamic derivatives used in the simulations [27,28].

Surge Force Derivatives		Lateral Force Derivatives		Yaw Moment Derivatives	
R_0'	0.022	Y_v'	−0.315	N_v'	−0.137
X_{vv}'	−0.040	Y_r'	0.083	N_r'	−0.049
X_{vr}'	0.002	Y_{vvv}'	−1.607	N_{vvv}'	−0.030
X_{rr}'	0.011	Y_{vvr}'	0.379	N_{vvr}'	−0.294
X_{vvvv}'	0.771	Y_{vrr}'	−0.391	N_{vrr}'	0.055
		Y_{rrr}	0.008	N_{rr}'	−0.013

Table 6. Comparison of overshoot angles.

Verification Condition	Overshoot Angle	Test (°)	MILS (°)	SVR (°)	MMG (°)
1	1st overshoot	12.17	7.89	7.51	10.71
	2nd overshoot	13.62	11.7	8.97	12.31
2	1st overshoot	12.03	8.97	7.75	16.88
	2nd overshoot	12.04	8.13	6.62	14.96

Table 7. Simulated and experimental error statistics of zigzag motions.

Verification Condition	Method	*RMSE* (°)	PCC
1	MILS	3.8414	0.9839
	SVR	5.9296	0.9639
	MMG	5.3710	0.9725
2	MILS	4.8138	0.9847
	SVR	6.2458	0.9733
	MMG	7.1617	0.9638

5. Conclusions and Prospects

Based on the parameter identification of the Nomoto model, the SVR method has been used to predict the parameters of ship maneuvering. Using a desktop computer (Intel® Core™ i7-10700F CPU @2.90 GHz, 40 GB RAM), the average training time of the SVR model was just 1.5 s. The scaled free-running ship model experiments show that the proposed method has good prediction accuracy and can quickly obtain the maneuvering model parameters. The specific conclusions are the following.

(1) The MILS and LS methods have good accuracy for parameter identification with the first-order non-linear Nomoto models. In general, the MILS method converges faster than the LS method. Thus, the MILS method was used to identify the parameters of the first-order Nomoto model with different maneuvering motions. The resulting dataset was trained using the SVR-based method to predict 15°/15° and 25°/25° zigzag motions. The results show that the SVR-based prediction method can obtain the Nomoto model parameters quickly, although the accuracy is slightly lower than that of the MILS method.

(2) The effects of rudder angle, engine speed, bow trim, stern trim, and load on the maneuvering coefficients were analyzed. K and T are larger in the bow trim condition than in the stern trim condition. In the full load condition, K and T are larger. K and T decrease as the rudder angle increases. Additionally, higher speeds lead to higher K values and lower T values. At an engine speed of 2000 rpm, α approaches zero as the rudder angle increases; however, this does not occur at an engine speed of 3000 rpm. The effective neutral rudder angle δ_r exhibits strong randomness under different conditions, making it difficult to determine the variation pattern of this parameter and its hydrodynamic action mechanism.

This study has presented a method for solving the problem of insufficient excitation of cargo ship navigation data. By considering the influence of engine speed, load, trim and rudder angle, the ship motion model under various load conditions in a daily voyage can be predicted. The main limitations of the proposed method are that the accuracy is slightly lower than that of the direct identification method (e.g., MILS), and large amounts of historical data of various navigation conditions are required for training. Although we have tried our best to understand and explain the hydrodynamics, it may be flawed in some aspects. In addition, the SVR model is trained based on zigzag experimental data and can only be used for zigzag-like maneuvers of ships. However, zigzag maneuvers seem not to be fully representative for anti-collision or track-change maneuvers in open sea. In future work, there should be a better balance between zigzag and turning circle tests in those fields of ship operations.

Author Contributions: Conceptualization, X.C.; methodology, M.Z.; validation, J.L. and S.D.; writing—review and editing, J.L. and M.Z.; funding acquisition, M.Z. All authors have read and agreed to the published version of the manuscript.

Funding: This research was funded by the National Natural Science Foundation of China (No. 52001243, 51920105014), the Ministry of Transport Project (No. SXHXGZ-2021-3), the Natural Science Project of Fujian Province (No. 2020J01860), and the Social Development Project of Fuzhou (No. 2021-S-236).

Informed Consent Statement: Not applicable.

Data Availability Statement: The data presented in this study are available in Appendix A.

Acknowledgments: Thank Zhu, T., Ma, Y. and Tian, G. for their assistance in the experiment.

Conflicts of Interest: The authors declare no conflict of interest.

Appendix A

Table A1. Training data of 2000 rpm.

Engine Speed (rpm)	Test Rudder Angle (°)	Bow Draft (cm)	Stern Draft (cm)	K (s^{-1})	T (s)	a (s/°^{2})	δ_r (°)	Weight
2000	10	7	5	7.49×10^{-1}	3.52×10^{0}	6.68×10^{-4}	2.59×10^{0}	6.24×10^{-1}
2000	10	7	5	9.19×10^{-1}	4.16×10^{0}	2.58×10^{-3}	1.61×10^{0}	4.68×10^{-1}
2000	10	7	5	9.48×10^{-1}	4.62×10^{0}	4.04×10^{-3}	2.69×10^{0}	1.45×10^{-1}
2000	10	7	5	1.03×10^{0}	4.59×10^{0}	5.21×10^{-3}	1.72×10^{0}	9.97×10^{-1}
2000	20	7	5	5.18×10^{-1}	2.16×10^{0}	1.85×10^{-4}	2.83×10^{0}	4.15×10^{-1}
2000	20	7	5	4.38×10^{-1}	1.88×10^{0}	-1.14×10^{-3}	2.80×10^{0}	7.69×10^{-1}
2000	20	7	5	4.99×10^{-1}	2.03×10^{0}	-2.11×10^{-4}	2.67×10^{0}	9.60×10^{-1}
2000	20	7	5	4.90×10^{-1}	2.14×10^{0}	8.27×10^{-6}	2.09×10^{0}	2.63×10^{-1}
2000	30	7	5	3.26×10^{-1}	1.45×10^{0}	-1.22×10^{-3}	3.55×10^{0}	5.14×10^{-1}
2000	10	7	7	5.91×10^{-1}	2.36×10^{0}	5.85×10^{-4}	1.69×10^{0}	4.74×10^{-1}
2000	10	7	7	6.49×10^{-1}	2.73×10^{0}	2.80×10^{-3}	2.33×10^{0}	9.43×10^{-1}
2000	10	7	7	7.04×10^{-1}	3.06×10^{0}	4.46×10^{-3}	2.27×10^{0}	9.79×10^{-1}
2000	20	7	7	4.46×10^{-1}	1.80×10^{0}	9.38×10^{-5}	2.97×10^{0}	9.83×10^{-1}
2000	20	7	7	4.53×10^{-1}	1.88×10^{0}	1.99×10^{-4}	2.96×10^{0}	9.34×10^{-1}
2000	20	7	7	4.05×10^{-1}	1.92×10^{0}	-8.31×10^{-4}	2.75×10^{0}	8.01×10^{-1}
2000	20	7	7	5.41×10^{-1}	2.21×10^{0}	2.24×10^{-3}	2.85×10^{0}	9.93×10^{-1}
2000	30	7	7	3.32×10^{-1}	1.50×10^{0}	-5.63×10^{-4}	3.90×10^{0}	7.37×10^{-1}
2000	30	7	7	3.87×10^{-1}	1.74×10^{0}	6.29×10^{-4}	4.03×10^{0}	9.11×10^{-1}
2000	10	9	9	1.24×10^{0}	7.94×10^{0}	2.89×10^{-2}	2.80×10^{0}	5.58×10^{-1}
2000	10	9	9	9.89×10^{-1}	5.31×10^{0}	1.20×10^{-2}	2.23×10^{0}	3.45×10^{-1}
2000	10	9	9	9.41×10^{-1}	5.50×10^{0}	1.48×10^{-2}	2.43×10^{0}	8.67×10^{-1}
2000	10	9	9	1.18×10^{0}	6.62×10^{0}	1.87×10^{-2}	2.71×10^{0}	5.57×10^{-1}
2000	20	9	9	8.54×10^{-1}	5.63×10^{0}	1.39×10^{-2}	2.83×10^{0}	1.00×10^{0}
2000	20	9	9	6.67×10^{-1}	4.00×10^{0}	5.29×10^{-3}	2.54×10^{0}	8.76×10^{-1}
2000	20	9	9	7.20×10^{-1}	4.41×10^{0}	9.80×10^{-3}	3.14×10^{0}	8.32×10^{-1}
2000	30	9	9	3.60×10^{-1}	2.23×10^{0}	7.08×10^{-5}	3.72×10^{0}	4.92×10^{-1}
2000	30	9	9	3.60×10^{-1}	2.23×10^{0}	7.08×10^{-5}	3.72×10^{0}	4.92×10^{-1}
2000	30	9	9	4.83×10^{-1}	3.33×10^{0}	4.99×10^{-3}	3.57×10^{0}	6.89×10^{-1}
2000	30	9	9	3.47×10^{-1}	2.30×10^{0}	-2.75×10^{-4}	3.46×10^{0}	8.10×10^{-1}
2000	10	5	7	3.04×10^{-1}	1.26×10^{0}	-6.21×10^{-3}	-3.25×10^{-1}	2.12×10^{-1}
2000	10	5	7	2.97×10^{-1}	7.04×10^{-1}	-6.71×10^{-3}	1.02×10^{0}	9.96×10^{-1}
2000	10	5	7	2.84×10^{-1}	8.30×10^{-1}	-1.08×10^{-2}	8.96×10^{-1}	8.57×10^{-1}
2000	10	5	7	3.10×10^{-1}	8.86×10^{-1}	-7.05×10^{-3}	1.66×10^{0}	8.57×10^{-1}
2000	20	5	7	3.27×10^{-1}	8.03×10^{-1}	-8.84×10^{-4}	1.28×10^{0}	6.36×10^{-1}
2000	20	5	7	2.95×10^{-1}	7.10×10^{-1}	-2.97×10^{-3}	1.76×10^{0}	8.17×10^{-1}
2000	30	5	7	2.66×10^{-1}	7.43×10^{-1}	-1.54×10^{-3}	1.73×10^{0}	8.48×10^{-1}
2000	30	5	7	2.86×10^{-1}	7.06×10^{-1}	-1.01×10^{-3}	3.07×10^{0}	8.70×10^{-1}
2000	30	5	7	3.20×10^{-1}	1.08×10^{0}	2.14×10^{-5}	3.48×10^{0}	7.23×10^{-1}

Table A2. Training data of 3000 rpm.

Engine Speed (rpm)	Test Rudder Angle (°)	Bow Draft (cm)	Stern Draft (cm)	K (s^{-1})	T (s)	a (s/°^{2})	δ_r (°)	Weight
3000	20	7	5	7.29×10^{-1}	1.39×10^{0}	1.01×10^{-4}	2.03×10^{0}	8.32×10^{-1}
3000	20	7	5	6.04×10^{-1}	1.17×10^{0}	-7.03×10^{-4}	1.98×10^{0}	6.30×10^{-1}
3000	20	7	5	6.84×10^{-1}	1.25×10^{0}	-2.13×10^{-4}	2.20×10^{0}	8.97×10^{-1}
3000	30	7	5	4.49×10^{-1}	9.40×10^{-1}	-6.16×10^{-4}	2.52×10^{0}	4.21×10^{-1}
3000	10	7	7	8.58×10^{-1}	1.58×10^{0}	6.65×10^{-4}	3.46×10^{0}	6.35×10^{-1}
3000	10	7	7	9.89×10^{-1}	2.06×10^{0}	1.55×10^{-3}	2.91×10^{0}	8.99×10^{-1}
3000	10	7	7	8.81×10^{-1}	1.99×10^{0}	1.39×10^{-3}	3.37×10^{0}	7.25×10^{-1}
3000	20	7	7	6.79×10^{-1}	1.39×10^{0}	-1.76×10^{-5}	2.36×10^{0}	2.71×10^{-1}
3000	20	7	7	5.76×10^{-1}	1.18×10^{0}	-5.17×10^{-4}	1.72×10^{0}	6.76×10^{-1}
3000	20	7	7	6.38×10^{-1}	1.20×10^{0}	-2.36×10^{-4}	2.10×10^{0}	8.99×10^{-1}
3000	30	7	7	4.04×10^{-1}	7.69×10^{-1}	-7.91×10^{-4}	2.27×10^{0}	4.87×10^{-1}
3000	30	7	7	4.35×10^{-1}	8.73×10^{-1}	-5.76×10^{-4}	3.15×10^{0}	7.23×10^{-1}
3000	10	9	9	1.25×10^{0}	3.70×10^{0}	2.30×10^{-3}	2.32×10^{0}	8.40×10^{-1}
3000	10	9	9	1.10×10^{0}	2.92×10^{0}	2.06×10^{-4}	2.34×10^{0}	7.43×10^{-1}
3000	10	9	9	1.14×10^{0}	3.53×10^{0}	1.86×10^{-3}	2.11×10^{0}	3.38×10^{-1}
3000	20	9	9	8.96×10^{-1}	2.48×10^{0}	1.48×10^{-3}	3.06×10^{0}	6.07×10^{-1}
3000	20	9	9	8.28×10^{-1}	2.39×10^{0}	1.62×10^{-3}	2.90×10^{0}	9.16×10^{-1}
3000	20	9	9	7.85×10^{-1}	2.08×10^{0}	5.39×10^{-4}	2.81×10^{0}	7.20×10^{-1}
3000	10	5	7	4.58×10^{-1}	1.16×10^{0}	-1.42×10^{-3}	1.32×10^{0}	9.43×10^{-1}
3000	10	5	7	4.77×10^{-1}	1.11×10^{0}	-1.73×10^{-3}	2.21×10^{0}	9.15×10^{-1}
3000	10	5	7	5.16×10^{-1}	1.71×10^{0}	3.40×10^{-3}	1.86×10^{0}	8.87×10^{-1}
3000	20	5	7	4.98×10^{-1}	1.21×10^{0}	1.19×10^{-5}	1.77×10^{0}	6.96×10^{-1}
3000	20	5	7	5.36×10^{-1}	1.68×10^{0}	8.87×10^{-4}	1.72×10^{0}	6.24×10^{-1}
3000	20	5	7	4.59×10^{-1}	8.25×10^{-1}	-6.71×10^{-4}	1.34×10^{0}	8.04×10^{-1}
3000	30	5	7	4.37×10^{-1}	1.05×10^{0}	-1.18×10^{-4}	1.81×10^{0}	5.79×10^{-1}
3000	30	5	7	5.05×10^{-1}	1.67×10^{0}	4.75×10^{-4}	2.22×10^{0}	4.12×10^{-1}

References

1. Zhang, M.; Kujala, P.; Hirdaris, S. A machine learning method for the evaluation of ship grounding risk in real operational conditions. *Reliab. Eng. Syst. Saf.* **2022**, *226*, 108697. [CrossRef]
2. Zhang, M.; Taimuri, G.; Zhang, J.; Hirdaris, S. A deep learning method for the prediction of 6-DoF ship motions in real conditions. *Proc. Inst. Mech. Eng. Part M J. Eng. Marit. Environ.* 2023, *online first*. [CrossRef]
3. Abkowitz, M.A. *Lectures on Ship Hydrodynamics–Steering and Manoeuvrability*; Technical Report Hy-5; Hydro-and Aerodynamic Laboratory: Lyngby, Denmark, 1964.
4. Norrbin, N.H. Theory and Observations on the Use of a Mathematical Model for Ship Manoeuvring in Deep and Confined Waters; Publication 68 of the Swedish State Shipbuilding Experimental Tank, Göteborg, Sweden. In Proceedings of the 8th Symposium on Naval Hydrodynamics, ONR, Pasadena, CA, USA, 24–28 August 1971; pp. E807–E905.
5. Ogawa, A.; Kasai, H. On the mathematical model of manoeuvring motion of ships. *Int. Shipbuild. Prog.* **1978**, *25*, 306–319. [illegible]
6. Nomoto, K.; Taguchi, K.; Honda, K.; Hirano, S. On the steering qualities of ships. [illegible] [PubMed]
7. Araki, M.; Sadat-Hosseini, H.; Sanada, Y.; Tanimoto, K.; Umeda, N.; Stern, F. Estimating maneuvering coefficients using system identification methods with experimental, system-based, and CFD free-running trial data. *Ocean Eng.* **2012**, *51*, 63–84. [CrossRef]
8. Carrillo, S.; Contreras, J. Obtaining first and second order Nomoto models of a fluvial support patrol using identification techniques. *Ship Sci. Technol.* **2018**, *11*, 19–28. [CrossRef]
9. Xu, H.; Hinostroza, M.; Soares, C.G. Estimation of hydrodynamic coefficients of a nonlinear manoeuvring mathematical model with free-running ship model tests. *Int. J. Marit. Eng.* **2018**, *160*, A3. [CrossRef]
10. Caccia, M.; Indiveri, G.; Veruggio, G. Modeling and identification of open-frame variable configuration unmanned underwater vehicles. *IEEE J. Ocean. Eng.* **2000**, *25*, 227–240. [CrossRef]
11. Luo, W.; Guedes Soares, C.; Zou, Z. Parameter identification of ship maneuvering model based on support vector machines and particle swarm optimization. *J. Offshore Mech. Arct. Eng.* **2016**, *138*, 031101. [CrossRef]
12. Zhu, M.; Hahn, A.; Wen, Y.; Bolles, A. Parameter identification of ship maneuvering models using recursive least square method based on support vector machines. *TransNav Int. J. Mar. Navig. Saf. Sea Transp.* **2017**, *11*, 23–29. [CrossRef]
13. Xue, Y.; Liu, Y.; Ji, C.; Xue, G. Hydrodynamic parameter identification for ship manoeuvring mathematical models using a Bayesian approach. *Ocean Eng.* **2020**, *195*, 106612. [CrossRef]
14. He, H.W.; Zou, Z.J. Black-box modeling of ship maneuvering motion using system identification method based on BP neural network. In Proceedings of the International Conference on Offshore Mechanics and Arctic Engineering, Virtual, 3–7 August 2020; Volume 84386, p. V06BT06A037.
15. Wang, T.; Li, G.; Wu, B.; Æsøy, V.; Zhang, H. Parameter identification of ship manoeuvring model under disturbance using support vector machine method. *Ships Offshore Struct.* **2021**, *16*, 13–21. [CrossRef]
16. Borkowski, P. Inference engine in an intelligent ship course-keeping system. *Comput. Intell. Neurosci.* **2017**, *2017*, 2561383. [CrossRef] [PubMed]

17. Himaya, A.N.; Sano, M.; Suzuki, T.; Shirai, M.; Hirata, N.; Matsuda, A.; Yasukawa, H. Effect of the loading conditions on the maneuverability of a container ship. *Ocean Eng.* **2022**, *247*, 109964. [CrossRef]
18. Zhang, M.; Conti, F.; Le Sourne, H.; Vassalos, D.; Kujala, P.; Lindroth, D.; Hirdaris, S. A method for the direct assessment of ship collision damage and flooding risk in real conditions. *Ocean Eng.* **2021**, *237*, 109605. [CrossRef]
19. Huajun, Z.; Xinchi, T.; Hang, G.; Shou, X. The parameter identification of the autonomous underwater vehicle based on multi-innovation least squares identification algorithm. *Int. J. Adv. Robot. Syst.* **2020**, *17*, 1729881420921016. [CrossRef]
20. Wang, S.; Wang, L.; Im, N.; Zhang, W.; Li, X. Real-time parameter identification of ship maneuvering response model based on nonlinear Gaussian Filter. *Ocean Eng.* **2022**, *247*, 110471. [CrossRef]
21. Tzeng, C.Y.; Chen, J.F. Fundamental properties of linear ship steering dynamic models. *J. Mar. Sci. Technol.* **2009**, *7*, 2. [CrossRef]
22. Hinostroza, M.; Xu, H.; Guedes Soares, C. *Experimental and Numerical Simulations of Zig-Zag Manoeuvres of a Self-Running Ship Model*; Taylor & Francis Group: London, UK, 2017; pp. E563–E570.
23. Xie, S.; Chu, X.; Liu, C.; Liu, J.; Mou, J. Parameter identification of ship motion model based on multi-innovation methods. *J. Mar. Sci. Technol.* **2020**, *25*, 162–184. [CrossRef]
24. Zhao, B.; Zhang, X.; Liang, C. A Novel Parameter Identification Algorithm for 3-DOF Ship Maneuvering Modelling Using Nonlinear Multi-Innovation. *J. Mar. Sci. Eng.* **2022**, *10*, 581. [CrossRef]
25. Cherkassky, V.; Ma, Y. Practical selection of SVM parameters and noise estimation for SVM regression. *Neural Netw.* **2004**, *17*, 113–126. [CrossRef]
26. Simman. Workshop on Verification and Validation of Ship Manoeuvring Simulation Methods. 2014. Available online: http://simman2014.dk (accessed on 1 March 2022).
27. Yasukawa, H.; Yoshimura, Y. Introduction of MMG standard method for ship maneuvering predictions. *J. Mar. Sci. Technol.* **2015**, *20*, 37–52. [CrossRef]
28. Yoshimura, Y.; Ueno, M.; Tsukada, Y. Analysis of steady hydrodynamic force components and prediction of manoeuvring ship motion with KVLCC1, KVLCC2 and KCS. In Proceedings of the Workshop Proceedings of SIMMAN2008, Copenhagen, Denmark, 7 April 2008; Volume 1, pp. E80–E86.

Journal of
Marine Science and Engineering

MDPI

Article

An Intelligent Algorithm for USVs Collision Avoidance Based on Deep Reinforcement Learning Approach with Navigation Characteristics

Zhe Sun [1,2], Yunsheng Fan [1,2,*] and Guofeng Wang [1,2]

1 College of Marine Electrical Engineering, Dalian Maritime University, Dalian 116026, China; rickey@dlmu.edu.cn (Z.S.); dmuwgf@dlmu.edu.cn (G.W.)
2 Key Laboratory of Technology and System for Intelligent Ships of Liaoning Province, Dalian 116026, China
* Correspondence: yunsheng@dlmu.edu.cn; Tel.: +86-155-2478-9899

Abstract: Many achievements toward unmanned surface vehicles have been made using artificial intelligence theory to assist the decisions of the navigator. In particular, there has been rapid development in autonomous collision avoidance techniques that employ the intelligent algorithm of deep reinforcement learning. A novel USV collision avoidance algorithm based on deep reinforcement learning theory for real-time maneuvering is proposed. Many improvements toward the autonomous learning framework are carried out to improve the performance of USV collision avoidance, including prioritized experience replay, noisy network, double learning, and dueling architecture, which can significantly enhance the training effect. Additionally, considering the characteristics of the USV collision avoidance problem, two effective methods to enhance training efficiency are proposed. For better training, considering the international regulations for preventing collisions at sea and USV maneuverability, a complete and reliable USV collision avoidance training system is established, demonstrating an efficient learning process in complex encounter situations. A reward signal system in line with the USV characteristics is designed. Based on the Unity maritime virtual simulation platform, an abundant simulation environment for training and testing is designed. Through detailed analysis, verification, and comparison, the improved algorithm outperforms the pre-improved algorithm in terms of stability, average reward, rules learning, and collision avoidance effect, reducing 26.60% more accumulated course deviation and saving 1.13% more time.

Keywords: unmanned surface vehicles; deep reinforcement learning; autonomous collision avoidance; COLREGs

Citation: Sun, Z.; Fan, Y.; Wang, G. An Intelligent Algorithm for USVs Collision Avoidance Based on Deep Reinforcement Learning Approach with Navigation Characteristics. *J. Mar. Sci. Eng.* **2023**, *11*, 812. https://doi.org/10.3390/jmse11040812

Academic Editor: Gerasimos Theotokatos

Received: 25 February 2023
Revised: 14 March 2023
Accepted: 20 March 2023
Published: 11 April 2023

1. Introduction

With the higher demand for unmanned surface vehicle (USV) intelligent technology, countries have taken measures to improve the effect of autonomous collision avoidance while safeguarding public life and property. Intelligence navigation refers to a USV that can use a sensing system to obtain current maritime navigation information and autonomously generate USV navigation decisions to achieve safe and reliable water navigation. Research on intelligent USV navigation is of great practical importance for the shipping industry and even human society.

USVs have been used for missions at sea due to their small size, high speed, low cost, and no risk of human casualties. In ocean mapping, USVs are used to carry GNSS and other equipment to assist in achieving accurate positioning and mapping [1]. The USV can efficiently achieve the detection of submarine geomorphology and underwater objects by carrying side scan sonar [2]. In hydrological monitoring, the detection area is covered by the navigation and control of the USVs, and the hydrological parameters are monitored by the sensor equipment [3,4]. In addition, many studies for path-tracking control [5], trajectory tracking [6], dynamic positioning [7], and collision avoidance [8] problems are a permanent basis for any application of USVs, and they can forcefully advance the development of the problem of autonomous USV navigation.

There are many ways of designing autonomous collision avoidance algorithms for USV, such as A*, artificial potential field, velocity obstacle, dynamic window, fast marching method, etc. Ren et al. [9] use the velocity obstacle method for collision avoidance algorithm design, optimizing the way of setting up the velocity obstacle region for multi-ship collision avoidance. Fan et al. [10] use an improved cuckoo search algorithm designed with an adaptive update step, which optimizes the global search capability and can plan a better collision avoidance strategy. Guan et al. [11] use the A* method and dynamic window method to design a collision avoidance algorithm for static and dynamic obstacles, optimizing the weight coefficient of the dynamic window algorithm by deep Q network. These methods have unique advantages in specific environments and are very effective in USV collision avoidance algorithms, but their disadvantages are also distinct. Firstly, the operational anthropomorphism degree of these algorithms is not sufficient. Secondly, the generalization ability of some algorithms is poor, and the tuning of parameters is complicated. Moreover, some algorithms simplify the actual situation when applied in training and are difficult to be applied in practice. However, the model-free deep reinforcement learning approach based on learning styles can effectively compensate for these shortcomings.

With the development of the deep reinforcement learning approach, it has achieved great results in many fields due to its outstanding perception and decision-making capabilities, such as Go [12], video games [13], autonomous navigation [14], and medicine [15]. It is based only on itself, without any human knowledge of collision avoidance beyond navigation rules. However, it can make excellent decisions in many challenging domains. Especially in USV collision avoidance, reliable samples for learning are hard to obtain and expensive. Therefore, independently, starting tabula rasa, deep reinforcement learning can fully compensate for these problems and complete collision avoidance tasks in complex USV encounter situations.

On the issues of USV collision avoidance, many pieces of research on autonomous collision avoidance based on the deep reinforcement learning approach have been developed. Most researchers focus on the model-free method because this direction is easy to implement, and for another reason, the model-based method is difficult. The model-free method used in the USVs collision avoidance algorithm is divided into value-function-based and policy-gradient-based [16]. The former is good at dealing with discrete action space. Chen et al. [17] provide a representative paradigm for the discrete ship movements and devise a unique way of training. Li et al. [18] employ the traditional artificial potential field (APF) method to optimize the reward signal of the DQN method, resulting in a more scientific reward signal. Shen et al. [19] improve the neural network framework for more efficient learning, obtaining a better collision avoidance effect than the algorithm before improvement. Zhou et al. [20] improve the collision algorithm to solve the sparse reward problem, using two networks to generate actions and evaluate behavior. Compared to the DQN algorithm, the training effect under the improved algorithm is better.

Research on the latter, the policy gradient method, is also widely studied. Du et al. [21] propose an improved algorithm based on the deep deterministic policy gradient (DDPG) algorithm [22] that combines with Douglas–Peucker (DP) algorithm to plan the path. The Long Short-Term Memory (LSTM) architecture and rich reward function are designed to improve the speed and stability of convergence. Xu et al. [23] also choose the DDPG algorithm and establish a risk assessment model, improving the network structure. Their algorithm has a good collision avoidance effect and real-time performance. Additionally, Chen et al. [24] propose a cooperative multi-agent deep reinforcement learning algorithm for ship collision avoidance, resulting in a good collision avoidance effect under simple typical ship encounter situations. Considering the results of this collision avoidance research, based on the reinforcement learning algorithm, there are some problems worth further consideration:

(1) In many pieces of research, the training environment in each episode is fixed, lacking practical significance, whether complex or not.
(2) Most researchers are not selecting more random seeds to verify the superiority and reliability of their algorithm.
(3) Some researchers are not considering the maneuverability characteristics of USVs adequately.

Deep reinforcement learning theory provides an extraordinary way for USVs' autonomous collision avoidance safety and efficiency. Compared with the traditional methods, it performed better in complex environments with simple designs. Furthermore, it does not require any prior knowledge from the expert navigator about avoiding the obstacle well; surprisingly, it is not even necessary to provide fully observed training environments, to accomplish the complex collision task.

Aiming at the above problems and characteristics, in this paper, a USV collision avoidance algorithm based on the deep reinforcement learning approach, DRLCA, is designed. The main contributions of this paper are as follows:

(1) This paper considers the restriction of maneuverability and international regulations for preventing collisions at sea (COLREGs) in the training process. A suitable training environment with stochasticity and complexity is designed based on the deep reinforcement learning approach. Additionally, considering the collision avoidance process for factors, a meticulous reward signal for USVs training is constructed, which makes training more practical.
(2) Double Q learning method is used to reduce overestimation, dueling neural network architecture to improve training effect, and prioritized experience replay to optimize sampling. The results of various improvements are analytically compared under an abundant training environment based on multiple random number seeds.
(3) Aiming at the hard-exploration problem caused by the training environment with strong randomness, the noisy network method is introduced, which can enhance the detection capability. Experimentally, the best way of noise adding in USV collision avoidance training is confirmed. Considering the characteristics of the USVs collision avoidance problem, the restriction of the dynamic area is introduced in training for calculation reduction and the clip of neural network state input for training effect improvement.

This paper is organized as follows. Kinematic parameters, COLREGs, ship domain, and USV maneuverability [illegible] in Section 2. Section 3 is about the deep reinforcement learning approach and its optimization methods. Section 4 describes the establishment of the training environment. In Section 5, the improved algorithm for USVs collision avoidance is tested in the Unity environment. The last section is the summary and prospect.

2. USV Collision Avoidance Parameters

2.1. USV Collision Avoidance Characteristics

As shown in Figure 1, $Y(N)$ and $X(E)$ point to the north and east. (x_U, y_U), (x_O, y_O), and (x_T, y_T) are the positions of our own USV, obstacle USV, and terminal. φ_U, φ_T, φ_O, ϕ, α_O, and θ are our USV course, absolute azimuth of the terminal and our own USV, obstacle USV course, relative azimuth of the terminal and our own USV, absolute azimuth of the obstacle USV and our own USV, and the relative azimuth of the obstacle USV and our own USV. The distance between obstacle USV and our own USV is d. The speed of our own USV and obstacle USV are V_U and V_O. To reflect the USV collision avoidance training characteristics, the following Norrbin ship mathematical model is selected [25],

$$\begin{cases} T\dot{\eta} + \eta + \alpha\eta^3 = K\delta \\ \eta = \dot{\varphi} \end{cases} \tag{1}$$

where, T, K, and α are related to USV maneuverability, and the course change caused by rudder angle change can be well calculated. Because the research object of this paper is the "Lan Xin" USV, which has a vector propulsion system, it is necessary to consider the influence of the characteristics of the steering gear. The vector propulsion system is a type of thruster that can change direction to achieve maximum speed propulsion in any direction and can obtain additional control torque to achieve control of ship attitude change. Therefore, the following equation, the second-order propulsion angle response model, is selected in this paper,

$$\ddot{\delta} + 2\zeta\omega_n\dot{\delta} + {\omega_n}^2\delta = k{\omega_n}^2\delta_r \quad (2)$$

where, ω_n, ζ, and k are intrinsic frequency, damping ratio and proportionality coefficient as $\omega_n = 0.958$, $\zeta = 0.811$ and $k = 0.923$. δ_r is the target rudder angle.

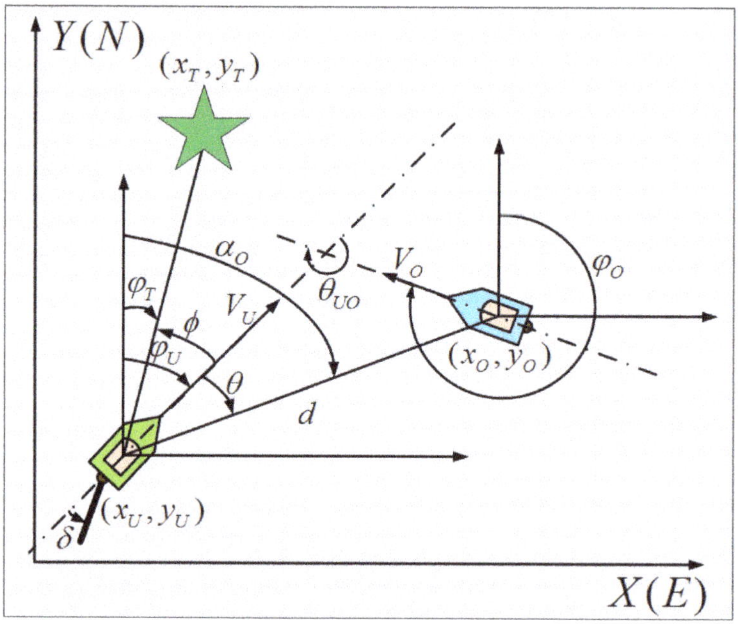

Figure 1. USV collision avoidance kinematic.

The accurate division of USV encounter situations is crucial for collision avoidance agent training, and it is divided into the following six conditions in this paper [26]:

(1) As shown in Figure 2a, when the obstacle USV_O and own USV_U are at the relative azimuth of $[355°, 360°) \cup [0°, 5°)$, it is the head-on encounter situation. According to the COLREGs, when there is a hazard of USV collision, both USVs have to avoid each other and should turn to the port side as they pass.

(2) As shown in Figure 2b, when USV_O is at the $[247.5°, 355°)$ relative azimuth of USV_U, and there is a risk of collision, it is the crossing-stand-on encounter situation. USV_U should stand on the course, and USV_O should turn to starboard.

(3) As shown in Figure 2c, when USV_O is at the $[5°, 112.5°)$ relative azimuth of USV_U, and there is a risk of collision, it is the crossing-give-way encounter situation. USV_U should turn to starboard, and USV_O should stand on the course.

(4) As shown in Figure 2d, when USV_U is at the $[112.5°, 247.5°)$ relative azimuth of USV_O, and there is a risk of collision, it is the overtaking encounter situation. USV_U should avoid the collision, and turns to starboard or port are allowed.

(5) As shown in Figure 2e, when USV_O is at the $[112.5°, 247.5°)$ relative azimuth of USV_U, and there is a risk of collision, it is the overtaking encounter situation. USV_U should stand on the course.

(6) Additionally, when the obstacle USV is in breach of rules, our own USV should avoid it proactively.

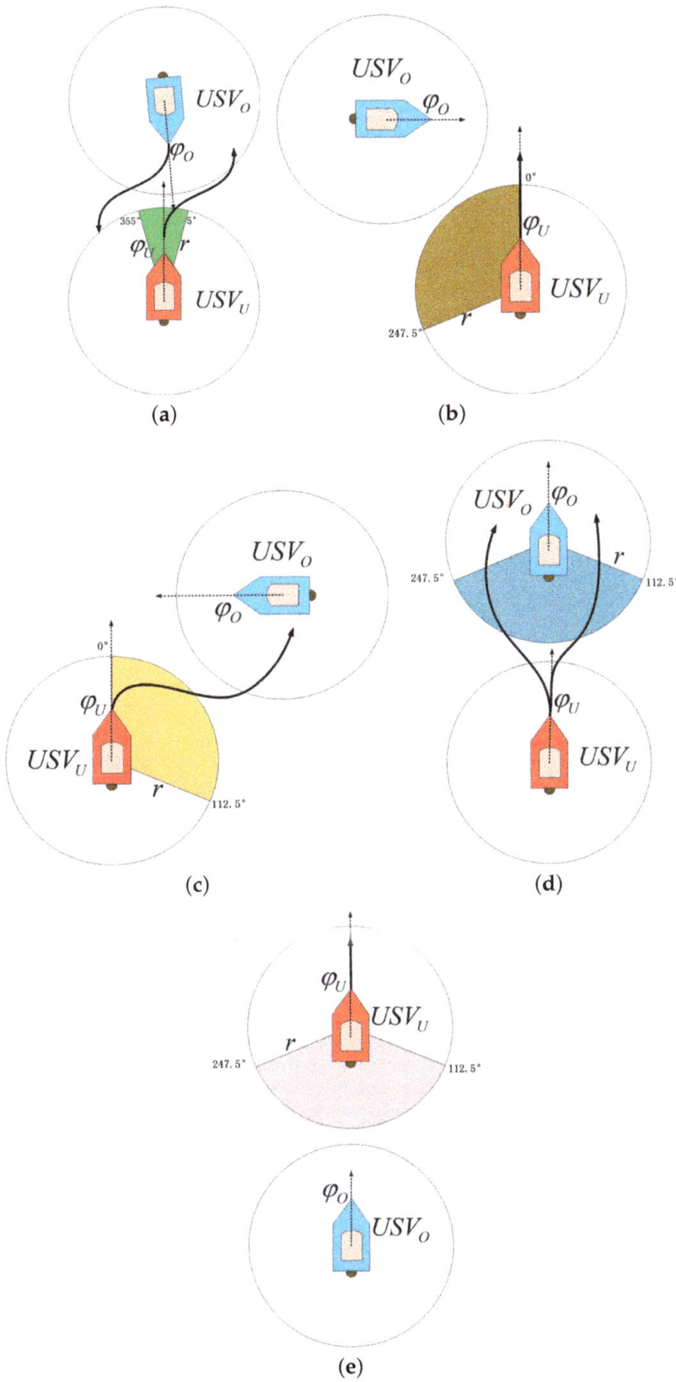

Figure 2. Encounter situation. (**a**) Head-on; (**b**) Crossing-stand-on; (**c**) Crossing-give-way; (**d**) Overtaking-give-way; (**e**) Overtaking-stand-on.

2.2. USV Collision Avoidance Characteristics

In Figure 2, our USV and the obstacles also have a region with the radius of r, called ship domain (SD) [27]. This domain is for each USV and is used to determine the collision when another USV invades. There are three main ways of defining the ship domain, theoretical analyses, experts' knowledge, and determined empirically. The first two are mainly based on a variable number of parameters to regulate the shape and size of the ship domain, which is complex and detailed. The third one lacks some details but is more concise. In this paper, the third way of a circular ship domain is chosen, which is a clean and practical ship domain shape. It is essential for deep reinforcement learning USV collision avoidance algorithm training. Another essential concept is the dynamic area (DA), recorded as R, planning a circular range around USV. When the obstacle USV is not in this range, there is no collision risk between the USVs. These two parameters are used to calculate the collision risk index (CRI), which can not only visualize the current risk for navigation intuitively but is also the key to guiding collision avoidance behavior. The distance at the closest point of approaching (DCPA) and time to the closest point of approaching (TCPA) are defined as shown in Figure 3, and they can be calculated as follows,

$$\begin{cases} DCPA = d_{OU}\sin(\lambda) \\ TCPA = d_{OU}\cos(\lambda)/V_{OU} \end{cases} . \tag{3}$$

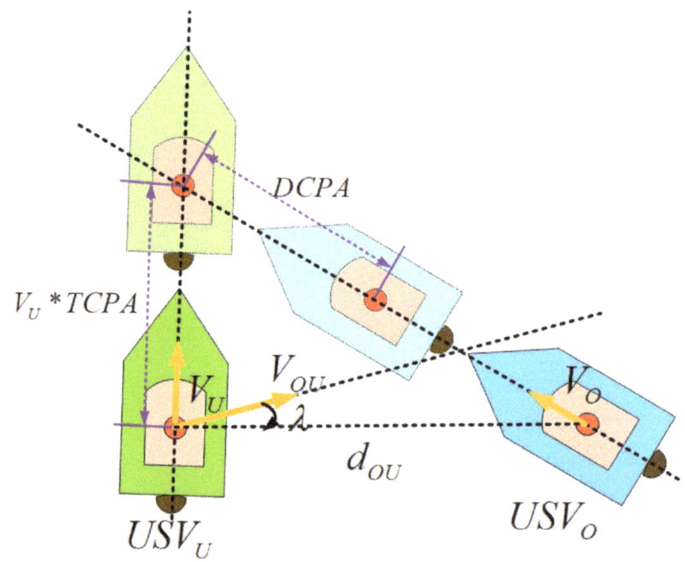

Figure 3. DCPA and TCPA.

Their membership functions u_D and u_T can be calculated as follows,

$$u_D = \begin{cases} 1, & |DCPA| \le r \\ 0.5 - 0.5\sin[\frac{\pi}{R-r} \times \frac{DCPA(R+r)}{2}], & r < |DCPA| \le R \\ 0, & |DCPA| > R \end{cases} \tag{4}$$

If TCPA > 0,

$$u_T = \begin{cases} 1, & TCPA \le T_1 \\ [\frac{T_2 - TCPA}{T_2 - T_1}]^2, & T_1 < TCPA \le T_2 \\ 0, & TCPA > T_2 \end{cases} \tag{5}$$

If TCPA ≤ 0,

$$u_T = \begin{cases} 1, & |TCPA| \leq T_1 \\ [\frac{T_2+TCPA}{T_2-T_1}]^2, & T_1 < |TCPA| \leq T_2 \\ 0, & |TCPA| > T_2 \end{cases} \tag{6}$$

T_1 and T_2 are expressed as follows,

$$T_1 = \begin{cases} \frac{\sqrt{D_1^2+DCPA^2}}{V_{OU}}, & D_1 \geq |DCPA| \\ 0, & D_1 < |DCPA| \end{cases} \tag{7}$$

$$T_2 = \begin{cases} \frac{\sqrt{D_2^2-DCPA^2}}{V_{OU}}, & D_2 \geq |DCPA| \\ 0, & D_2 < |DCPA| \end{cases} \tag{8}$$

Therefore, the USV collision risk u_{CRI} can be calculated as follows,

$$u_{CRI} = \begin{cases} 0, & u_D = 0 \\ 0, & u_D \neq 0, u_T = 0 \\ \max(u_D, u_T), & u_D \neq 0, u_T \neq 0 \end{cases} \tag{9}$$

3. Deep Reinforcement Learning

Deep reinforcement learning theory focuses on the interaction in learning, which addresses how an agent can maximize their reward through learning conspicuous behavior in different states under a specific environment. It is worthwhile to note that it requires only a small amount of prior knowledge provided by humans, but it can complete many challenging problems. Such as path planning in a grid [28], imitating humans playing video games [29], and controlling the movement of vehicles [30]. The deep reinforcement learning approach does not need to investigate internal connection and hidden architecture to the object. Through trial-and-error and delay reward, it can perform control and environment identification tasks simultaneously [31]. The reinforcement learning theory with strong decision-making ability is very suitable for use in the research of USVs. Based on the USV [illegible] and training maritime environment, this algorithm can accomplish the task of USV collision avoidance well in a complex environment.

3.1. Deep Q Learning

Reinforcement learning theory is very suitable for use in unstructured and intricate environments because it is almost impossible for other algorithms to accurately describe the implicit relationship between the environment and the agent. In this way, the component that can make decisions is called the agent, and all the others are the environment. All the frames in reinforcement learning theory are around the constant interaction between these two components [32]. The core of agent-environment interaction consists of the following four elements [33]. The first element is the policy π. It describes the mapping relationship from state to action, similar to a functional relationship, and determines the selected action in the current state. The second element is reward signal G_t. It is a scalar form of feedback from the environment to the agent's behavior, and it defines the long-standing goal of the agent in reinforcement learning processing. The third element is the value function v. It is similar to the reward signal, a quantitative description of advantages based on the latest state. The value function analyzes the better choice from a longer perspective. Finally is the model, which reflects whether the agent has the ability to react to the external environment. The method of driving the interaction process is through the Markov decision process (MDP) [34], which consists of the following three parts. The first part is the state S, which describes the agent and environment at the current moment t. The second part is the action A, which represents all possible ways an agent can influence the environment. The third part is the reward function R, which describes the value of the reward for taking a particular action in a given state. The last part is transition probability $p(s',r|s,a)$, which selects a specific probability distribution for each state and action. In every step t, the agent influences the environment by its action A_t, basing the state S_{t-1} and reward R_{t-1} of the last step, and then getting a new state S_t and reward R_t. Through such a learning process, the reinforcement learning agent can continue understanding the environment deeper and achieve clever solutions to complex control problems.

Q learning is an algorithm [35] that not only contains the bootstrapped learning idea from Dynamic Programming algorithm [36], learning without waiting for the end of an episode but also allows continuous interaction with the environment without modeling, as in the Monte Carlo method [37]. The Q learning algorithm is updated using the following functions,

$$Q(S_t, A_t) \leftarrow Q(S_t, A_t) + \alpha[R_{t+1} + \gamma \max_a Q(S_{t+1}, a) - Q(S_t, A_t)] \tag{10}$$

Though Q learning has performed well in many control problems, there are still many limitations because of the updating form of the value function based on the Q table. In many complex control problems, the reinforcement learning approach is used to deal with problems that are common, complex, and high-dimensional, such as tasks with huge state space or a complex combination of forms, whereas traditional Q learning is inadequate. Therefore, the Q table is fitted using a specific neural network. The gradient descent technique is used instead of the Q table to update the neural network, which perfectly makes up for the defects of traditional Q learning [38].

Figure 4 shows an update schematic diagram of the DQN algorithm. At every training step t, the agent interacts with the environment, constantly enriches its knowledge, and improves their behavior. The state S is fed into the neural network θ and influences the environment by maximizing the action value obtained at this training step t. Then, the state S changes to S' as the environment changes, and the agent can obtain the corresponding rewards signal. Whereafter, the information of interaction (s, a, r, s') is saved in the experience replace buffer for sample and learning. The evaluating Q network output $Q(s, a; \theta)$, and target Q network output $Q(s', a'; \theta^-)$. In this way, all the parameters of the loss function needed for neural network training can be obtained, and the gradient descent method is used to update the evaluating Q network. Finally, the parameters of evaluating the Q network are copied to the target network every N step to achieve the real policy update.

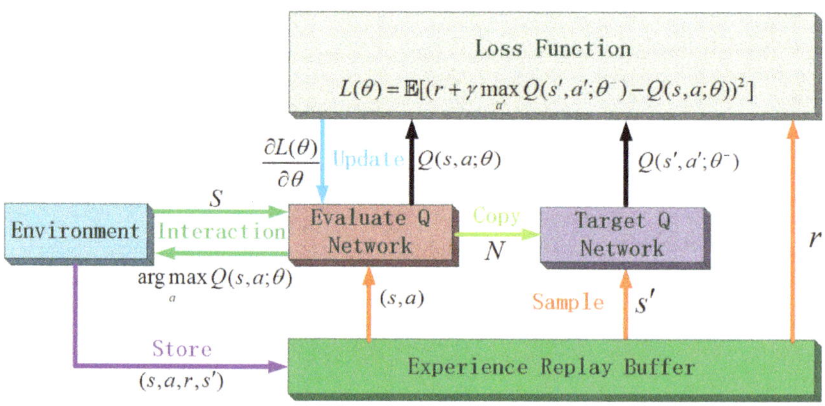

Figure 4. DQN algorithm architecture.

3.2. Collision Avoidance Algorithm for USV

When building a USV collision avoidance training framework with the deep reinforcement learning approach, a complete set of state, action, and reward signals is essential and can facilitate efficient training.

The first part is the design of state space. The USVs autonomous collision avoidance system must be able to face complex and time-varying maritime conditions. With reliable sensors, USVs can perceive the real-time information of obstacles within a range and can be used for USV collision avoidance behavior training. So, the following state space is designed,

$$S = \{\varphi_U, \dot{\varphi}_U, \delta_U, \dot{\delta}_U, \vartheta_t, d_t, d_{O1}, \vartheta_{O1}, \varphi_{O1}, \ldots, d_{Om}, \vartheta_{Om}, \varphi_{Om}, d_{S1}, \vartheta_{S1}, \ldots, d_{Sn}, \vartheta_{Sn}\} \quad (11)$$

The state space can be divided into four parts. The first part is the state of our own USV, which reflects the navigation information of the USV. It contains its own USV course φ_U, the change rate of our USV course $\dot{\varphi}_U$, its own USV rudder angle δ_U, and the change rate of the USV's rudder angle $\dot{\delta}_U$. The second part is the terminal state, which describes the navigation destination of each episode. It contains the absolute azimuth of the terminal and USV_U, and the distance between USV_U and the terminal. The third part is the state of obstacle USVs, which reflects the navigation information of obstacle USVs near our USV, and for m obstacle USVs, there are m groups. It contains the absolute azimuth of USV_O and USV_O, the distance between our own USV and the obstacle USVs, and the obstacle USVs' course. The fourth part is the state of static obstacles with n sets of information for n obstacles USV. It contains the absolute azimuth of static obstacles and USV_O, and the distance between our USV and static obstacles.

The second part is the building of the action space. It is appropriate to design some actions in action space as the change of rudder angle because the USV changes its navigation state by rudder changes, and it can be designed as the following action space,

$$A = \{\Delta\delta_1, \Delta\delta_2, \Delta\delta_3, \Delta\delta_4, \ldots, \Delta\delta_k\} \quad (12)$$

After selecting the action, the change of target rudder angel $\Delta\delta_k$ is obtained, and the new target rudder angle is as follows,

$$\delta_r \leftarrow \delta_r + \Delta\delta_k \quad (13)$$

In this paper, 11 different sizes of actions are designed in the action space to enable the USV to adopt various behaviors, such as not changing the rudder angle, slightly changing, and hard changing. Therefore, the specific designs of action space are as follows,

$$A = \{-5^\circ, -4^\circ, -3^\circ, -2^\circ, -1^\circ, +0^\circ, +1^\circ, +2^\circ, +3^\circ, +4^\circ, +5^\circ\} \quad (14)$$

The third part is the design of the reward signal, which evaluates the various USV behaviors at each training step. The training is to continuously learn about the new environmental conditions and maximize their estimation of future benefits, but this estimation is derived from the design of reliable reward signals. The reward design of this paper is divided into the following two parts,

(1) The reward for goal
   (a) Terminal reward
   The terminal is where the end of USV navigation is on each training episode. The design of the terminal reward can encourage this good behavior and affect the whole training environment through bootstrap. When $\sqrt{(x_U - x_t)^2 + (y_U - y_t)^2} < r_{\min} + r_t$, it is considered that reaching the terminal, and getting the reward, R_t.
   (b) Collision reward
   Avoiding obstacles is another important goal in training. Punishment for collision can teach the trained USV to keep a safe distance from obstacles. When

$\sqrt{(x_U - x_O)^2 + (y_U - y_O)^2} < r$ or $\sqrt{(x_U - x_O)^2 + (y_U - y_O)^2} < r_O$, it is considered that colliding the obstacle USV. The collision reward obtained is R_O.

When $\sqrt{(x_U - x_O)^2 + (y_U - y_O)^2} < r + r_S$, it is considered that colliding the static obstacle. The collision reward is R_S.

(c) COLREGs reward

COLREGs provide a constraint for USV behaviors. Integrating COLREGs into the training process in a reward signal can endow the trained USV agent with regularized avoidance behavior. When $E \in \{E_3, E_4\}$ and $a \notin \{0°, +1°, +2°, +3°, +4°, +5°\}$, the reward signal $R_C = k_C u_{CRI}$ can be obtained. The more dangerous the moment of breaking the COLREGs, the higher the penalty for USV. When $E \in \{E_3, E_4\}$ and $a \in \{0°, +1°, +2°, +3°, +4°, +5°\}$, or $E \in \{E_1, E_2, E_5, E_6\}$, there are the conditions that the our USV should go straight or turn left or right. The designed reward signal is 0.

(d) Seamanship reward

When there are no obstacles or no duty to give way, our USV should keep straight as far as possible. Therefore, the following seamanship reward is designed to restrain the navigation behavior of the USV: When $a \notin \{0°\}$ and $u_{CRI} = 0$, the reward is R_δ.

(2) Guiding reward

The guiding reward can enrich the reward signal in a training environment and avoid the training difficulty caused by the sparse rewards problem.

(a) Course reward

The course that points more toward the terminal is considered to be a better state, so the course reward signal is designed as follows,

$$R_\varphi = \mathrm{k}_\varphi(\varphi_\mathrm{k} - |\varphi_U - \varphi_T|) \tag{15}$$

where φ_k is the critical value of the positive or negative reward.

(b) Course better reward

The agent's behavior is positive if it makes the course more pointed toward the terminal after an action, so the course better reward signal is designed as follows,

$$R_{\Delta\varphi} = \begin{cases} r_s & ,\text{if } \phi \text{ smaller} \\ r_b & ,\text{if } \phi \text{ bigger} \\ r_{e1} & ,\text{if } \phi \text{ doesn't change, and } \phi = 0 \\ r_{e2} & ,\text{if } \phi \text{ doesn't change, and } \phi \neq 0 \end{cases} \tag{16}$$

Thus, the complete reward signal function can be expressed as,

$$R = R_t + R_O + R_S + R_C + R_\delta + R_\varphi + R_{\Delta\varphi} \tag{17}$$

After designing the state space, action space, and reward signal, the training system is completed. Figure 5 shows the complete training architecture. At each step, the state information is input into the neural network, then the value of all actions based on the current network parameters and state is obtained through the neural network. Then the selected action is obtained, resulting in the environment update.

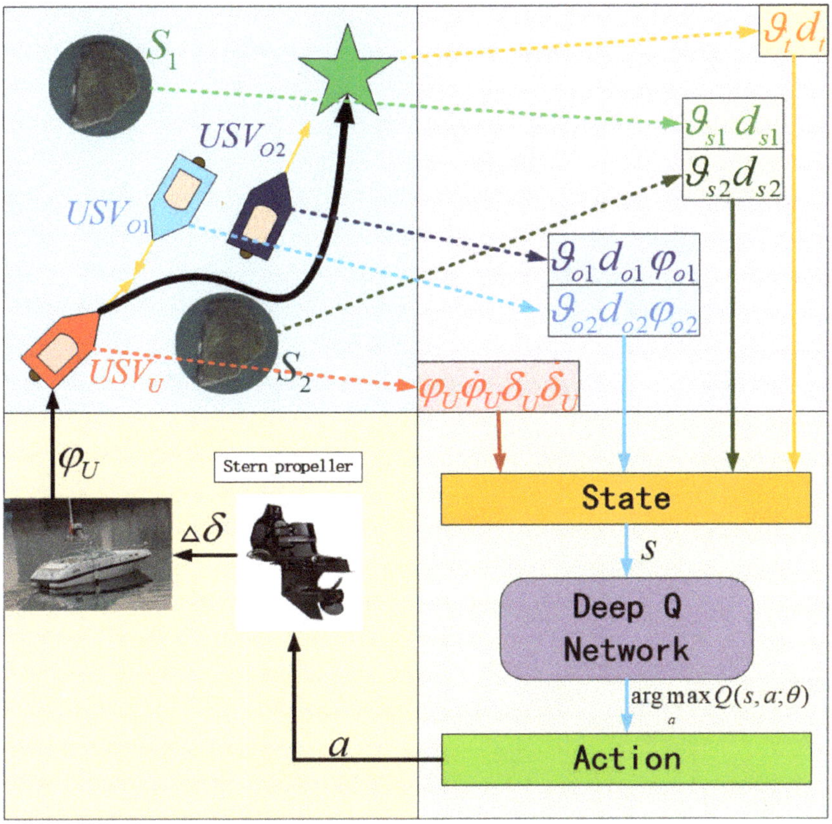

Figure 5. Training Architecture.

4. Improvement for USV Collision Avoidance Algorithm

4.1. Double DQN

In the training process of the DQN algorithm, uncontrollable overestimation of the action value generally exists, which is caused by the unknown of the real action value in the learning process, resulting in the performance of the training being affected and even falling into local optimal [39]. The DQN algorithm uses a greedy policy based on the action with the maximum action value in the action space, which will introduce significant maximization bias. This kind of overestimation is common. However, the influence of overestimation on the optimal policy can be reduced as much as possible by Double Q learning. In the traditional DQN algorithm, as shown in Equation (18), the action is chosen through the target network, while the value estimation output is also. This operation is the root cause of the overvaluation. Therefore, it can be decoupled by two neural network outputs. The action is selected by the evaluated network rather than based on the target network to reduce the impact of the overestimation problem [40]. A new calculation method can be obtained as shown in Equation (19).

$$Y_t^{DQN} = r + \gamma \max_{a'} Q(s', a'; \theta^-) \tag{18}$$

$$Y_t^{Double\_DQN} = r + \gamma Q(s', \arg\max_{a'} Q(s', a'; \theta); \theta^-) \tag{19}$$

4.2. Dueling DQN

Dueling neural network architecture is an outstanding method [41]. As shown in Equation (20), the action value function is divided into a state value function and an advantage function. The former describes the worth of a state, while the latter describes the relative importance of each action. It can distinguish the value of different states and actions, leading to more robust training.

$$Q(s,a;\theta,\alpha,\beta) = V(s;\theta,\beta) + A(s,a;\theta,\alpha) \tag{20}$$

4.3. Prioritized Experience Replay

As a pivotal part of the DQN algorithm, the experience replay buffer can reduce sample correlation and improve sample utilization. The traditional DQN algorithm uniformly samples the experience from the replay buffer after storing samples through action policy. However, this form may not be optimal. In different training conditions and steps, the value of each interaction for network training is distinct. If interactions have different sampling probability weights, the interactions that are more valuable to the current USV training will be assigned higher weights, which will be more conducive to agent learning [42]. Here a significant index, TD-error, forms an essential part of the loss function and a basis for gradient descent. This index provides easily accessible and valid evidence for the definition of priority, making the interactions in the experience replay buffer with a certain tilt. This random-priority method makes learning from the experience more robust. TD-error is as follows,

$$TD = r + \gamma \max_{a'} Q(s',a';\theta^-) - Q(s,a;\theta) \tag{21}$$

Interactions with high sampling weights are also not guaranteed to be sampled at any step, while those with low sampling weights are not necessarily not sampled. The priority only makes the samples that need to be chosen much easier. The probability of sampling is as follows,

$$P(i) = \frac{p_i^\alpha}{\sum_m p_m^\alpha} \tag{22}$$

where $P(i)$ is the probability of sampling for each interaction, and there are m interactions in the experience replay buffer. α is used to adjust the effect of priority. When $\alpha \to 0$, it is uniform sampling, and the higher α is, the more prominent the effect of priority on sampling. This paper adopts proportional prioritization as follows,

$$p_i = |TD_i| + \varepsilon \tag{23}$$

where, TD_i is the TD-error after normalization of $TD_i = \frac{TD_i}{TD_{\max}}$, and ε is a parameter to avoid the problem of zero denominators.

The off-policy method requires importance sampling weights (ISW) to correct the bias of estimation based on different samples. However, the transfer probability of the Q-learning algorithm is not dependent on policy but on the environment. Therefore, it does not need ISW. Nevertheless, the prioritized experience replay method breaks this advantage because the unbiasedness of the expected value estimation depends on the sample with the same distribution as the expected value. The prioritized experience replay changes the distribution of the samples in an uncontrolled way. Therefore, it is necessary to compensate for this bias by adding ISW, which is calculated as follows,

$$\omega_i = \left(\frac{1}{N}\frac{1}{P(i)}\right)^\beta \tag{24}$$

This normalized ISW can make the updating process more stable. β can adjust the impact of prioritized experience replay. When $\beta \to 1$, it cancels out the inconsistent probability effect of prioritized experience replay. So, its loss function is shown as follows,

$$L(\theta) = \mathbb{E}\left[\omega_i TD_i^{\,2}\right]. \tag{25}$$

4.4. Noisy Network

Because the training environment considered in this paper is multivariate, it places high demands on the exploratory capabilities of the algorithm. Although promoting the exploration with ε greedy policy is easy to implement and their effect is acceptable in most cases, it is not a reasonable enough choice because the agent acting up to the greedy-action with a specific probability value is too aimless. So, it can drive exploration by adding a learnable noise instead of ε greedy policy [43], which has not been used in USV collision avoidance. The weight and bias of the neural network become uncertain due to the noise parameters, which will increase the uncertainty of neural network output and promote agent exploration. After adding noise, the neural network can be expressed as follows,

$$y \doteq (\mu^w + \sigma^w \odot \varepsilon^w)x + \mu^b + \sigma^b \odot \varepsilon^b \tag{26}$$

where, weight w is divided into two parts, the weight μ^w without noise and the weight σ^w with noise. ε^w is the noise parameter, and $\odot$ represents element-wise multiplication for adding the noise to each weight σ^w in the neural network. Similarly, μ^b is the bias without noise, and σ^b is the bias with the noise parameter of ε^w. The noisy network forms a way of exploration by reasonably controlling the noise added to each network parameter, which can meet the different exploration needs in different training conditions. Factorized Gaussian noise is chosen to construct the noise in the neural network, which has a lower computational cost, that is suitable for the DQN algorithm. The noise can be constructed as follows,

$$\begin{cases} \varepsilon^w_{i,j} = \zeta(\varepsilon_i)\zeta(\varepsilon_j) \\ \varepsilon^b_j = \zeta(\varepsilon_j) \end{cases} \tag{27}$$

where, $\zeta(a) = \mathrm{sgn}(a)\sqrt{|a|}$ is a function for construction of noise. All the value of parameters ε_i and ε_j are obeyed Gaussian distribution as $\varepsilon \sim N(0, k)$. The variance k limits the noise size, and the higher variance means a greater ability to explore. The mean of zero means introducing noise parameters will not bias the original policy. Introducing a noisy network instead of the traditional neural network is more beneficial for environment exploration. The noise drives the exploration making the algorithm more flexible, reasonable, and efficient. So, its loss function is as follows,

$$L'(\zeta)_{Noisy} = \mathbb{E}_{\varepsilon,\varepsilon^-}\left[\mathbb{E}_{(s,a,r,s')}\left[r + \gamma \max_{a'} Q(s', a', \varepsilon^-; \zeta^-) - Q(s, a, \varepsilon; \zeta)\right]^2\right]. \tag{28}$$

4.5. Improvements with USV Characteristics

Since the input to the neural network in the form of USV states contains scalar values with many differences in the order of magnitude, such as rudder angle values, direction values, and distance values, it is not reasonable to put them directly into training. USV state clipping is a benefit for improving the network training efficiency. By normalizing the input, the potential problem of the large descent gradient caused by the differential input can be avoided as much as possible, which makes neural network training more robust.

If the obstacles are too far from our own USV, the changes in the distance between the USVs do not affect the navigation of the USV. Therefore, it is not worth wasting more computational resources on learning these conditions. In this paper, a simplified way is designed, using the collision avoidance parameter of DA, the distances between USVs and obstacles or the terminal more than the DA value R are limited to R, which can effectively improve the training effect.

The code for this Algorithm 1 is as follows,

Algorithm 1 DRLCA algorithm code

Initialize USV training environment
Initialize experience replay buffer H to capacity of C
Initialize evaluation neural network in θ
Initialize target network in $\theta^- = \theta$
Initialize variance of noise in k
For episode = 1, n do
    Initialize initial states of each USV and static obstacles
    Initialize speed of our USV and obtain ship domain size r in current episode
    While true
        Update the USV collision avoidance training environment
        Generate ε_i and ε_j
        Get noise parameters
        Select the action with $a_m = \arg\max_a Q(s', \arg\max_a Q(s', a; \theta), \varepsilon^-; \zeta^-)$
        Changing rudder angel by execute action $a^* = a_m$ in environment, and obtain s_{m+1}
        Obtain collision avoidance reward signal $R = R_t + R_O + R_S + R_C + R_\delta + R_\varphi + R_{\Delta\varphi}$
        Store current transition (s_m, a_m, r_m, s_{m+1}) in experience replay buffer H
        Assign current transition to highest priority $p_{\max}$
        By priority for each transition $P(i)$, sample the random minibatch of transitions (s_k, a_k, r_k, s_{k+1}) from H for learning
        Caculate ISW ω_i for each transition in minibatch
        Caculate TD-error TD_i.
        Obtain $y_i = \begin{cases} r_j, & \textit{if } \mathrm{j+1} \textit{ is the terminal} \\ r_j + \gamma \max_{a'} Q(s_j, a', \varepsilon^-; \zeta^-), & \textit{otherwise} \end{cases}$
        Using gradient descent with ISW, Update evaluate network parameters θ
        Update $p_i = |TD_i| + \varepsilon$ for all samples
        If it is target network updating step xN
            update the weight $\theta^- = \theta$
        End if
        The number of steps counted plus 1
    End while
End for
Return the weight $\theta^* = \theta^-$ of target network

5. Experiments

5.1. Training Environment

Figure 6 shows a collision avoidance training environment designed with a single obstacle USV. The left part of this figure shows how the obstacle USV is generated, with 360 initial positions, spaced the degree of $\eta = 1^\circ$ evenly distributed around the circumference. The right part of this picture shows how the static obstacles are generated, with eight potential locations for each initial location of obstacle USV. Six of these locations are evenly distributed around the circumference of the circle, and the other two are at the intersection of line P1P3, line P4P6, and line P2P5. In each episode, randomly select two locations from these eight locations to generate static obstacles. It constitutes a simulation environment with 360 × 8 × 7 = 20,160 random combinations of obstacles. In addition, the simulation in this paper is based on the Unity virtual marine simulation training and testing platform.

This experiment is based on 2× RTX 2080Ti GPUs, Xeon Gold 5218 CPU, Python 3.6, and Tensorflow 1.15 for environment building, algorithm training, and collision avoidance testing.

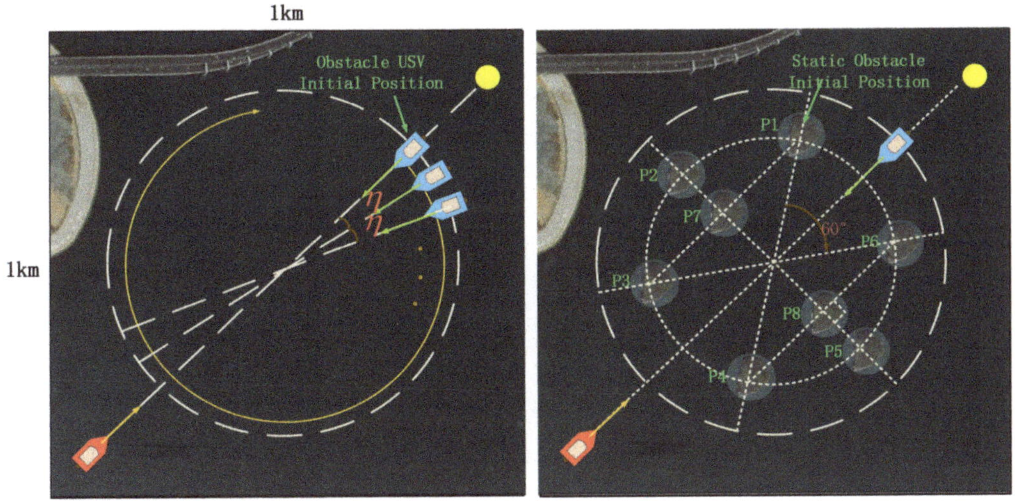

Figure 6. Training environment.

5.2. Framework for Training

Based on the various improvements of the algorithm structure and simulation environment, the training hyperparameters selected in this paper are shown in Table 1. For more stable training, the learning rate is designed to be very low and the discount factor very high, 0.0001 and 0.99, respectively. The update frequency of the target network is 4096, and the sampling number is 600. The noise variance is 0.1, and the mean value is 0, which ensures a suitable exploration capability while avoiding the bias caused by noise. The ISW is initially 0.5, and the priority experience replay factor is 0.4. For better training, start training on the 2000th step. The experience replay memory size of 1,000,000 ensures that no experience is dropped.

Table 1. Hyperparameters for training.

Hyperparameter	Value
Learning Rate	0.0001
Discount Factor	0.99
Target Network Update Frequency	4096
Replay Memory Size	1,000,000
Batch Size	600
Noise Variance	0.1
Noise Mean	0.0
Greedy Value	1.0
Importance Sampling	0.5
Linearly Anneal of Importance Sampling	1.25×10^{-6}
Priority Experience Replay	0.4
Replay Start Size	2000

Figure 7 shows the neural network architecture in this paper. With 900 neurons in each layer, the green part is the traditional layer of the Q network, the yellow is the noise layer, the blue is the dueling network architect, the orange is the addition operation, the red is state input, and the black is the action output. The optimizer is Adam. The activation function is Lecky_ReLU.

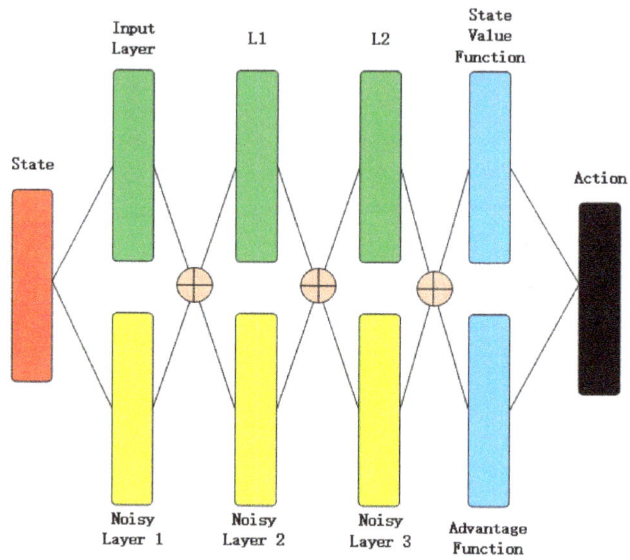

Figure 7. Neural network architecture.

5.3. Training

The average reward in training is shown in Figure 8. The same ten sets of random number seeds are selected for all algorithms for ten simulations, and the average reward graph containing confidence regions is plotted. In ten simulations, the average reward per 40 episodes is averaged to one value, forming ten values, and the point on the curve is the average of these ten values. The shaded part in this figure shows the confidence region, whose upper and lower bounds are the maximum and minimum values among these ten values. The means of the average reward value of each algorithm over different training stages is shown in Table 2. An individual improvement is limited, but the algorithm proposed in this paper, DRLCA, combining all improvement methods, has a very significant collision avoidance performance. The average reward of DRLCA rises very fast in the early stage, especially the first point on the curve, which is already much higher than others and smooths out at a higher average reward value position in the later stage of training.

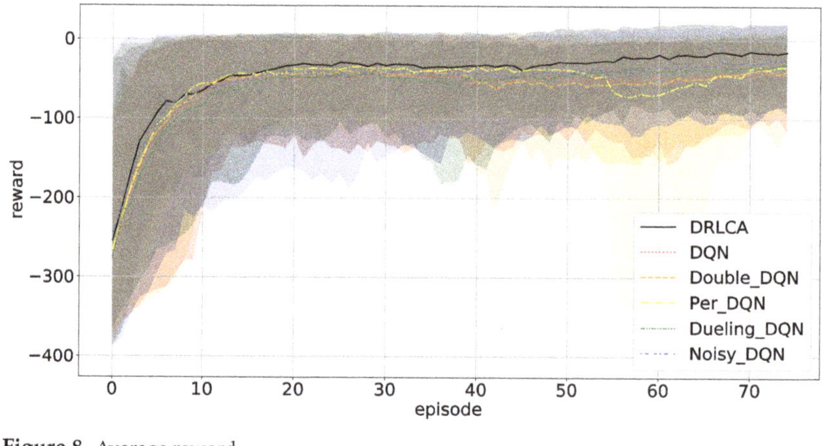

Figure 8. Average reward.

Table 2. Mean of rewards at different stages of training.

Alogrithm	First Third of Training	Middle Third of Training	Last Third of Training
DRLCA	−75.86	−32.61	−20.83
DQN	−91.05	−48.37	−47.26
Double DQN	−85.33	−47.62	−49.55
Per DQN	−81.92	−37.46	−51.17
Dueling DRLCA	−96.42	−48.82	−38.18
Noisy DQN	−81.97	−43.11	−34.09

As Figure 9a shows the condition of USV training in the first episode, the USV will keep rotation because the agent has no knowledge about this environment and is basing its movement on random exploration. Figure 9b shows the training effect in the 9th episode, where the USV tries to collect more experience. Figure 9c shows the training effect in the 20th episode. The USV has tried more behaviors to explore this environment. Figure 9d shows the training effect in the 46th episode, where the USV reaches the terminal for the first time, which is very important for the training and proves that the guidance reward designed in this paper is very effective. Figure 9e shows the effect in the 124th episode, where the failed collision experience with the obstacle USV is crucial for better learning. As shown in Figure 9f, the effect in the 358th episode, the USV constantly optimizes its behavior, and within the next thousands of training episodes, the optimal collision avoidance policy is approached continuously.

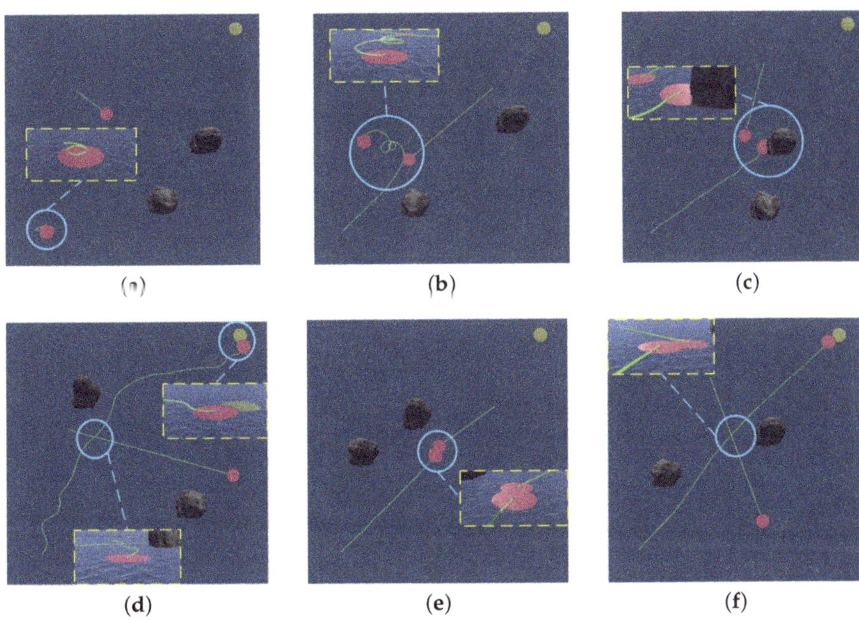

Figure 9. Training. (**a**) First episode; (**b**) 9th episode; (**c**) 20th episode; (**d**) 46th episode; (**e**) 124th episode; (**f**) 358th episode.

To verify whether all the improvements have positive effects and whether removing one leads to better training, an ablation study is performed. The average reward of the ablation study is shown in Figure 10. The means of the average reward value of each algorithm over different ablation study stages is shown in Table 3. The average reward height and increase rate of the algorithms with any one improvement removed are lower than the DRLCA, verifying all the improvement methods are complementary.

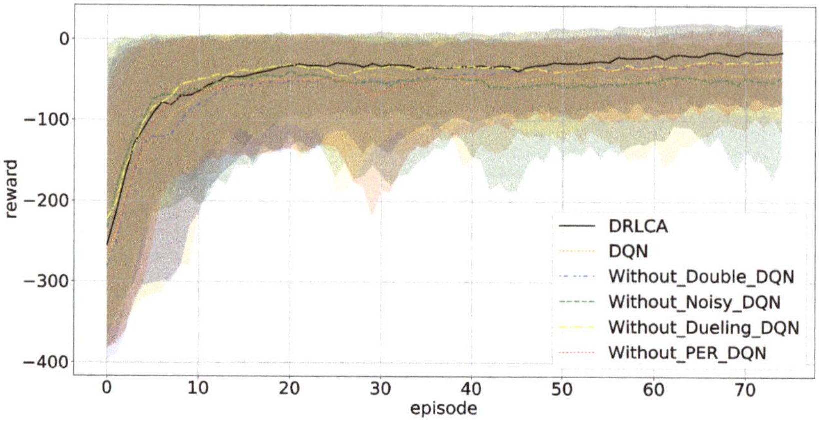

Figure 10. Ablation study.

Table 3. Mean of rewards at different stages of ablation study.

Alogrithm	First Third of Training	Middle Third of Training	Last Third of Training
DRLCA	−75.86	−32.61	−20.83
DQN	−91.05	−48.37	−47.26
Without Double DQN	−97.36	−43.94	−31.61
Without Per DQN	−88.91	−49.68	−33.53
Without Dueling DRLCA	−69.30	−35.92	−31.44
Without Noisy DQN	−73.84	−52.57	−52.04

5.4. Test

The first test environment is shown in Figure 11, where our USV is in the No. 6 encounter situation with the USV_O, and USV_U should avoid the USV_O and steers both to port and starboard are allowed. Concurrently, the two static obstacles do not obviously block the navigation of our own USV that can be used to test whether the USV collision avoidance agent has learned to ignore the non-hazardous obstacles. Figure 11a shows the effect of collision avoidance of the DRLCA algorithm and Figure 11b DQN algorithm. The initial position of the USV_O, static obstacles 1 and 2 are $(402.33, 783.66)$, $(216.34, 402.33)$, and $(442.76, 205.51)$. The course of obstacle USV is $161°$. Figure 11c shows the change of rudder angle and course of collision avoidance of the DRLCA, Figure 11d DQN algorithm. Figure 11e shows the changes in the distance of collision avoidance of the DRLCA algorithm, Figure 11f DQN algorithm. By adding up $180 - \phi$ (if $\phi > 180°$) or ϕ (if $\phi \leq 180°$) in each second, the accumulated course deviation can be obtained, and the DRLCA is $1730.28°$and DQN is $3781.80°$. The closest distances to the three obstacles are 43.89 m, 136.66 m, and 162.22 m for DRLCA, and 129.81 m, 169.04 m, and 104.55 m for DQN.

The second test environment is shown in Figure 12, where it is in the No. 1 encounter situation, and USV_U should avoid the USV_O and steers both to port and starboard are allowed. In addition, there are two static obstacles, one of which does not affect our USV's navigation, and the other does. It tests the collision avoidance ability when encountering static obstacles and USV at the same time. Figure 12a,b show the effect of collision avoidance of DRLCA and DQN. The initial position of the USV_O, static obstacles 1 and 2 are $(654.51, 757.15)$, $(505.24, 799.95)$, and $(628.56, 422.74)$. The course of obstacle USV is $211°$. Figure 12c,d shows the change of rudder angle and course of collision avoidance of DRLCA and DQN. Figure 12e,f shows the distance in collision avoidance of DRLCA and DQN. In the index of accumulated course deviation, the DRLCA is $707.51°$, and DQN is $2357.59°$.

The closest distances to the three obstacles are 44.32 m, 245.9 m, and 99.51 m for DRLCA and 57.6 m, 274.29 m, and 66.81 m for DQN.

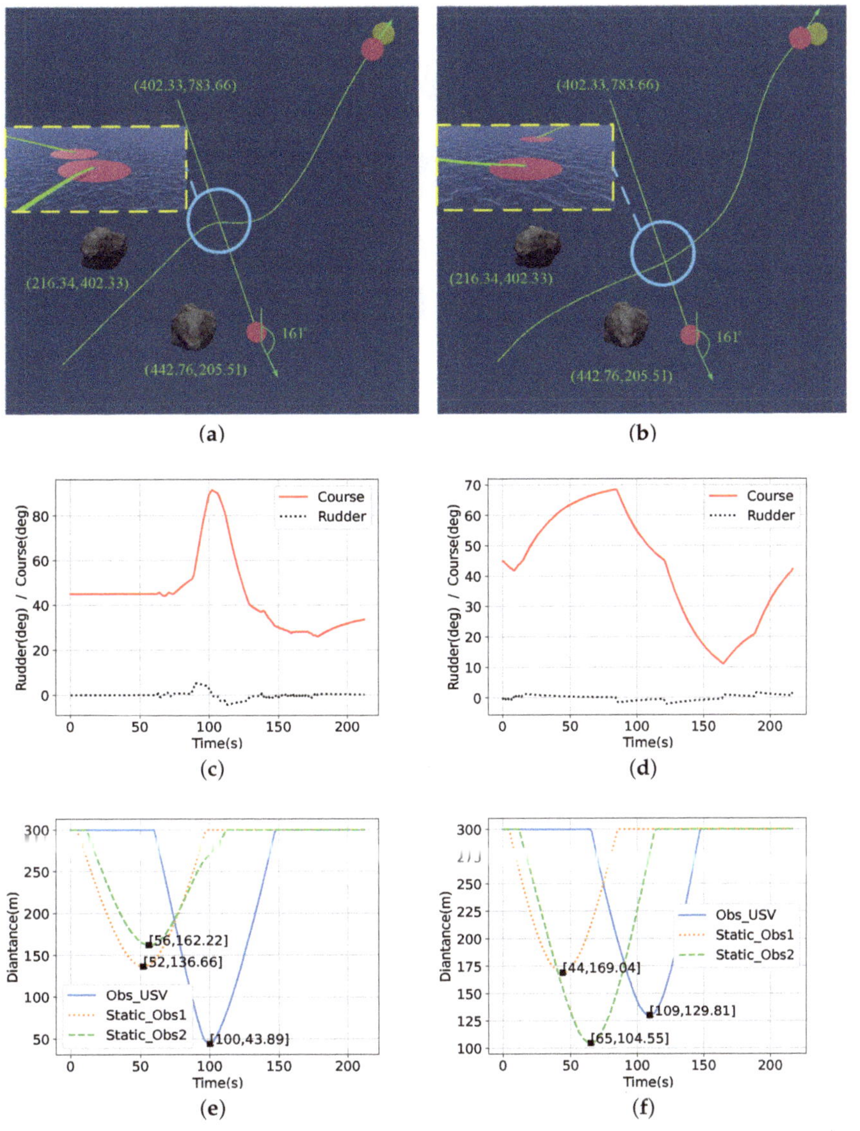

Figure 11. Encounter situation 1. (**a**) Path planned by DRLCA; (**b**) Path planned by DQN; (**c**) Course and rudder for DRLCA; (**d**) Course and rudder for DQN; (**e**) Distance for DRLCA; (**f**) Distance for DQN.

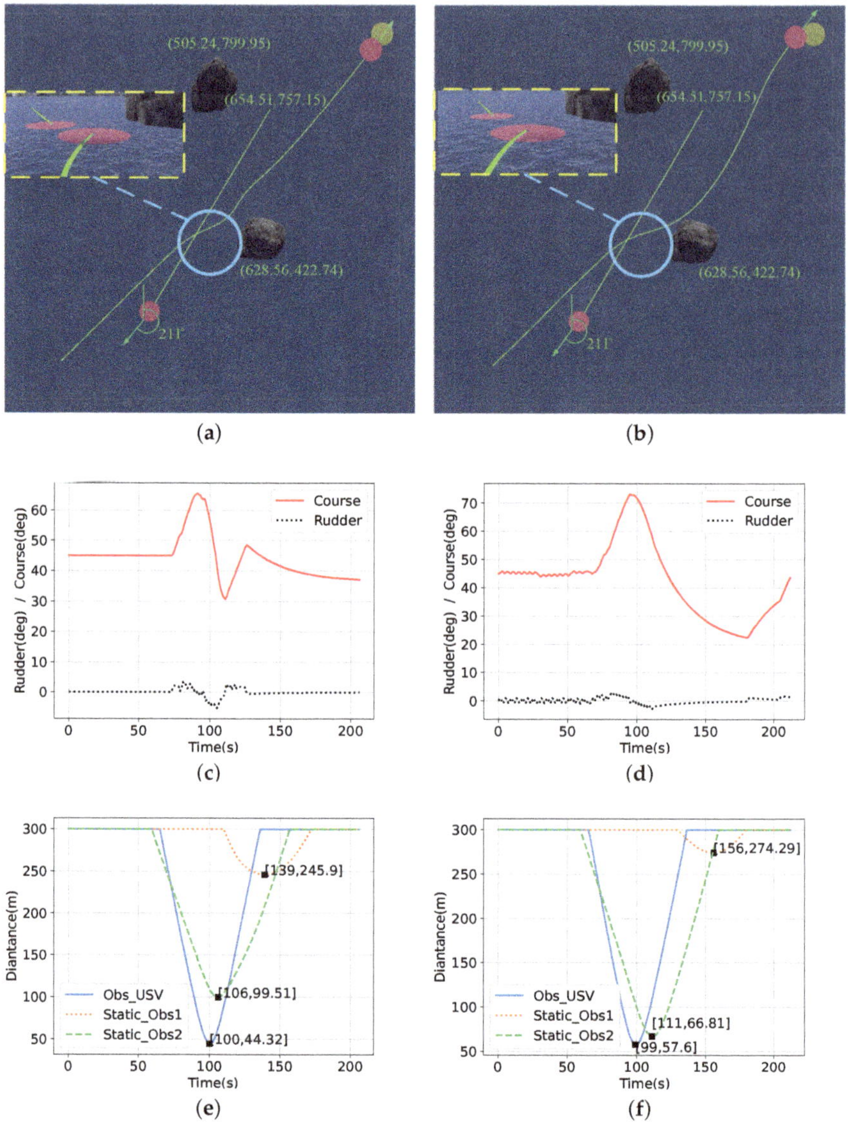

Figure 12. Encounter situation 2. (**a**) Path planned by DRLCA; (**b**) Path planned by DQN; (**c**) Course and rudder for DRLCA; (**d**) Course and rudder for DQN; (**e**) Distance for DRLCA; (**f**) Distance for DQN.

The third test environment is shown in Figure 13. The interference from the obstacle USV is not significant, which can verify the ability to avoid static obstacles. Figure 13a,b show the effect of collision avoidance of DRLCA and DQN. The initial position of the USV_O, static obstacles 1 and 2 are $(789.77, 422.35)$, $(538.82, 644.89)$, and $(461.18, 355.11)$. The course of obstacle USV is 285. Figure 13c,d shows the change of rudder angle and course of collision avoidance of DRLCA and DQN. Figure 13e,f shows the distance in collision avoidance of DRLCA and DQN. The accumulated course deviation is 3371.51° for DRLCA, and DQN could not compare because of the incomplete navigation. The closest distances to the three obstacles are 136.02 m, 100.57 m, and 138.48 m for DRLCA and 36.45 m, 65.79 m, and 103.96 m for DQN.

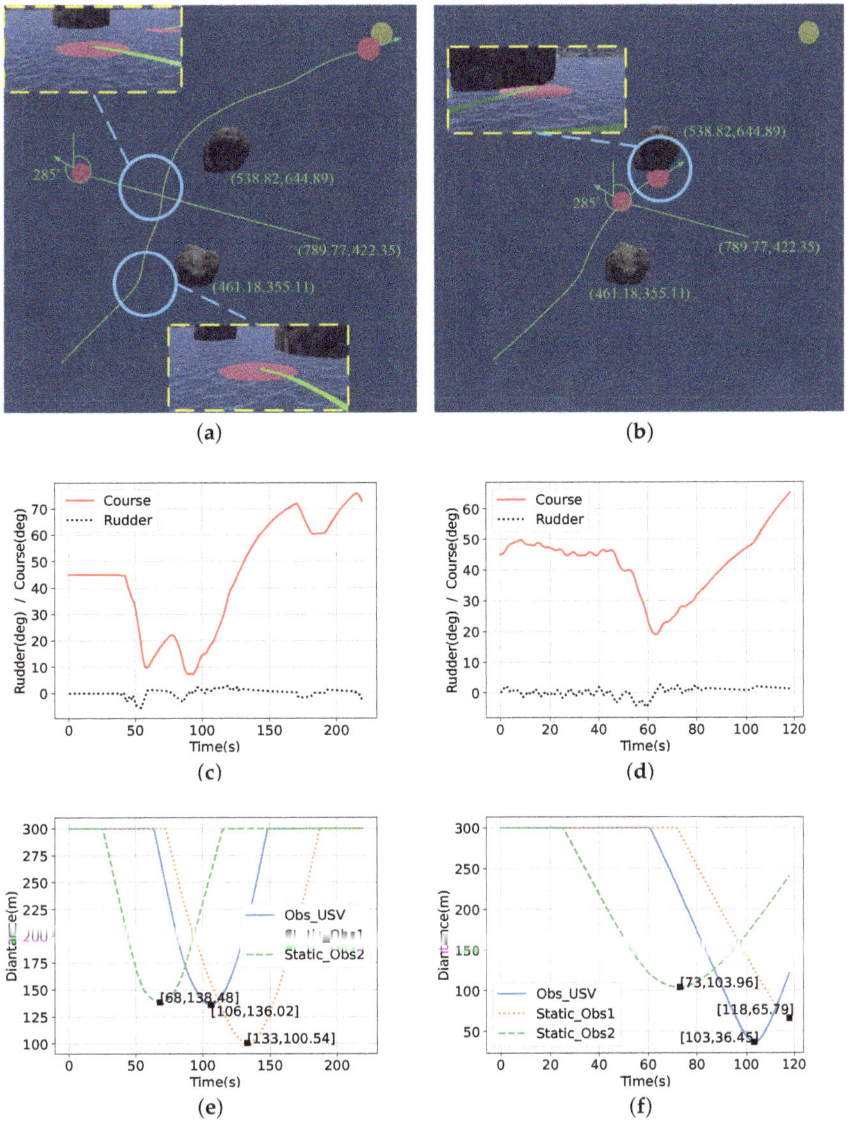

Figure 13. Encounter situation 3. (**a**) Path planned by DRLCA; (**b**) Path planned by DQN; (**c**) Course and rudder for DRLCA; (**d**) Course and rudder for DQN; (**e**) Distance for DRLCA; (**f**) Distance for DQN.

The fourth test environment is shown in Figure 14. Our USV is in the No. 4 encounter situation with the obstacle USV, and there is no interference from the static obstacles. Figure 14a,b show the effect of collision avoidance of DRLCA and DQN. The initial position of the USV_O, static obstacles 1 and 2 are $(363.80, 232.70)$, $(232.70, 636.20)$, and $(633.65, 431.90)$. The course of obstacle USV is 27°. Figure 14c,d shows the change of rudder angle and course of collision avoidance of DRLCA and DQN. Figure 14e,f shows the distance in collision avoidance of DRLCA and DQN. The accumulated course deviation is 909.19° and 2074.92° for DRLCA and DQN. The closest distances to the three obstacles are 55.11 m, more than 300 m, and 88.98 m for DRLCA and 33.16 m, 265.91 m, and 183.41 m for DQN.

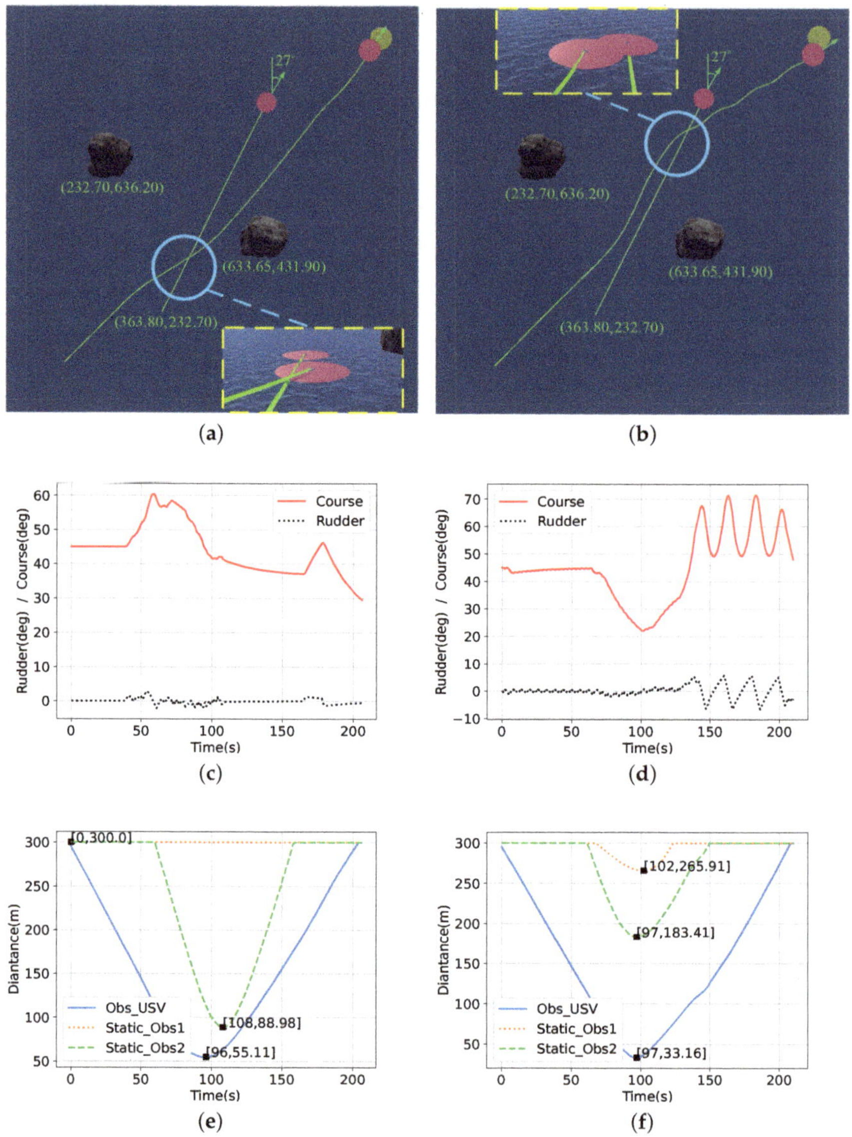

Figure 14. Encounter situation 4. (**a**) Path planned by DRLCA; (**b**) Path planned by DQN; (**c**) Course and rudder for DRLCA; (**d**) Course and rudder for DQN; (**e**) Distance for DRLCA; (**f**) Distance for DQN.

Table 4 shows the comparison results for four experiments. In groups 1, 2, and 4 experiments, the DRLCA reduced the total course deviation by 54.25%, 70.00%, and 56.18%, respectively, compared with the DQN, with an average improvement of 60.14%.

In total, 100 experiments of collision avoidance under the same random number seed are carried out. As shown in Table 5, the results of these experiments are recorded. The number of successful arrivals, out-of-bounds, and collisions are also recorded, and the success rate of the DRLCA is much higher. Because the failed collision avoidance will affect the result of the accumulated course deviation, only the experiments that reach the terminal are used for the calculation. The accumulated course deviation for DRLCA is 2150.02°, and DQN is 2929.36°, improving 26.60%. Finally, the average time per experiment is compared for DRLCA and DQN, which are 211.875 and 214.304 s, improving by 1.13%. The improved algorithm performs better on collision avoidance problems in this environment.

Table 4. Results of four experiments.

Experiment	Result	Course Deviation
Test environment 1, DRLCA	arrival	1730.28°
Test environment 1, DQN	arrival	3781.80°
Test environment 2, DRLCA	arrival	707.51°
Test environment 2, DQN	arrival	2357.59°
Test environment 3, DRLCA	arrival	3371.51°
Test environment 3, DQN	collision	/
Test environment 4, DRLCA	arrival	909.19°
Test environment 4, DQN	arrival	2074.92°

Table 5. Results of 100 times experiments.

Algorithm	Successful Arrival	Out of Bound	Collision	Average Accumulated Deviation of Course	Average Time
DRLCA	97	3	0	2150.02°	211.875 s
DQN	56	30	14	2929.36°	214.304 s

5.5. Multi-Obstacle USV Collision Avoidance

A test effect diagram of multi-obstacle USV collision avoidance is shown in Figure 15. As shown in Figure 15a, the first stage shows the initial condition of a test environment. The obstacle USVs are at the position of $(250, 600)$, $(570, 480)$, and $(800, 200)$. The static obstacles are at $(250, 270)$ and $(400, 500)$. The course of obstacle USVs are 90°, 225°, and 315°. As shown in Figure 15b, stage 2 is a condition when our own USV avoids a static obstacle. As shown in Figure 15c, our USV encounters USV_{O2} in the No. 1 encounter situation, turns to the starboard, and successfully avoids the obstacle USV. As shown in Figure 15d, our USV encounters the USV_{O3} in the No. 3 encounter situation. Our USV successfully turned to the starboard to avoid the obstacle USV according to the COLREGs. As shown in Figure 15e, stage 5, the USV_U is close to the USV_{O1}, but there is no hazard of collision. Therefore, our USV continued to navigate to the terminal. Finally, as shown in Figure 15f, our USV arrives at the terminal.

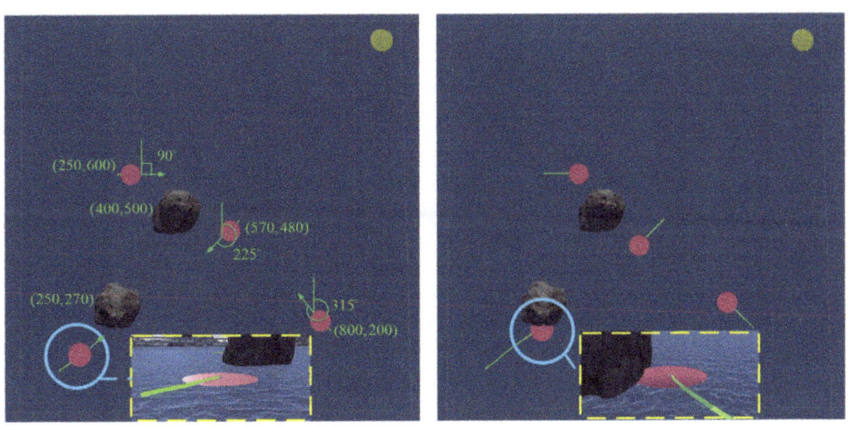

(a) (b)

Figure 15. *Cont.*

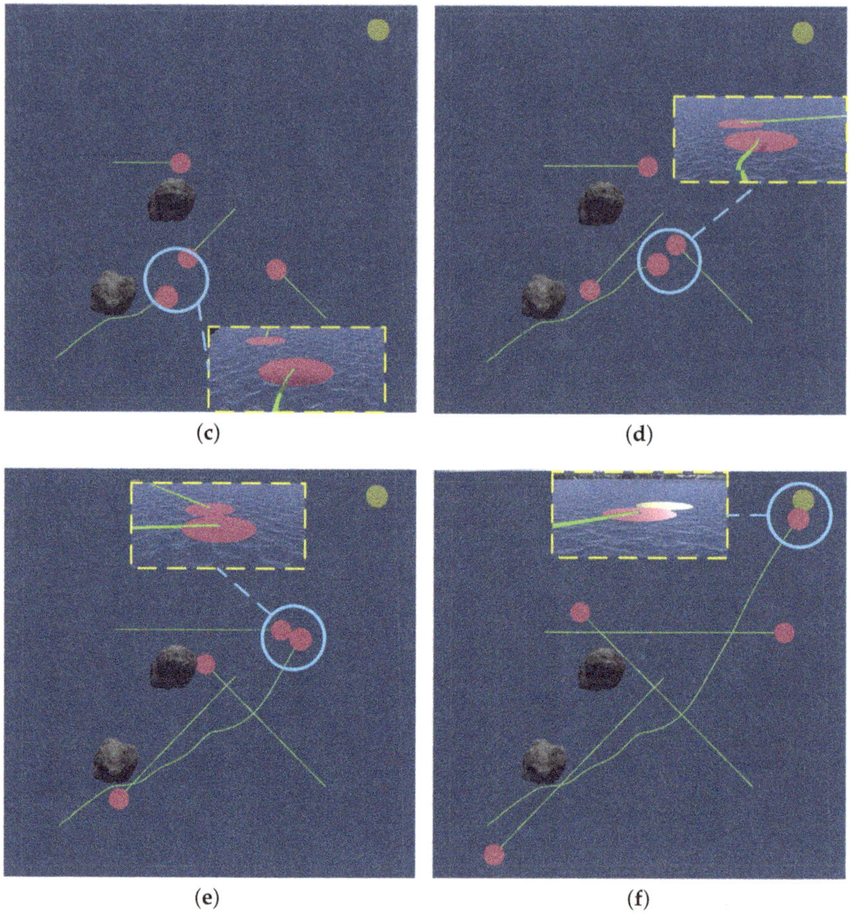

Figure 15. Multi-USVs collision avoidance environment. (**a**) Test stage 1; (**b**) Test stage 2; (**c**) Test stage 3; (**d**) Test stage 4; (**e**) Test stage 5; (**f**) Test stage 6.

6. Conclusions

This paper proposes an autonomous USV collision avoidance framework, DRLCA, which can be applied to USV navigation. The collision avoidance characteristics and maneuverability of USV are considered, and an efficient method for collision avoidance agent training is designed accordingly. A dueling architecture and a double learning method are used to improve training efficiency. Prioritize experience replay method is used instead of the uniform sampling method to improve sample utilization. The noisy network method, which has not been applied to the USV collision avoidance problem, is used to increase the exploration capability in USV training, verifying the feasibility of this method. Combining the characteristics of USV collision avoidance, two effective improvement methods are proposed in this paper, namely USV state clipping and DA distance restriction. Combined with the Unity virtual marine platform, which has realistic physical characteristics, the effect of the DRLCA is reliably verified and compared. The result shows that the improved collision avoidance algorithm proposed in this paper has a superior USV collision avoidance effect.

In the future, the ship domain will be replaced by an ellipse that can vary with speed to achieve a more accurate simulation of realistic collision avoidance situations. At the same time, the reward signal will be designed to be more detailed and associated with different information, such as distance, speed, and angle. How the stability and generalization ability of the algorithm can be further improved will be investigated to cope with the situation that about 3% of the DRLCA algorithm proposed in this study still does not arrive at the terminal.

Author Contributions: Funding acquisition, G.W.; Writing—original draft, Z.S.; Review and editing, Y.F. All authors have read and agreed to the published version of the manuscript.

Funding: This research was funded by "National Natural Science Foundation of China" (Grant number 61976033), "Key Development Guidance Program of Liaoning Province of China" (Grant number 2019JH8/10100100), "Pilot Base Construction and Pilot Verification Plan Program of Liaoning Province of China" (Grant number 2022JH24/10200029), "China Postdoctoral Science Foundation" (Grant number 2022M710569).

Institutional Review Board Statement: Not applicable.

Informed Consent Statement: Informed consent was obtained from all subjects involved in the study.

Data Availability Statement: The data that support the findings of this study are available from the corresponding author, [Yunsheng Fan], upon reasonable request.

Conflicts of Interest: The authors declare no conflict of interest.

Abbreviations

The following abbreviations are used in this manuscript:

USV	unmanned surface vehicle
DQN	deep Q network
DRL	deep reinforcement learning
COLREGs	international regulations for preventing collisions at sea
DRLCA	deep reinforcement learning collision avoidance
DDPG	deep deterministicpolicy gradient
LSTM	long short-term memory
SD	ship domain
DA	dynamic area
CRI	collision risk index
DCPA	the distance at the closest point of approaching
TCPA	time to the closest point of approaching

References

1. Dabrowski, P.S.; Specht, C.; Specht, M. Integration of multi-source geospatial data from GNSS receivers, terrestrial laser scanners, and unmanned aerial vehicles. *Can. J. Remote Sens.* **2021**, *47*, 621–634. [CrossRef]
2. Kurowski, M.; Thal, J.; Damerius, R. Automated survey in very shallow water using an unmanned surface vehicle. *IFAC-PapersOnLine* **2019**, *52*, 146–151. [CrossRef]
3. Li, C.; Jiang, J.; Duan, F. Modeling and experimental testing of an unmanned surface vehicle with rudderless double thrusters. *Sensors* **2019**, *19*, 2051. [CrossRef]
4. Luis, S.Y.; Reina, D.G.; Marín, S.L.T. A multiagent deep reinforcement learning approach for path planning in autonomous surface vehicles: The Ypacaraí lake patrolling case. *IEEE Access* **2021**, *9*, 17084–17099. [CrossRef]
5. Mu, D.; Wang, G.; Fan, Y. Adaptive trajectory tracking control for underactuated unmanned surface vehicle subject to unknown dynamics and time-varing disturbances. *Appl. Sci.* **2018**, *8*, 547. [CrossRef]
6. Stateczny, A.; Specht, C.; Specht, M. Study on the positioning accuracy of GNSS/INS systems supported by DGPS and RTK receivers for hydrographic surveys. *Energies* **2021**, *14*, 7413. [CrossRef]
7. Gao, S.; Liu, C.; Tuo, Y. Augmented model-based dynamic positioning predictive control for underactuated unmanned surface vessels with dual-propellers. *Ocean Eng.* **2022**, *266*, 112885. [CrossRef]
8. Li, Y.; Zhang, H. Collision Avoidance Decision Method for Unmanned Surface Vehicle Based on an Improved Velocity Obstacle Algorithm. *J. Mar. Sci. Eng.* **2022**, *10*, 1047. [CrossRef]
9. Ren, J.; Zhang, J.; Cui, Y. Autonomous obstacle avoidance algorithm for unmanned surface vehicles based on an improved velocity obstacle method. *ISPRS Int. J. Geo-Inf.* **2021**, *10*, 618. [CrossRef]

10. Fan, Y.; Sun, X.; Wang, G. Collision avoidance controller for unmanned surface vehicle based on improved cuckoo search algorithm. *Appl. Sci.* **2021**, *11*, 9741. [CrossRef]
11. Guan, W.; Wang, K. Autonomous Collision Avoidance of Unmanned Surface Vehicles Based on Improved A-Star and Dynamic Window Approach Algorithms. *IEEE Intell. Transp. Syst. Mag.* **2023**, 2–17. [CrossRef]
12. Silver, D.; Schrittwieser, J.; Simonyan, K. Mastering the game of go without human knowledge. *Nature* **2017**, *550*, 354–359. [CrossRef] [PubMed]
13. Vinyals, O.; Babuschkin, I.; Czarnecki, W.M. Grandmaster level in StarCraft II using multi-agent reinforcement learning. *Nature* **2019**, *575*, 350–354. [CrossRef] [PubMed]
14. Wang, N.; Gao, Y.; Zhao, H. Reinforcement learning-based optimal tracking control of an unknown unmanned surface vehicle. *IEEE Trans. Neural Netw. Learn. Syst.* **2020**, *32*, 3034–3045. [CrossRef]
15. Bastani, H.; Drakopoulos, K.; Gupta, V. Efficient and targeted COVID-19 border testing via reinforcement learning. *Nature* **2021**, *599*, 108–113. [CrossRef]
16. Kiran, B.R.; Sobh, I.; Talpaert, V. Deep reinforcement learning for autonomous driving: A survey. *IEEE Trans. Intell. Transp. Syst.* **2021**, *23*, 4909–4926. [CrossRef]
17. Chen, C.; Chen, X.Q.; Ma, F. A knowledge-free path planning approach for smart ships based on reinforcement learning. *Ocean. Eng.* **2019**, *189*, 106299. [CrossRef]
18. Li, L.; Wu, D.; Huang, Y. A path planning strategy unified with a COLREGS collision avoidance function based on deep reinforcement learning and artificial potential field. *Appl. Ocean. Res.* **2021**, *113*, 102759. [CrossRef]
19. Shen, H.; Hashimoto, H.; Matsuda, A. Automatic collision avoidance of multiple ships based on deep Q-learning. *Appl. Ocean. Res.* **2019**, *86*, 268–288. [CrossRef]
20. Zhou, C.; Wang, Y.; Wang, L. Obstacle avoidance strategy for an autonomous surface vessel based on modified deep deterministic policy gradient. *Ocean Eng.* **2022**, *243*, 110166. [CrossRef]
21. Du, Y.; Zhang, X.; Cao, Z. An Optimized Path Planning Method for Coastal Ships Based on Improved DDPG and DP. *J. Adv. Transp.* **2021**, *2021*, 7765130. [CrossRef]
22. Lillicrap, T.P.; Hunt, J.J.; Pritzel, A. Continuous control with deep reinforcement learning. *arXiv* **2015**, arXiv:1509.02971.
23. Xu, X.; Cai, P.; Ahmed, Z. Path planning and dynamic collision avoidance algorithm under COLREGs via deep reinforcement learning. *Neurocomputing* **2022**, *468*, 181–197. [CrossRef]
24. Chen, C.; Ma, F.; Xu, X. A Novel Ship Collision Avoidance Awareness Approach for Cooperating Ships Using Multi-Agent Deep Reinforcement Learning. *J. Mar. Sci. Eng.* **2021**, *9*, 1056. [CrossRef]
25. Norrbin, N.H. Theory and observations on the use of a mathematical model for ship manoeuvring in deep and confined waters. In *Publication 68 of the Swedish State Shipbuilding Experimental Tank, Proceedings of the 8th Symposium on Naval Hydrodynamics, Pasadena, CA, USA, 24–28 August 1970*; Elanders Boktryckeri Aktiebolag: Göteborg, Sweden, 1971; pp. 807–905.
26. Fan, Y.; Sun, Z.; Wang, G. A Novel Reinforcement Learning Collision Avoidance Algorithm for USVs Based on Maneuvering Characteristics and COLREGs. *Sensors* **2022**, *22*, 2099. [CrossRef]
27. Fujii, Y.; Tanaka, K. Traffic capacity. *J. Navig.* **1971**, *24*, 543–552. [CrossRef]
28. Piray, P.; Daw, N.D. Linear reinforcement learning in planning, grid fields, and cognitive control. *Nat. Commun.* **2021**, *12*, 4942. [CrossRef]
29. Aytar, Y.; Pfaff, T.; Budden, D. Playing hard exploration games by watching YouTube. In Proceedings of the 32nd International Conference on Neural Information Processing Systems, Montreal, QC, Canada, 3–8 December 2018; pp. 2935–2945.
30. Bellemare, M.G.; Candido, S.; Castro, P.S. Autonomous navigation of stratospheric balloons using reinforcement learning. *Nature* **2020**, *588*, 77–82. [CrossRef]
31. Sutton, R.S.; Batro, A.G. *Reinforcement Learning: An Introduction*; MIT Press: Cambridge, MA, USA, 2018.
32. Dabney, W.; Kurth-Nelson, Z.; Uchida, N. A distributional code for value in dopamine-based reinforcement learning. *Nature* **2020**, *577*, 671–675. [CrossRef]
33. Bain, A. *The Emotions and the Will*; John W. Parker and Son: London, UK, 1859.
34. Alagoz, O.; Hsu, H.; Schaefer, A.J. Markov decision processes: A tool for sequential decision making under uncertainty. *Med. Decis. Mak.* **2010**, *30*, 474–483. [CrossRef]
35. Watkins, C.J.; Dayan, P. Q-learning. *Mach. Learn.* **1992**, *8*, 279–292. [CrossRef]
36. Bellman, R. Dynamic programming. *Science* **1966**, *153*, 34–37. [CrossRef]
37. Metropolis, N.; Ulam, S. The monte carlo method. *J. Am. Stat. Assoc.* **1949**, *44*, 335–341. [CrossRef] [PubMed]
38. Mnih, V.; Kavukcuoglu, K.; Silver, D. Playing atari with deep reinforcement learning. *arXiv* **2013**, arXiv:1312.5602.
39. Hasselt, H. Double Q-learning. In Proceedings of the 23rd International Conference on Neural Information Processing Systems, Vancouver, BC, Canada, 6–9 December 2010; pp. 2613–2621.
40. Van Hasselt, H.; Guez, A.; Silver, D. Deep reinforcement learning with double q-learning. In Proceedings of the 13th AAAI Conference on Artificial Intelligence, Phoenix, AZ, USA, 12–17 February 2016; Volume 30.
41. Wang, Z.; Schaul, T.; Hessel, M. Dueling network architectures for deep reinforcement learning. In Proceedings of the 33rd International Conference on Machine Learning, New York, NY, USA, 19–24 June 2016; pp. 1995–2003.

42. Schaul, T.; Quan, J.; Antonoglou, I. Prioritized experience replay. *arXiv* **2015**, arXiv:1511.05952.
43. Fortunato, M.; Azar, M.G.; Piot, B. Noisy networks for exploration. *arXiv* **2017**, arXiv:1706.10295.

Journal of
Marine Science and Engineering

MDPI

Article

Neural Network, Nonlinear-Fitting, Sliding Mode, Event-Triggered Control under Abnormal Input for Port Artificial Intelligence Transportation Robots

Yaping Zhu [1], Qiang Zhang [1,2,*], Yang Liu [3], Yancai Hu [1] and Sihang Zhang [1]

1 School of Navigation and Shipping, Shandong Jiaotong University, 1508 Hexing Road, Huancui District, Weihai 264209, China
2 Shandong Intelligent Transportation Key Laboratory of Shandong Jiaotong University, 5001 Haitang Road, Changqing University Science Park, Jinan 250357, China
3 Department of Maritime Transportation, Mokpo National Maritime University, 91 Haeyangdaehang-ro, Mokpo City 58628, Republic of Korea
* Correspondence: zq20060054@163.com

Abstract: A new control algorithm was designed to solve the problems of actuator physical failure, remote network attack, and sudden change in trajectory curvature when a port's artificial intelligence-based transportation robots track transportation in a freight yard. First of all, the nonlinear, redundant, saturated sliding surface was designed based on the redundant information of sliding mode control caused by the finite nature of control performance; the dynamic acceleration characteristic of super-twisted sliding mode reaching law was considered to optimize the control high frequency change caused by trajectory mutation; and an improved super-twist reaching law was designed. Then, a nonlinear factor was designed to construct a nonlinear, fault-tolerant filtering mechanism to compensate for the abnormal part of the unknown input that cannot be executed by adaptive neural network reconstruction. On this basis, the finite-time technology and parameter-event-triggered mechanism were combined to reduce the dependence on communication resources. As a result, the design underwent simulation verification to verify its effectiveness and superiority. In the comparative simulation, under a consistent probability of a network attack, the tracking accuracy of the algorithm proposed in this paper was 22.65%, 12.69% and 11.48% higher those that of the traditional algorithms.

Keywords: nonlinear-fitting redundant sliding mode; event triggered; abnormal input; neural network; artificial intelligence transportation robots; track tracking

Citation: Zhu, Y.; Zhang, Q.; Liu, Y.; Hu, Y.; Zhang, S. Neural Network, Nonlinear-Fitting, Sliding Mode Event-Triggered Control under Abnormal Input for Port Artificial Intelligence Transportation Robots. *J. Mar. Sci. Eng.* **2023**, *11*, 659. https://doi.org/10.3390/jmse11030659

Academic Editor: Kamal Djidjeli

Received: 16 February 2023
Revised: 12 March 2023
Accepted: 16 March 2023
Published: 21 March 2023

1. Introduction

With the advent of the era of Industry 4.0, artificial intelligence transportation robots have gradually matured and have become widely used in ports, logistics, and other freight-related situations [1]. Tianjin Port uses cutting-edge technologies, such as unmanned driving, artificial intelligence, and big data, to replace traditional transportation equipment. There are artificial intelligence transportation robots and a high level of automation for terminal operations, and it serves as a reference for the construction and development of domestic container automation terminals. In 2020, Hefei Port introduced artificial intelligence transportation robots, which continuously improve and expand their perception capabilities in a real operating environment through fusion algorithms, and realize the trajectory prediction of surrounding traffic participants. They not only realize safe and stable operation, but also take into account operational efficiency to ensure efficient and smooth operation. Other port yards have gradually introduced artificial intelligence transportation robots, as shown in Figure 1. However, unknown control anomalies caused by dynamic uncertainty, control signal transmission noise, network attacks, and program faults in the real environment make it impossible for artificial intelligence transportation robots to achieve precise control. In order to ensure the efficient operation of artificial intelligence transportation robots in ports and terminals, high-precision trajectory tracking control is an urgent problem to be solved.

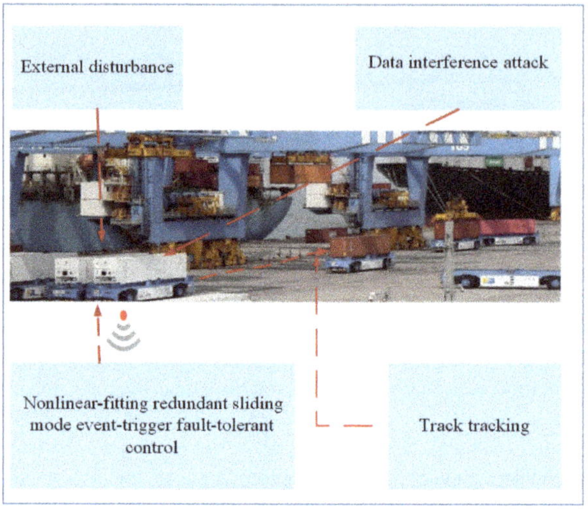

Figure 1. Artificial intelligence transportation robots: operational diagram of the Qingdao Port Freight Yard.

Nowadays, the robustness of mature control algorithms such as PID [2] and linear feedback [3] cannot meet the actual needs of jobs in complex scenes. Sliding mode control [4] is recognized by the control community more and more because of its good robustness and simple structure. It is worth noting that how to effectively reduce the chattering effect of the algorithm itself has become a main premise of practical engineering applications. Therefore, under the premise of ensuring robustness, some scholars used the high-dimensional sliding surface [5] to optimize the buffeting threshold, and others used the dynamic characteristics of the sliding mode to design the variable-speed reaching law [6–10] to slow the speed of the near sliding surface to reduce chattering.

It is worth noting that reducing chattering will reduce the robust performance of sliding mode control to a certain extent. The interference caused by dynamic uncertainty especially will mean the robustness of the algorithm cannot be fully brought into play. Therefore, Baek et al. [11] established a stochastic stability judgment mechanism for uncertainty, and used a time-delay estimation scheme combined with adaptive technology to achieve good asymptotic stochastic stability. Niu et al. [12] used neural network approximation to realize on-line robust sliding mode adaptive control. However, adaptive conservatism will waste control resources, and intelligent algorithms require high performance of the controller and need to be further optimized. Zhang et al. [13] proposed a high-order, fast, non-singular, terminal sliding mode controller based on a double-disturbance observer, which effectively weakens the chattering phenomenon of the system. Wang et al. [14] proposed an adaptive, proportional-integral-derivative, fractional-order, non-singular, terminal sliding mode control method based on time-delay estimation, which realizes timely and accurate adjustment of the control gain of the robust term. Shao et al. [15] proposed an adaptive, recursive terminal sliding mode controller. The fast, non-singular end sliding function, and the recursive integral end sliding function were designed by using the recursive structure, so that the sliding surface reaches continuously, which significantly improves the tracking-error-convergence speed and the anti-interference ability speed-wise.

In practical engineering, robot tasks are required to be timely, and most of the above algorithms are asymptotically stable. Therefore, how to complete the tracking movement within a specified time is an urgent need for the project. Therefore, finite-time technology was proposed and applied in control systems [16–18]. Wang et al. [19] used a variable division technique and fuzzy control. Wang et al. [20] took advantage of the approximation of fuzzy logic systems. Fang et al. [21] gave sufficient conditions for practical fixed-time stability, as did Ba et al. [22], with the help of a neural network and backstepping technology. Zuo [23] carried out finite-time control design for second-order nonlinear systems with uncertainties and disturbances. However, the convergence time of the system with finite-time stability increases with the initial error, which greatly weakens the convergence performance of the system. At the same time, in actual engineering, there is are artificial intelligence transportation robots that have physical limits in the initial stage, and the error is mostly the maximum error of the control task, but this characteristic has not been paid attention. The use of limited control resources to complete the control an abnormal environment has not been further considered.

When transporting in a port's cargo yard, due to the automatic operation of multi-frequency scanning for a long time, the equipment will be worn out, which greatly reduces the control accuracy. Academia mainly studies problems of this kind from the point of view of being with or without detectors [24–27]. Although sliding mode control can be passively fault-tolerant without a detector [28], its fault-tolerant response is not sensitive enough, and its reconstruction accuracy is not high. Therefore, it is often combined with adaptivity, an observer, and other technologies. There is also a zero-order hold mechanism for event-triggering to compensate for faults while reducing communication resources [29,30]. In addition, the remote control signal of the robot is calculated and sent by the wireless upper control terminal, but the host may have some problems, such as Trojan horse implantation, redundant data interference, illegal attack, transmission noise, and so on. This will lead to matching interference in the transmission information [31]. However, for when these effects lead to abnormal control, how to better carry out the soft compensation of the control side is the focus of this paper.

To sum up, it can be known that reducing the parameter-tuning complexity of sliding mode reaching-law control to reduce chattering and ensure reaching efficiency is a direction that needs to be improved in the research of super-twisted reaching laws. The physical limitations of the robot lead to the integration-performance redundancy of unutilized residual errors on the traditional sliding mode surface, which is also challenging. In addition, how to better compensate for the abnormal control problems caused by signal interference, network attacks, and faults is also very important. Therefore, a new type of sliding surface was designed. The dynamic acceleration characteristics of the sliding mode are considered to improve the super-distortion reaching law, the nonlinear saturated filtering fault-tolerant mechanism is used to fit the abnormal information, and the adaptive neural network technology is used to fit and compensate. Finally, the stable control of artificial intelligence transportation robots was realized and event-triggered. The main innovations of this paper are as follows:

1. The integral processing easily produces the problem of stable error, but the error state is bounded; that is, the redundant information can be said to be bounded by the state, the reference trajectory, and information beyond the limit. The elimination of redundant information is limited to the bounded range, which effectively reduces the problem of stable error. Using the function of eliminating redundant information of residual error by integral term, the integral saturation mechanism is designed. While avoiding integral saturation, the redundant information of residual error, which can be offset by the maximum control performance, is removed. Compared with the ordinary integral sliding mode control, it will improve the controllable stability of sliding mode control in the case of fault tolerance and saturation.

2. The speed of traditional super-twist near the sliding mode surface is larger than that of the sliding mode dynamic method in this paper. From the angle of approaching dynamics, a better approaching state can be obtained from a lower speed, so there is a better buffeting suppression effect than traditional super-twist, and the parameter adjustment

is also simpler. The control method overcomes the shortcomings of the traditional super-twist reaching law, such as the complex adjustment of the parameters and the tendency to increase the buffeting when approaching the instantaneous mutation. Considering the acceleration dynamic characteristics of the reaching law, the nonlinear, variable damping reaching law is designed to reduce the change rate of the switching interval between the reaching stage and the sliding stage of the sliding mode, thereby improving the buffeting weakening ability and the reaching efficiency.

3. When solving the problem of the control signal being attacked by data interference and the partial failure of the actuator, the low fitting accuracy of the control abnormal information and the difficulty of signal-data interference-attack peeling in the literature [30–32] are overcome. The nonlinear fitting factor was designed based on the virtual hypothesis of abnormal information, and the nonlinear saturation-fault-tolerant filtering mechanism was designed for the dynamic information of system state.

The rest of this paper is divided into four sections: Section 2 establishes the kinematic and dynamic motion models of the tracked underwater vehicle and sets out the preliminary knowledge; Section 3 contains four parts: The first part proposes a new, nonlinear, projection redundant, feedforward sliding mode surface and a new sliding mode reaching law. It also provides the theoretical comparison and proof of the advantages of the method. The second part describes the event-triggered mechanism. Then, in the third part, the new nonlinear fault-tolerant subsystem is proposed, and its effectiveness is demonstrated. In the fourth part, the kinematics and dynamics controller are designed. In the Section 4, the Simulink simulation is compared with the control system using a traditional sliding mode approach law to verify the effectiveness of the control scheme proposed in this paper. Section 5 gives the conclusion of this paper.

2. Model and Preliminaries

2.1. Artificial Intelligence Transportation Robot Model

The stability and safety of artificial intelligence transportation robots in container cargo transportation are important, so differential mobile robots are often used. On the other hand, the differential mobile robots have the characteristic of a nonholonomic constraint. According to reference [33], the kinematic and dynamic models of robot motion plane are as follows:

$$\dot{q}(k) = S(q(k))u(k) \tag{1}$$

$$\dot{u}(k) = \bar{M}^{-1}(q)[\bar{B}(q)\tau_l - F_m(\dot{q}) - \bar{\tau}_d] \tag{2}$$

where $\bar{M}(q) = S^T(q)M(q)S(q)$, $\bar{\tau}_d = S^T(q)\tau_d$, $\bar{B}(q) = S^T(q)B(q)$, and $\tau_{\max} \geq \|\tau\|$. $F_m(\dot{q})$ is dynamic uncertainty of robot model. $M(q) = \begin{bmatrix} \mathrm{m} & 0 & -\mathrm{md}\sin\phi \\ 0 & \mathrm{m} & \mathrm{md}c\cos\phi \\ -\mathrm{md}\sin\phi & \mathrm{m}\cos\phi & \mathrm{md}^2 + J \end{bmatrix}$; $C(q,\dot{q}) = \begin{bmatrix} 0 & 0 & -\dot{\mathrm{m}}\mathrm{mdcos}\phi \\ 0 & 0 & -\dot{\mathrm{m}}\mathrm{md}\sin\phi \\ 0 & 0 & 0 \end{bmatrix}$; $M(q) = \begin{bmatrix} \mathrm{m} & 0 & -\mathrm{md}\sin\phi \\ 0 & \mathrm{m} & \mathrm{md}c\cos\phi \\ -\mathrm{md}\sin\phi & \mathrm{m}\cos\phi & \mathrm{md}^2 + J \end{bmatrix}$; $C(q,\dot{q}) = \begin{bmatrix} 0 & 0 & -\dot{\mathrm{m}}\mathrm{mdcos}\phi \\ 0 & 0 & -\dot{\mathrm{m}}\mathrm{md}\sin\phi \\ 0 & 0 & 0 \end{bmatrix}$; $T = \begin{bmatrix} T_L \\ T_R \end{bmatrix}$; $B(q) = \begin{bmatrix} \frac{\cos\phi}{\mathrm{r}} & \frac{\sin\phi}{\mathrm{r}} & -\frac{\mathrm{b}}{\mathrm{r}} \\ \frac{\cos\phi}{\mathrm{r}} & \frac{\sin\phi}{\mathrm{r}} & \frac{\mathrm{b}}{\mathrm{r}} \end{bmatrix}$; $A^T(q) = \begin{bmatrix} \sin\phi \\ -\cos\phi \\ -\mathrm{d} \end{bmatrix}$; $\lambda = -m(\dot{x}\cos\phi + \dot{y}\sin\phi)\dot{\phi}$; $\tau_d = \begin{bmatrix} T_{\mathrm{d}1} \\ T_{\mathrm{d}2} \\ T_{\mathrm{d}3} \end{bmatrix}$. λ is the Lagrange dynamics' dykoll coordinates to kinetic multipliers of generalized coordinates. The linear velocity and angular velocity matrix is $u = \begin{bmatrix} v \\ w \end{bmatrix}$. The kinematic model's coefficient matrix $s(q) = \begin{bmatrix} \cos\theta & d\sin\theta \\ \sin\theta & -d\sin\theta \\ 0 & 1 \end{bmatrix}$.

2.2. Mathematical Model of Abnormal Control

The data transmission network from controller to actuator is vulnerable to random noise interference and data-injection interference. In addition, the actuator of the artificial intelligence transportation robot will have a physical saturation limitation, so it is necessary to carry out saturation fitting in advance. Here, the input nonlinear fitting model is introduced. From Equation (2), the true actuator input of using hyperbolic tangent function to fit the saturation characteristics of robot actuators has the following form:

$$\tau_l(k) = A_\tau \tau_{\max} \tanh(\tau(k) + A_D) \tag{3}$$

where the $\tau(k)$ is the controller-calculated online signal. A_r is an input partial fault with coupling characteristics. A_D is a kind of data interference network attack with concealment characteristics. $A_D = \Im A_{\Im}, \Im$ has the characteristics of an independently distributed Bernoulli sequence with a value of $\{0\}$ or $\{1\}$. The $A_{\Im}$ is a virtual interference data value of the control signal caused by the attack.

2.3. RBF Neural Network Fitter

This is inspired by the paper [34]. The radial basis function neural network (RBFNN) approximation [35] is cited. As shown in Figure 2, a RBF neural network is a three-layer neural network because of the nonlinear characteristics of system uncertainty. If there exists an m-dimensional compact set $\Xi^{\mathrm{m}} \subseteq \mathrm{R}^{\mathrm{m}} \to \mathrm{R}$ and there is an unknown nonlinear function $f(Q)$ with initial value 0 defined on Ξ^{m}, the RBF approximator (4) is used to fit the dynamic values of $f(Q)$.

$$f(Q) = W^{*\,\mathrm{T}} Z(Q) + e_Z(Q), \quad \forall Q \in \Xi^{\mathrm{m}} \tag{4}$$

where $e_z(Q)$ is the bounded RBF fitting error, which is defined on the compact set Ξ^{m}. $|e_z(Q)| \le \bar{e}_z$, and $\bar{e}_z$ is the maximum nuclear distance. To improve the nonlinear local approximation ability, the Gaussian function $Z(Q) = \exp((Q-\kappa)^{\mathrm{T}}(Q-\kappa)/-l^2)$ is selected as the smooth kernel function. The κ is approaching the center column distance vector. The l is a varying constant value. * is the order-m dimensional weight row vector, which is optimally fitted: as

$$W^* = \arg\left(\min_{\hat{W}}\left\{\sup_{Q \in E^m} \left|\hat{W}^{-} Z(Q) - f(z)\right|\right\}\right) \tag{5}$$

where $\hat{W}$ is the minimum estimate of W^* that optimizes $E_f(Q)$.

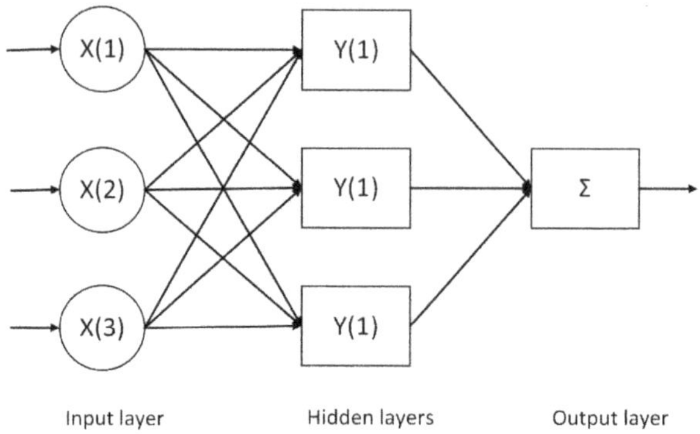

Figure 2. RBF neural network diagram.

The weight parameters in the online approximation process of the neural network approximator have been identified using parameter-adaptive technology, which is different from other neural networks [36] that generally need offline training. This method is based on the adaptive RBF neural network method in the paper [37], combined with the minimum parameter-learning method to perturbate the model's parameters caused by the disturbance and the neural network's weight parameters for online adaptation, so the weight matrix adjustment is automatically adjusted by the adaptive law. There are differences in the use of control processes in other areas.

The RBF neural network used in this paper is based on the ideas in the paper [35], and it is used as an online universal approximator, as a regression fitter, and it is fitted online with the data, so it is different from the general neural network, and this article adds the minimum parameter-learning method, as described in the paper [37]. The complexity and calculation time are adjusted with adaptive changes in the control process, and there is no need to train in advance when the uncertainty approximator of the control system, the activation function, can be set and the weight matrix can be determined to achieve universal approximation [35]. Therefore, taking advantage of the universal approximation property of RBF and the absence of a need for training in advance, the uncertainty caused by model dynamics and network attacks can be approximated nonlinearly.

2.4. Preliminaries

Lemma 1. *For the system (1), when $x^* \in \mathbb{R}$ and $x^* \neq 0$, if Lyapunov function $V(x^*) > 0$ exists, Lyapunov condition of the finite-time stability can be given as [30]*

$$\dot{V}(x^*) + \beta_1 V(x^*) + \beta_2 V^*(x^*) \leq 0 \tag{6}$$

where $\beta_1 > 0, \beta_2 > 0$, and $0 < \kappa < 1$, so the system is globally finite-time stable, and the stable time depending on the initial state e_0 is given as

$$T_V = \ln\Big(\Big(\beta_2 \cdot V^{1-\kappa}(e_0) + \beta\Big)/\beta_2\Big)/(\beta_1 - \beta_1\kappa) \tag{7}$$

The relevant proof is shown in reference [30].

Lemma 2. *According to the Cauchy–Schwarz inequality, for any number a_i and $b_i (i = 1, 2, \ldots, n)$, we can know*

$$\left(\sum_{i=1}^{n} a_i b_i\right)^2 \leq \left(\sum_{i=1}^{n} a_i^2\right)\left(\sum_{i=1}^{n} b_i^2\right) \tag{8}$$

and any $0 < l < 1$, *there exists*

$$\left(\sum_{i=1}^{n}|a_i|\right)^{t} \leq \sum_{i=1}^{n}|a_i|^{t} \tag{9}$$

Lemma 3. *For any* $b_c > 0$ *and* $z_t \in \mathbb{R}$, $\tanh(\cdot)$ *has the following property:*

$$0 \leq |z_t| - z_t \tanh(z_t/b_c) \leq 0.2785 b_c \tag{10}$$

Assumption 1. *The unknown bounded low frequency time-varying disturbance* $\bar{\tau}_d$ *is* $\|\bar{\tau}_d\| \leq \hat{\tau}_d$. *The initial system state errors,* $u_e(0)$ *and* $q_e(0)$, *are defined on a compact set, and it is assumed as* $\|u_e(t)\| \leq \bar{u}_e$ *and* $\|q_e(t)\| \leq \hat{q}_e$.

Assumption 2. *To the limited range of the freight yard, the range of motion and the desirably reference trajectory of the robot are bounded . The desirably reference trajectory* $q_r = \begin{bmatrix} x_r & y_r & \theta_r \end{bmatrix}^T$, *and reference positive scalar speed* $u_r = \begin{bmatrix} v_r & w_r \end{bmatrix}^T$. *Their derivatives are smooth and bounded.*

Assumption 3. *The system (6) is a controllable system that satisfies Lemma 2. For facilitate matrix operation, all constant terms are in the form of a diagonal matrix.*

3. Controller Design

The virtual kinematic subsystem is designed to analyze velocity state in this section. It can obtain tracking position of artificial intelligence transportation robots. Based on characteristics of contaminated velocity state signals, the fault-tolerant filtering subsystem is designed. Then, the nonlinear sliding mode surface is designed to improve the robustness of the controller by establishing a nonlinear bounded state space and combined with the reaching law to reduce the vulnerability of faults and contaminated communications. Figure 3 is the schematic diagram of artificial intelligence transportation robot trajectory-tracking control flow.

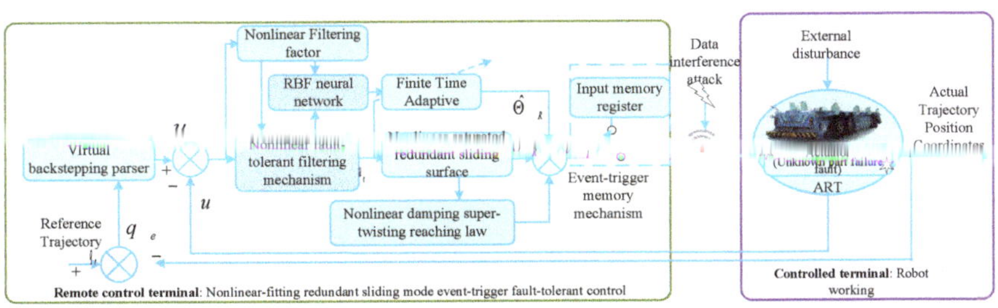

Figure 3. Schematic diagram of artificial intelligence transportation robots with a trajectory-tracking control flow.

3.1. Nonlinear Saturation Fault-Tolerant Filtering Mechanism

According to the system (1), the kinematic position error is defined as

$$q_e = \begin{bmatrix} x_e \\ y_e \\ \theta_e \end{bmatrix} = \begin{bmatrix} \cos\theta & \sin\theta & 0 \\ -\sin\theta & \cos\theta & 0 \\ 0 & 0 & 1 \end{bmatrix} \cdot \begin{bmatrix} x_r - x \\ y_r - y \\ \theta_r - \theta \end{bmatrix} \tag{11}$$

According to the characteristics of virtual backstepping and kinematic spatial, the method of velocity motion control at low latitudes of dynamics is used to design a virtual-kinematics control law as the dynamic desire value (12).

$$u_d = \begin{bmatrix} v_r \cos\theta_e - w\theta_e + \lambda_2(x_e - d + d\cos\theta_e) \\ w_r + \lambda_1^{-1} v_r(\vartheta_1(y_e + \theta_e) + (d\vartheta_1 + \lambda_1\vartheta_2)\sin\theta_e) \end{bmatrix} \tag{12}$$

where $u_r = [v_r, w_r]^T$ is the velocity of the reference trajectory.

Combined with the system state information, the error is designed as

$$U_e = \begin{bmatrix} v_e \\ \theta_e \end{bmatrix} = u_d - u \tag{13}$$

Remark 1. *The system input is contaminated by GPS sensor failure, unreliable signal, and actuator failure in the system, which will result in an unknown input. After this kind of input is executed, it will not be able to achieve stability control. Therefore, the nonlinear factor* $\eta_\tau = \frac{A_\tau}{1+\tanh(\tau)\tanh(A_d)}$ *is involved in constructing the saturation-fault-tolerant mechanism. The nonlinear saturation-fault-tolerant filtering mechanism* $\eta_f(k) = 1 - \eta_\tau + \eta_\tau \tanh(A_d)$ *denotes that the decoupling form about the unknown influence of the signal is dealt with nonlinearly. According to error Equation (2), the fault-tolerant saturation filtering dynamic subsystem is designed as* $\dot{u} = \bar{M}^{-1}(q(k))[\bar{B}(q(k))\tau_l(k) - \bar{F}_m(\dot{q}(k)) - \bar{\tau}_d(k)] + \eta_f(k)$. *The* $\bar{F}_m(\dot{q}(k)) = F_m(\dot{q}(k)) + \tau_{\max}\bar{M}(q(k))\eta_f(k)$.

This mechanism makes use of unknown information of attack and fault loss to form a nonlinear virtual hypothesis. According to this hypothesis, signal attack and fault features can be extracted better, and the fault-tolerant and adaptive ability can be improved further according to saturation analysis method in the literature [32]. The filtering dynamic error is

$$\dot{E}_R(k) = \dot{u}_d(k) - \left(\bar{M}^{-1}(q(k))[\bar{B}(q(k))\tau_l - \bar{F}_m(\dot{q}(k)) - \bar{\tau}_d]\right) - \eta_f(k) \tag{14}$$

3.2. Design of Nonlinear-Fitting, Redundant, Sliding Mode, Event-Trigger Fault-Tolerant Control

Step 1. A new type of a nonlinear, saturated, redundant sliding surface (NSRSMS). The NSRSMS is denoted as:

$$S_R(k) = \dot{E}_R(k) + \gamma_a E_R(k) + B_0(E_R(k)) - B_l(E_R(k)) \tag{15}$$

where $B_l(E_R(k)) = \int \tanh(E_R(t)) \ln\left[\frac{B_\psi(E_R(l))\exp(1)+1}{1+\exp(1)}\right] dl$, and $B_\psi(E_R(k)) = \beta_a E_R(k)\tanh(E_R(k))$.

The $\beta_a = \mathrm{diag}\{\beta_{a1}, \beta_{a2}\}$ is positive permanent diagonal function. The $B_0(E_R(k)) = -[\gamma_a E_R(0) - l(E_R(k))]$ is initial global approach term to ensure the global mode of SMC.

Remark 2. *When the error is designed on a SMS, the maximum error that can be eliminated at a single time has the characteristic of saturation because of the physical limitation of the control ability of the controlled robot. Therefore, the redundant information of saturation is used to give full play to the control performance, prevent the control performance from overshoot, and design a nonlinear mechanism* $B_l(E_R(k))$ *to improve the integral saturation. The characteristics of NSRSMS and linear SMS* $S_I(k) = \dot{E}_R(k) + \int E_R(l)dl + E_R(k) + B_0(E_R(k))$ *are shown in Figure 4, where* $\int E_R(l)dl$ *is y-axis and* $E_R(k)$ *is the x-axis.*

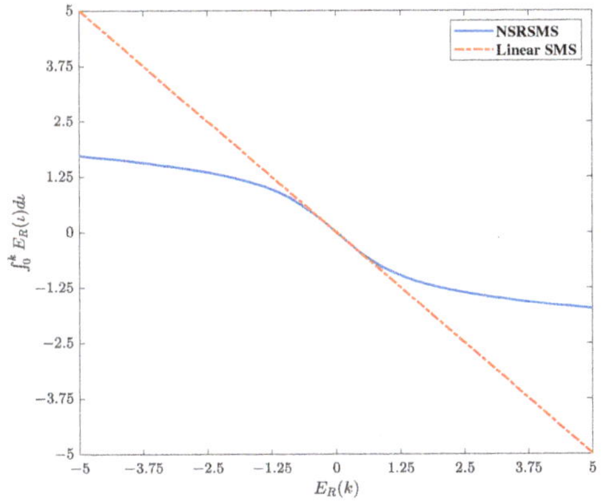

Figure 4. Comparison of NSRSMS and linear sliding mode surface.

Remark 3. *The control performance error struggles to meet the error overshoot when the error is large; that is to say, the error of the expected performance is relatively easy to control, and if the error exceeds a certain expected region, it will need to be compensated step by step. It can be seen in Figure 4 that the saturation term is designed and applied to the sliding surface, the saturated information is regarded as redundant information, the filtered saturated information is filtered directly, and the expectation of a certain error is used as the fault-tolerant information control point. It will be more conducive to the realization of fault tolerant control and reduce the waste of control performance.*

In addition, to avoid control instability caused by singularity in practical engineering, the singularity of the sliding surface is verified through Equation (16).

$$\dot{S}_R = \ddot{E}_R(k) + \gamma_a \dot{E}_R(k) + \dot{B}_0(E_R(k)) - B_\gamma \tag{16}$$

where $B_\gamma = \tanh(E_R(t))\left[\ln\left(B_\psi(E_R(t))\exp(1) + 1\right) + \ln(1 + \exp(1))\right]$, *and we can know that NSRSMS does not contain singularities in the control process.*

Step 2. Nonlinear-damping, super twisting reaching law (NDSTRL) considering acceleration.

Although the redundant information of the integral saturated sliding mode surface can be used to improve the control accuracy and reduce the residual error, if there is a sudden instantaneous error, the buffeting problem can not be ignored, so the NDSTRL is designed according to the traditional STRL. The NSTRL is denoted as

$$\dot{S}_r = -\gamma_r \chi(k)\,\mathrm{sign}\left(S_R^T\right) - \gamma_k S_R \tag{17}$$

where $\chi(k) = \tanh\left(\|S_R(k)\|^{\frac{3}{2}}\right)\|S_R(k)\|^{\frac{1}{2}}$ shows nonlinear time-varying gain. The STRL denotes $\dot{S}_s = -\gamma_{rs}\|S_R(k)\|^{\frac{1}{2}}\,\mathrm{sign}(S_R^T) - \gamma_{ks}S_R$, and ERL is defined as $\dot{S}_E = -\gamma_{rE}\,\mathrm{sign}(S_R^T) - \gamma_{kE}S_R$. Figure 5 shows NDSTRL, STRL, and ERL with coefficients equal to one.

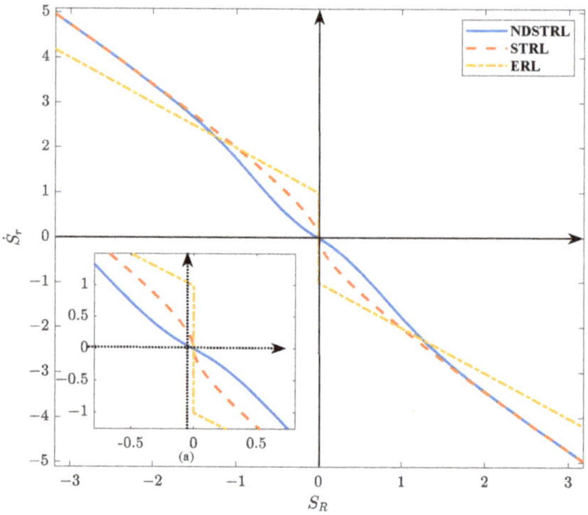

Figure 5. Comparison of NDSTRL and STRL.

Remark 4. *From Figure 5, we can see that the STRL can effectively regulate the reaching speed according to the system state value. When the reaching speed reaches the SMS, the reaching speed will be decelerated smoothly. The sudden deceleration at the break point of ERL (see the ERL in Figure 5a), which can not avoid chattering, will increase the chattering amplitude. Figure 5a shows that the acceleration of the NDSTRL decreases gradually, and the speed is adjusted gradually. The acceleration of SMC reaching dynamics are better controlled, which is more conducive to reducing buffeting. Additionally, even better, in the case of a large error, the speed of sliding mode will not be reduced.*

Remark 5. *In addition, we can see in Figure 6 that the NDSTRL acceleration gain has the effect of adjusting the acceleration trend. The status is closer to that of the SMS; the gain is more, which will slow down the velocity trend (see Figure 6a). Moreover, the reaching speed can be dynamically accelerated when the error is large, and speed gain does not need to be adjusted (see Figure 6b).*

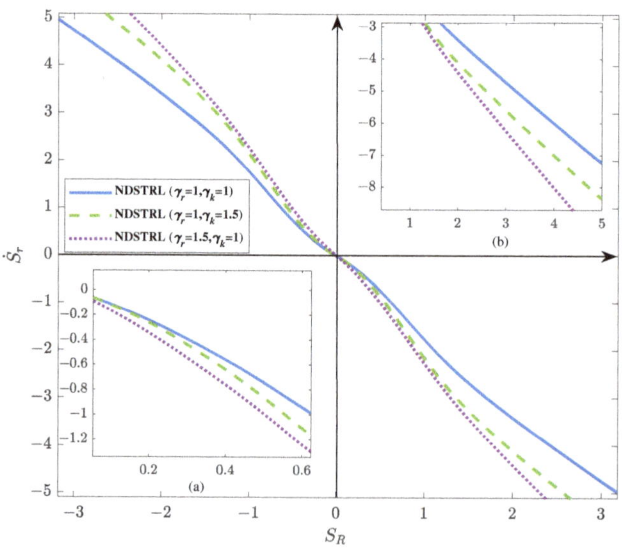

Figure 6. Comparison of NSTRL and STRL.

Step 3. Controller based on event-triggered memory input mechanism.

Under the influence of unexpected situations such as network attacks and faults, saving communication resources can reduce the instability of network control to a certain extent, so an online memory input event-triggered mechanism was designed (shown in Equation (18)). This mechanism allows the controlled robot to complete stable control without achieving the ideal control precision, which can also help to achieve stable control with a certain level of precision under network attacks.

$$\|T_l(k)\| \geqslant \lambda \|\tau_l(k_l)\| \tag{18}$$

where $\vartheta|\tau_l(t_k)|$ is the memory input with trigger weight gain, and $\tau_l(t_k)$ is the input that meets the trigger condition. $T_l(k) = \tau_l(k) - \tau_l(k_l)$ shows the dynamic characteristics and the input, and contains the state information of the robot affected by faults and attacks. When the input value of online calculation satisfies Equation (18), $\tau_l(k) = \tau_l(k_l), \forall k \in [k_l, k_{l+1})$, it denotes the input affected by the zero-order retention effect of the robot installed in advance.

Using Equation (17) and Equations (21)–(39), one can obtain the control law (Equation (19)) and adaptive law (Equation (20)).

$$\tau_l(k) = \tau_c(k) + \tau_\varsigma(k) \tag{19}$$

$$\dot{\hat{\Theta}}_R = \frac{1}{4r_s}\tanh\left(\frac{S_R}{D}\right)S_R - b\hat{\Theta}_R + r_s r_\Theta \|S_R\|^2 \varphi^4(Z_R) \tag{20}$$

where $\tau_\zeta(k)$ denotes an adaptive nonlinear fault-tolerant filter control mechanism, and $\tau_c(k)$ is the NDSTRL distance. $\tau_c(k) = -\left(\gamma_a \tau_{\max}\tilde{B}\right)^{-1} G_c \dot{S}_r + \gamma_a \dot{u}_d$ and $\tau_\varsigma(k) = \left(\gamma_a \tau_{\max}\tilde{B}\right)^{-1}\Big(\frac{1}{4r_s}\tanh\left(\frac{S_R}{D}\right) + r_s r_\Theta S_R \varphi^4(Z_R)\Big)\hat{\Theta}_R$. $r_\Theta > 0, r_s > 0, Z_1 = \left[H_1^{\mathrm{T}}, \dot{u}_d^{\mathrm{T}}, \varsigma^{\mathrm{T}}\right]^{\mathrm{T}}, \Theta_R = \max\Big\{\|\tau_{\max}E_r\|, \|\psi^T\|, \left\|\tau_{\max}E_f\right\|, \|\bar{\tau}_D\|\Big\}, H_1 = \ddot{E}_R(k) + \dot{B}_0(E_R(k))$, and $\varphi(Z_R) = \|D_r\|\Big(2\|\tilde{B}^T(q)\| + \|\tilde{B}^T(q)\tau_{\max}\phi(u)\| + \|D_r\|^{-1}\|\ddot{E}_R(k) + \dot{B}_0(E_R(k))\|\Big)$.

3.3. Theoretical Proof

To verify the stability of the controller, the effectiveness of the kinematic virtual controller, dynamic controller and event-triggered controller will be verified theoretically.

Proof. Theoretical proof of the virtual-kinematics control law.

Next, take the Lyapunov function to prove its stability:

$$V_q = \frac{1}{2}{S_x}^2 + \frac{1}{2}{S_y}^2 + S_\theta \tag{21}$$

where $S_x = x_e + d\cos\theta_e + d, S_y = y_e + d\sin\theta_e + \theta_e, S_\theta = k_1(1 - \cos\theta_e), k_1$ and k_2 are positive parameters. After the first order guidance, one can obtain

$$\dot{V}_q = S_x\dot{S}_x + S_y\dot{S}_y + \dot{S}_\theta \tag{22}$$

so we obtain

$$\dot{V}_q = -\lambda_2(x_e - d + d\cos\theta_e)^2 - \alpha\frac{v_r}{\lambda_1}(y_e + d\sin\theta_e + \theta_e)^2 \leq 0 \tag{23}$$

As (23) is negative, the system tends to be stable, and it is also proved in [26].

$$u_d = \begin{bmatrix} v_r\cos\theta_e - w\theta_e + \lambda_2(x_e - d + d\cos\theta_e) \\ w_r + \lambda_1^{-1}v_r(\vartheta_1(y_e + \theta_e) + (d\vartheta_1 + \lambda_1\vartheta_2)\sin\theta_e) \end{bmatrix} \tag{24}$$

□

Proof. Theoretical proof of the dynamic controller.

The Lyapunov function $V_{R1} = \frac{1}{2}S_R^T S_R$ is designed, and the first derivative of V_{R1} with sampling time can be obtained:

$$\begin{aligned} \dot{V}_{R1} &= S_R^{\mathrm{T}} \dot{S}_R \\ &= S_R^{\mathrm{T}} \big[D_r \big(\dot{u}_d - \tilde{B}(q) \eta_\tau \tau_{\max} (\tanh(\tau) + \tanh(A_D)) \\ &\quad + \bar{M}^{-1}(q) \bar{F}_m(\dot{q}) - \eta_f + \bar{\tau}_D \big) + \ddot{E}_R(k) + \dot{B}_0(E_R(k)) - B_\gamma \big] \end{aligned} \tag{25}$$

where $\gamma_a = D_r, \tilde{B}(q) = \bar{M}^{-1}(q)\bar{B}(q), \bar{\tau}_D = \bar{M}^1(q)\bar{\tau}_d$.

By using the saturation analysis mechanism (see Remark 1), the multiplicative fault is nonlinearized and projected, and one can obtain:

$$\begin{aligned} \dot{V}_{R1} &\le S_R^{\mathrm{T}} \big[D_r \big(\dot{u}_d - \tilde{B}(q) \tau_{\max} (\tanh(\tau) + (1 - \eta_\tau + \eta_\tau \tanh(A_D))) \\ &\quad + \bar{M}^{-1}(q) \bar{F}_m(\dot{q}) - \eta_f + \bar{\tau}_D \big) + \ddot{E}_R(k) + \dot{B}_0(E_R(k)) - B_\gamma \big] \end{aligned} \tag{26}$$

According to $\eta_f = 1 - \eta_\tau + \eta_\tau \tanh(A_d)$, the nonlinear fitting characteristics of neural network are used to fit the fault, attack, and dynamic nonlinearity. One can obtain

$$\begin{aligned} \dot{V}_{R1} &\le S_R^{\mathrm{T}} \big[D_r \big(\dot{u}_d - \tilde{B}(q) \tau_{\max} (\tau - E_\tau) + \dot{B}_0(E_R(k)) - B_\gamma \\ &\quad + \tau_{\max} \tilde{B}^T(q) \left(\psi^T \phi(u) + E_f \right) + \bar{\tau}_D \big) + \ddot{E}_R(k) \big] \end{aligned} \tag{27}$$

where $E_\tau = \tau - \tanh(\tau), F_D(\dot{q}) = \tau_{\max}^{-1} \bar{B}^{-1}(q) F_m(\dot{q}) + \eta_f$. According to the above formula, we can obtain:

$$\begin{aligned} &\left\| S_R^{\mathrm{T}} \right\| \left(\| \tau_{\max} E_\tau \| \left\| D_r \tilde{B}^T(q) \right\| + \left\| \psi^T \right\| \left\| D_r \tilde{B}^T(q) \tau_{\max} \phi(u) \right\| + \left\| \tau_{\max} E_f \right\| \left\| D_r \tilde{B}^T(q) \right\| \right. \\ &\left. + \| \bar{\tau}_D \| \| D_r \| + \| \ddot{E}_R(k) + \dot{B}_0(E_R(k) \| \right) \le \| S_R^{\mathrm{T}} \| \Theta_R \varphi(Z_R) \end{aligned} \tag{28}$$

From this, we can obtain

$$\dot{V}_{R1} \le \left\| S_R^{\mathrm{T}} \right\| \Theta_R \varphi(Z_R) - S_R^{\mathrm{T}} D_r \left[\dot{u}_d - \tilde{B}(q) \tau_{\max} \tau \right] \tag{29}$$

The global Lyapunov function is denoted as

$$V_R = V_{R1} + \frac{1}{2} \tilde{\Theta}_R^2 \tag{30}$$

where $\tilde{\Theta}_R = \Theta_R - \hat{\Theta}_R$, and $\dot{\Theta}_R = -\dot{\Theta}_R$. The time derivative of V_u is

$$\dot{V}_u = \dot{V}_{R1} - \tilde{\Theta}_R^T \dot{\Theta}_R \le \left\| S_R^{\mathrm{T}} \right\| \Theta_R \varphi(Z_R) - S_R^{\mathrm{T}} D_r \tau_{\mathrm{ma}} \tilde{B}^T(q) (\tau_c(k) + \tau_\xi(k)) - \tilde{\Theta}_R^T \dot{\hat{\Theta}}_R \tag{31}$$

By substituting Equations (19)–(20) into Equation (31), one can obtain

$$\begin{aligned} \dot{V}_R &\le \left\| S_R^{\mathrm{T}} \right\| \Theta_1 \varphi(Z_R) + S_R^{\mathrm{T}} \dot{S}_r + S_R^{\mathrm{T}} \left[-\frac{1}{4r_s} \tanh\left(\frac{S_R}{D} \right) \hat{\Theta}_R \right. \\ &\left. - r_s r_\Theta S_R \hat{\Theta}_1 \varphi^4(Z_R) \right] - \tilde{\Theta}_1^T \left[\frac{1}{4r_s} \tanh\left(\frac{S_u}{D} \right) S_u - b \hat{\Theta}_R \right. \\ &\left. + r_s r_\Theta \| S_R \|^2 \varphi^4(Z_R) \right] \end{aligned} \tag{32}$$

According to Young's inequality, we can obtain the following:

$$\left\|S_R^{\mathrm{T}}\right\|\Theta_R\varphi(Z_R) \leq r_s\left\|S_R^{\mathrm{T}}\right\|\Theta_R\varphi^2(Z_R) + \frac{\|S_R^{\mathrm{T}}\|\Theta_R}{4r_s} \tag{33}$$

Substitute Equation (33) into Equation (32) as

$$\begin{aligned}\dot{V}_R \leq\ & S_R^{\mathrm{T}}\dot{S}_r - \frac{1}{4r_s}S_R^{\mathrm{T}}\tanh\left(\frac{S_R}{D}\right)\Theta_R + r_s\left\|S_R^{\mathrm{T}}\right\|\Theta_R\varphi^2(Z_R) \\ & - \tilde{\Theta}_R^T\left(-b\hat{\Theta}_R + r_s r_\Theta\|S_R\|^2\varphi^4(Z_R)\right) + \frac{\|S_R^{\mathrm{T}}\|\Theta_R}{4r_s} \\ & - r_s r_\Theta\|S_R\|^2\hat{\Theta}_R\varphi^4(Z_R)\end{aligned} \tag{34}$$

Using Young's inequality, we can obtain

$$r_s\Theta_R\left\|S_R^{\mathrm{T}}\right\|\varphi^2(Z_R) \leq r_s r_\Theta\Theta_R\left\|S_R^{\mathrm{T}}\right\|^2\varphi^4(Z_R) + \frac{r_s\Theta_R}{4r_\Theta} \tag{35}$$

Further, using Equation (35), $\tilde{\Theta}_1 = \Theta_1 - \hat{\Theta}_1$ can be rewritten as

$$\dot{V}_R \leq S_R^{\mathrm{T}}\dot{S}_r - \frac{1}{4r_s}S_R^{\mathrm{T}}\tanh\left(\frac{S_R}{D}\right)\Theta_R + \frac{\Theta_R}{4r_s}\left\|S_R^{\mathrm{T}}\right\| + b\tilde{\Theta}_R^T\hat{\Theta}_R + \frac{r_s\Theta_R}{4r_\Theta} \tag{36}$$

According to Lemma 3 and $\tilde{\Theta}_R^T\hat{\Theta}_R \leq \tilde{\Theta}_R^T(\Theta_R - \tilde{\Theta}_R) \leq \frac{1}{2}\Theta_R^2 - \frac{1}{2}\tilde{\Theta}_R^T\tilde{\Theta}_R$, we have the following inequality:

$$\begin{aligned}\dot{V}_R &\leq S_R^{\mathrm{T}}\dot{S}_r - \frac{1}{4r_s}S_R^{\mathrm{T}}\tanh\left(\frac{S_R}{D}\right)\Theta_R + \frac{\Theta_R}{4r_s}\left\|S_R^{\mathrm{T}}\right\| + b\tilde{\Theta}_R^T\hat{\Theta}_R + \frac{r_s\Theta_R}{4r_\Theta} \\ &\leq S_R^{\mathrm{T}}\dot{S}_r - \frac{b}{2}\tilde{\Theta}_R^T\tilde{\Theta}_R + \frac{1}{4r_s}0.2785D\Theta_R + \frac{r_s\Theta_R}{4r_\Theta} + \frac{b}{2}\Theta_R^2\end{aligned} \tag{37}$$

Using $\frac{1}{4}\|\tilde{\Theta}_R\| \leq \frac{b}{4}\|\tilde{\Theta}_R\|^2 + \frac{1}{16b}$, we can obtain

$$\begin{aligned}\dot{V}_R \leq\ & S_R^{T}\dot{S}_r - \frac{1}{4}\left(\tilde{\Theta}_R^T\tilde{\Theta}_R\right)^{\frac{1}{2}} - \frac{b}{4}\tilde{\Theta}_R^T\tilde{\Theta}_R + \frac{1}{4r_s}0.2785D\Theta_R \\ & + \frac{r_s\Theta_R}{4r_\Theta} + \frac{b}{2}\Theta_R^2 + \frac{1}{16b}\end{aligned} \tag{38}$$

According to Equation (38), we can obtain

$$\begin{aligned}\dot{V}_R \leq\ & -\left(S_R^{\mathrm{T}}\left(\sqrt{2}\gamma_r(g_s - I_E)\right)^2 S_R\right)^{\frac{1}{2}} - S_R^{\mathrm{T}}\left(\gamma_r + \gamma_k - \frac{1}{2}I_E\right)S_R \\ & - \frac{1}{4}\left(\tilde{\Theta}_R^T\tilde{\Theta}_R\right)^{\frac{1}{2}} - \frac{b}{4}\tilde{\Theta}_R^T\tilde{\Theta}_R + \circlearrowright \\ \leq\ & -\eta_1 V_R - \eta_2 V_R^{\frac{1}{2}} + \circlearrowright\end{aligned} \tag{39}$$

where $\eta_1\left\{\lambda_{\min}\left(\gamma_r + \gamma_k - \frac{1}{2}I_E\right), \frac{\lambda_{\min}(b)}{4}\right\}$, $\eta_2 = \left\{\lambda_{\min}\left(\sqrt{2}\gamma_r(g_s - I_E)\right), \frac{1}{4}\right\}$, and $\circlearrowright = \frac{1}{4r_s}0.2785D\Theta_R + \frac{r_s\Theta_R}{4r_\Theta} + \frac{b}{2}\Theta_R^2 + \frac{1}{16b}$. □

4. Simulation Results and Analysis

To verify the effectiveness and superiority of this control algorithm, a comparative simulation, SISMAEFC, was established by using a conventional integral SMC combined with the super-twisted sliding mode reaching law [38], and the algorithm was verified and analyzed by error-simulation results, sliding surface simulation results, and double-actuator-input simulation results (see Figures 7–10). In addition, because SISMAEFC does not use the improved nonlinear fault tolerance mechanism and does not reconstruct faults and attacks, it is compensated directly by adaptive algorithms, which is demonstrated by adaptive laws (Figure 11). All simulations in this section adopt control parameters as: $\gamma_a = \text{diag}\{45, 15.5\}$, $\beta_a = \text{diag}\{0.01, 0.01\}, G_c = \text{diag}\{1, 1\}, \tau_{\max} = \text{diag}\{10, 10\}$, $\gamma_r = \text{diag}\{10, 40\}, \gamma_k = \text{diag}\{0.01, 0.1\}$, $r_s = \text{diag}\{0.02, 0.04\}, r_\Theta = \text{diag}\{0.01, 0.01\}, b = \text{diag}\{10.5, 10.5\}, D = \text{diag}\{1, 1\}$. According to the experimental model of the nonholonomic underactuated robot in the laboratory as the controlled object, the relevant parameters are: $m = 15$ kg, $r = 0.05$ m, $b = 0.5$ m, $J = 5\ \text{kg} \cdot \text{m}^2, d = 0.05$ m, $r = 0.05$ m. The RBF · NNs for H(Z) contain 15 nodes with centers evenly spaced in the range $[-3, 3] \times \ldots \times [-3, 3]$ and widths $\omega_l = 1.8 (l = 1, \ldots, 15)$. The simulation interval is designed according to the sampling rate 100hz of the main control chip of the experimental robot. The total simulation time was 150 s. To verify the tracking control performances on different trajectories and the control stability under attack, the expected trajectory used a trapezoidal line with a combination of a straight line and a curve [39].

The position error and angular velocity error of the robot under the two algorithms change over time, and the position error under the algorithm designed in this paper can converge to zero quickly. From Figure 7, it can be observed that x_e stabilized after about 4 s at the earliest stage, and ϕ_e stabilized at the latest at around 13 s. The whole system's position error could converge in about 15 s, and the stability of the curve was smooth relatively after the change in curvature after convergence. Although the error curve of SISMAEFC can converge, the curve shows obvious jitter, and SISMAEFC recovered to a stable state slowly when numerical fluctuation occurred, which was affected greatly by model uncertainty, and the convergence was not as good as that of the algorithm designed in this paper. Additionally, due to the network attack (the attack frequency is 30%), it can be seen from the local detail diagram that the SISMAEFC jitter is more obvious and the attack has a greater impact. It is worth noting that the error state under the NRSMEFC attack and affected by the fault is better, so we can see the attack and fault in this paper.

As shown in Table 1, the two parameters of the algorithm in this paper are smaller, and the MIAC is 18% lower than the minimum value and 45% lower than the maximum value of the comparison algorithm. The MISE is 22.6% lower than the minimum value and 11.4% lower than the maximum value of the comparison algorithm. Additionally, under the same attack, we can see from Figures 9 and 10 that the proposed algorithm is better.

Table 1. Quantitative analysis of the controller's control effect.

Evaluation Criteria	MIAC	MISE
Algorithm in this paper	[0.90, 2.03]	[8.307, 8.844, 3.16]
Comparison algorithm	[1.107, 3.75]	[10.74, 10.13, 3.57]

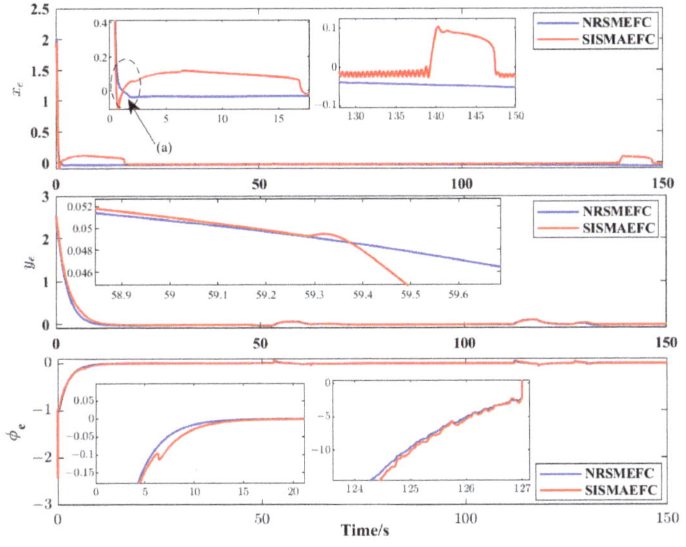

Figure 7. Comparison of system error.

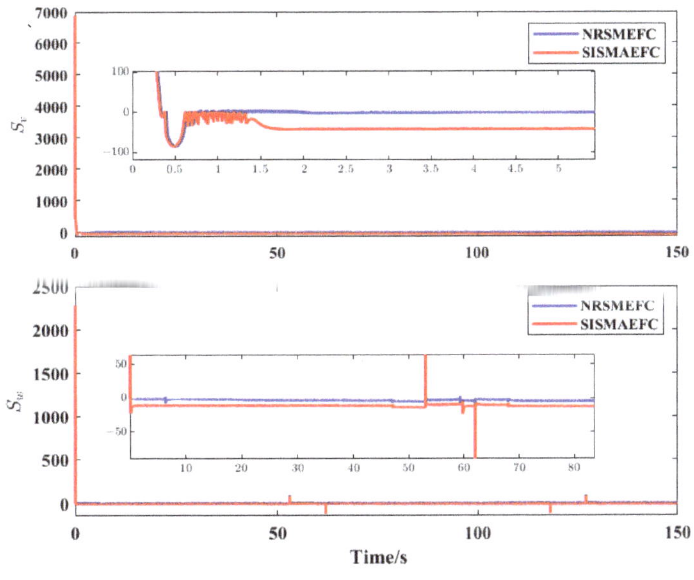

Figure 8. Sliding surface S_R comparison.

As shown in Figure 8, when the surface of the traditional sliding surface changes abruptly, the influence of nonlinear terms will increase significantly near the sliding surface. The system's response changes, and the system's buffeting changes significantly. Due to the accelerated dynamic characteristics of the reaching law, the residual redundancy will appear when the system's error reaches the sliding mode surface. In this paper, the sliding surface is designed to make up for the deficiency of the traditional sliding surface under the integral saturation mechanism and nonlinear variable damping reaching law. It reduces the sharp change in the curvature of the sliding surface, thereby reducing the probability of surface mutation. Thus, it eliminates the chattering phenomenon, makes up for the error redundancy, and improves the stability rate.

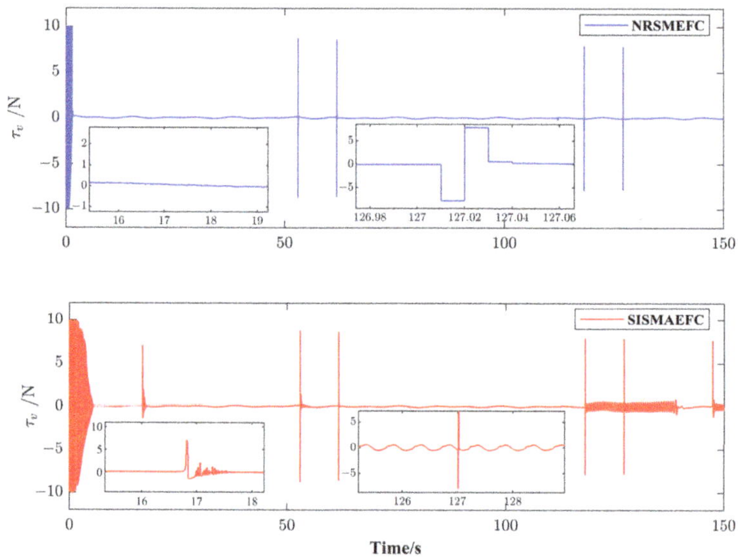

Figure 9. Differential coupling input τ_v comparison.

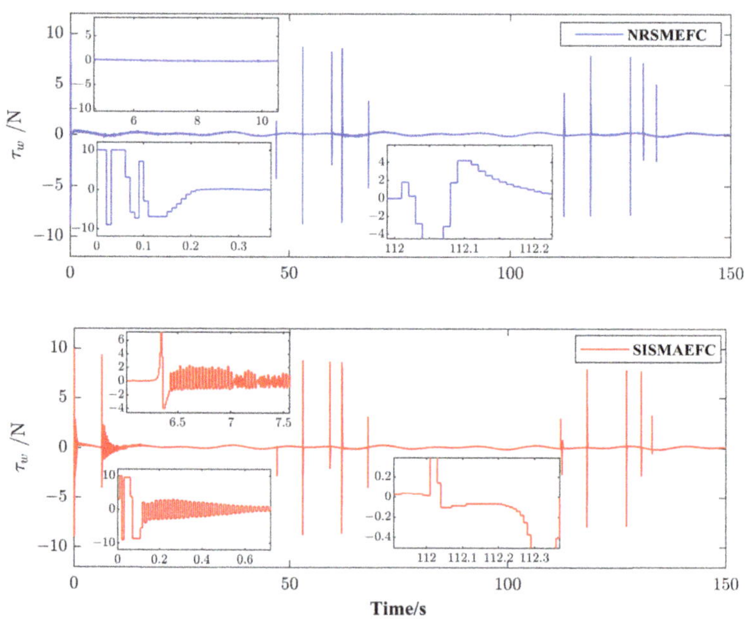

Figure 10. Differential coupling input τ_w comparison.

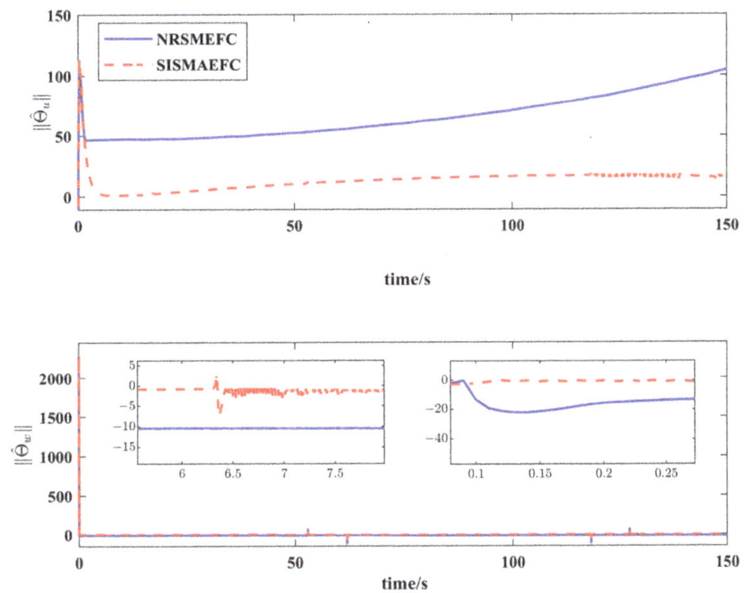

Figure 11. Comparison of compensation values of adaptive law.

From Figures 9 and 10, we can see that the control input of the algorithm designed in this paper does not fluctuate obviously under the attack and continues to be stable over time, but the control scheme shows severe chattering when the attack occurs. In Figure 11, we can see that the system cannot read the input signal correctly, which will affect the control effect. It can be seen in trigger times that the dynamic effect is better for the scheme of nonlinear fitting, followed by adaptive compensation by the controller in this paper. As the system's input may be subject to data errors, security attacks, system failures, and other problems, the system's input accuracy cannot be guaranteed. As can be seen in Figure 10, under the algorithm in this paper, the effect of event trigger is obviously better. The active transmission frequencies of the dual controller output channel of the proposed method and the comparison method are (8671, 6563) and (10,[illegible], [illegible]), respectively, which shows that the method in this paper has better control performance and is more stable state in cases of attacks and failures.

5. Conclusions

In this paper, a neural network, nonlinear-fitting, redundant, sliding mode event-trigger control system affected by abnormal input was designed. Firstly, according to the dynamic saturation input characteristics, the nonlinear redundant sliding surface was designed by using the nonlinear fitting function. Then, to reduce the chattering problem caused by the system, improve the input, and improve the approaching efficiency of a sliding mode surface with a large error, a nonlinear-damping, super-torsion reaching law was designed to improve the robust response efficiency of a system with a large error. For the problem of matching input interference and signal noise in the process of cable-based signal transmission, the input anomaly is non-linearly fitted by the fault-tolerant mechanism of the saturation filter, then stripped by saturation analysis, and then fitted by the nonlinear neural network. A set of nonlinear fault-tolerant subsystems was designed, which is controlled by an event-trigger mechanism. It improves the tracking accuracy of an intelligent robot in the cases of physical failure of the actuator, remote network attacks, and trajectory curvature mutation. In the comparison of simulation experiments, the pose error of this algorithm was improved by 11.48% at least. It can effectively improve the work efficiency of the freight yard and save on work costs. It has certain application prospects for engineering. In addition, with the development of artificial intelligence research, a class of algorithms for recognition of human activities has emerged. These include semi-supervised recurrent convolutional attention model algorithms [40], adaptive semi-supervised feature analysis algorithms [41], and convolutional neural network and recurrent neural network algorithms [42]. Consider applying such algorithms to artificial intelligence transport robots. The algorithm proposed in this paper needs to be improved, and the next step will be to investigate this problem.

Author Contributions: Conceptualization, Y.Z. and Q.Z.; methodology, Y.Z.; software, Y.Z.; validation, Y.Z.; formal analysis, Y.Z.; investigation, Q.Z.; resources, Q.Z.; data curation, Y.L.; writing—original draft preparation, Y.Z.; writing—review and editing, Y.Z. and S.Z.; visualization, Y.H.; supervision, Q.Z.; project administration, Q.Z. All authors have read and agreed to the published version of the manuscript.

Funding: This research was funded by project ZR2022ME087 and the supported by Shandong Provincial Natural Science Foundation, National Natural Science Foundation of China (51911540478).

Institutional Review Board Statement: Not applicable.

Informed Consent Statement: Not applicable.

Data Availability Statement: Not applicable.

Conflicts of Interest: The authors declare no conflict of interest.

Abbreviations

The following abbreviations are used in this manuscript:

NSRSMS	New type of nonlinear, saturated, redundant sliding surface
NDSTRL	Nonlinear-damping, super-twisting reaching law
MIAC	Mean integration absolute control
MISE	Mean integration square error

References

1. Yoerger, D.R.; Jakuba, M.; Bradley, A.M.; Bingham, B. Techniques for Deep Sea Near Bottom Survey Using an Autonomous Underwater Vehicle. *Int. J. Robot. Res.* **2007**, *26*, 41–54. [CrossRef]
2. Yang, J.; Sun, R.; Cui, J.; Ding, X. Application of composite fuzzy-PID algorithm to suspension system of Maglev train. In Proceedings of the 30th Annual Conference of IEEE Industrial Electronics Society, Busan, Republic of Korea, 2–6 November 2004.
3. Jiang, G.P.; Wei, X.Z. An LMI criterion for linear-state-feedback based chaos synchronization of a class of chaotic systems . *Chaos Solitons Fractals* **2005**, *26*, 142–149. [CrossRef]
4. Zhao, X.; Yang, H.; Xia, W.; Wang, X. Adaptive Fuzzy Hierarchical Sliding-Mode Control for a Class of MIMO Nonlinear Time-Delay Systems With Input Saturation. *J. IEEE Trans. Fuzzy Syst.* **2016**, *25*, 1061–1077. [CrossRef]
5. Levant, A. Quasi-Continuous High-Order Sliding-Mode Controllers. In Proceedings of the 42nd IEEE Conference on Decision and Control Maui, Maui, HI, USA, 9–12 December 2003.
6. Bartoszewicz, A. A new reaching law for sliding mode control of continuous time systems with constraints. *J. Trans. Inst. Meas. Control* **2014**, *37*, 515–521. [CrossRef]
7. Ma, H.; Wu, J.; Xiong, Z.A. Novel Exponential Reaching Law of Discrete-Time Sliding-Mode Control. *J. IEEE Trans. Ind. Electron.* **2017**, *64*, 3840–3850. [CrossRef]
8. Fallaha, C.J.; Saad, M.; Kanaan, H.Y.; Al-Haddad, K. Sliding-Mode Robot Control With Exponential Reaching Law. *J. IEEE Trans. Ind. Electron.* **2011**, *58*, 600–610. [CrossRef]
9. Bartoszewicz, A. Pawe Latosiński. Discrete time sliding mode control with reduced switching—A new reaching law approach. *J. Int. J. Robust Nonlinear Control* **2016**, *26*, 47–68. [CrossRef]
10. Chakrabarty, S.; Bandyopadhyay, B. A generalized reaching law with different convergence rates. *J. Autom. Oxf.* **2016**, *63*, 34–37. [CrossRef]
11. Baek, J.; Jin, M.; Han, S. A New Adaptive Sliding Mode Control Scheme for Application to Robot Manipulators. *J. IEEE Trans. Ind. Electron.* **2016**, *63*, 3628–3637. [CrossRef]

12. Niu, Y.; Lam, J.; Wang, X.; Ho, D.W. Neural adaptive sliding mode control for a class of nonlinear neutral delay systems. *J. Dyn. Syst. Meas. Control* **2008**, *130*, 758–767. [CrossRef]
13. Kang, Z.; Limei, W.; Xin, F. High-order fast nonsingular terminal sliding mode control of permanent magnet linear motor based on double disturbance observer. *IEEE Trans. Ind. Appl.* **2022**, *58*, 3696–3705.
14. Wang, Y.; Peng, J.; Zhu, K.; Chen, B.; Wu, H. Adaptive PID-fractional-order nonsingular terminal sliding mode control for cable-driven manipulators using time-delay estimation. *Int. J. Syst. Sci.* **2020**, *51*, 3118–3133. [CrossRef]
15. Shao, K.; Zheng, J.; Huang, K.; Wang, H.; Man, Z.; Fu, M. Finite-Time Control of a Linear Motor Positioner Using Adaptive Recursive Terminal Sliding Mode. *IEEE Trans. Ind. Electron.* **2020**, *67*, 6659–6668. [CrossRef]
16. Wang, Y.; Zhu, B.; Zhang, H.; Zheng, W.X. Functional observer-based finite-time adaptive ISMC for continuous systems with unknown nonlinear function. *J. Autom.* **2021**, *125*, 109468. [CrossRef]
17. Sui, S.; Tong, S.; Chen, C. Finite-Time Filter Decentralized Control for Nonstrict-Feedback Nonlinear Large-Scale Systems. *J. IEEE Trans. Fuzzy Syst.* **2018**, *26*, 3289–3300. [CrossRef]
18. Chen, M.; Wang, H.; Liu, X.; Hayat, T.; Alsaadi, F.E. Adaptive finite-time dynamic surface tracking control of nonaffine nonlinear systems with dead zone. *J. Neurocomputing* **2019**, *366*, 66–73. [CrossRef]
19. Wang, H.; Liu, P.X.; Zhao, X.; Liu, X. Adaptive Fuzzy Finite-Time Control of Nonlinear Systems with Actuator Faults. *J. IEEE Trans. Cybern.* **2019**, *99*, 1–12.
20. Wang, F.; Zhang, X. Adaptive Finite Time Control of Nonlinear Systems Under Time-Varying Actuator Failures. *J. IEEE Trans. Syst. Man, Cybern.* **2019**, *49*, 1845–1852. [CrossRef]
21. Fang, W.A.; Gl, B. Fixed-time control design for nonlinear uncertain systems via adaptive method. *J. Syst. Control Lett.* **2020**, *140*, 104704.
22. Ba, D.; Li, Y.X.; Tong, S. Fixed-time adaptive neural tracking control for a class of uncertain nonstrict nonlinear systems. *J. Neurocomputing* **2019**, *363*, 273–280. [CrossRef]
23. Zuo, Z. Non-singular fixed-time terminal sliding mode control of non-linear systems. *J. IET Control Theory Appl.* **2014**, *9*, 545–552. [CrossRef]
24. Liu, W.; Jian, T.; Yang, H.; Li, S.; Wang, H.; Wang, Y. Tunable Multichannel Adaptive Detector for Mismatched Subspace Signals. *J. J. Electron. Inf. Technol.* **2016**, *38*, 3011–3017.
25. Liu, W.; Liu, J.; Huang, L.; Yang, Z.; Yang, H.; Wang, Y.L. Distributed Target Detectors With Capabilities of Mismatched Subspace Signal Rejection. *J. Abbr.* **2017**, *53*, 888–900. [CrossRef]
26. Yang, F.; Zhang, H.; Jiang, B.; Liu, X. Adaptive reconfigurable control of systems with time-varying delay against unknown actuator faults. *J. Int. J. Adapt. Control Signal Process.* **2013**, *28*, 1206–1226. [CrossRef]
27. Li, L.; Luo, H.; Ding, S.X.; Yang, Y.; Peng, K. Performance-based fault detection and fault-tolerant control for automatic control systems. *J. Autom.* **2019**, *99*, 308–316. [CrossRef]
28. Zuo, Z.; Ho, D.; Wang, Y. Fault tolerant control for singular systems with actuator saturation and nonlinear perturbation. *J. Autom.* **2010**, *46*, 569–576. [CrossRef]
29. Zhang, G.; Gao, S.; Li, J.; Zhang, W. Adaptive neural fault-tolerant control for course tracking of unmanned surface vehicle with event-triggered input. *J. Proc. Inst. Mech. Eng. Part I J. Syst. Control Eng.* **2021**, *235*, 1594–1604. [CrossRef]
30. Zhang, X.M.; Han, Q.L. Event-triggered dynamic output feedback control for networked control systems. *J. IET Control Theory Appl.* **2014**, *8*, 226–234. [CrossRef]
31. Domingo, M.C. Securing underwater wireless communication networks. *J. IEEE Wirel. Commun.* **2011**, *18*, 22–28. [CrossRef]
32. Chen, Q.; Zhang, Q.; Hu, Y.; Liu, Y.; Wu, H. Euclidean distance damping–based adaptive sliding mode fault-tolerant event-triggered trajectory-tracking control . *Proc. Inst. Mech. Eng. Part I J. Syst. Control. Eng.* **2022**, . [CrossRef]
33. Zhang, C.; Su, J.; Zhang, W.; Zhou, J. Design of Crawler Mobile Car with Infrared Remote Control. In Proceedings of the 2020 6th International Conference on Control, Automation and Robotics, Singapore, 20–23 April 2020.
34. Najafi, B.; Ardabili, S.F. Application of ANFIS, ANN, and logistic methods in estimating biogas production from spent mushroom compost (SMC). *Resour. Conserv. Recycl.* **2018**, *133*, 169–178. [CrossRef]
35. Sun, M.; Luan, T.; Liang, L. RBF neural network compensation-based adaptive control for lift-feedback system of ship fin stabilizers to improve anti-rolling effect. *Ocean Eng.* **2018**, *163*, 307–321. [CrossRef]
36. Vinod, J.; Sarkar, B.K. Francis turbine electrohydraulic inlet guide vane control by artificial neural network 2 degree-of-freedom PID controller with actuator fault. *Proc. Inst. Mech. Eng.* **2021**, *235*, 1494–1509.
37. Zhu, G.; Ma, Y.; Li, Z.; Malekian, R.; Sotelo, M. Adaptive neural output feedback control for MSVs with predefined performance. *IEEE Trans. Veh. Technol.* **2021**, *70*, 2994–3006. [CrossRef]
38. Mobayen, S. Adaptive global sliding mode control of underactuated systems using a super-twisting scheme: An experimental study. *J. Vib. Control.* **2019**, *25*, 2215–2224. [CrossRef]
39. Chen, Q.; Hu, Y.; Zhang, Q.; Jiang, J.; Chi, M.; Zhu, Y. Dynamic Damping-Based Terminal Sliding Mode Event-Triggered Fault-Tolerant Pre-Compensation Stochastic Control for Tracked ROV. *J. Mar. Sci. Eng.* **2022**, *10*, 1228. [CrossRef]
40. Chen, K.; Yao, L.; Zhang, D.; Wang, X.; Chang, X.; Nie, F. A Semisupervised Recurrent Convolutional Attention Model for Human Activity Recognition. *IEEE Trans. Neural Networks Learn. Syst.* **2020**, *31*, 1747–1756. [CrossRef]

41. Luo, M.; Chang, X.; Nie, L.; Yang, Y.; Hauptmann, A.G.; Zheng, Q. An Adaptive Semisupervised Feature Analysis for Video Semantic Recognition. *IEEE Trans. Cybern.* **2018**, *48*, 648–660. [CrossRef]
42. Zhang, D.; Yao, L.; Chen, K.; Wang, S.; Chang, X.; Liu, Y. Making Sense of Spatio-Temporal Preserving Representations for EEG-Based Human Intention Recognition. *IEEE Trans. Cybern.* **2020**, *50*, 3033–3044. [CrossRef]

MDPI
St. Alban-Anlage 66
4052 Basel
Switzerland
www.mdpi.com

Journal of Marine Science and Engineering Editorial Office
E-mail: jmse@mdpi.com
www.mdpi.com/journal/jmse

www.ingramcontent.com/pod-product-compliance
Lightning Source LLC
Chambersburg PA
CBHW040751250926
51850CB00007BB/1464
* 9 7 8 3 0 3 9 2 8 6 2 3 2 *